Particle Physics: Theory and Applications (Volume I)

Particle Physics: Theory and Applications (Volume I)

Zachary Walter

Larsen & Keller
www.larsen-keller.com

Particle Physics: Theory and Applications (Volume I)
Zachary Walter
ISBN: 979-8-88836-014-9 (Hardback)

Larsen & Keller
Published by Larsen and Keller Education,
5 Penn Plaza,
19th Floor,
New York, NY 10001, USA

Cataloging-in-Publication Data

Particle physics : theory and applications (Volume I) / Zachary Walter.
p. cm.
Includes bibliographical references and index.
ISBN 979-8-88836-014-9
1. Particles (Nuclear physics). 2. Nuclear physics. 3. Physics. I. Walter Zachary.
QC793.2 .P37 2023
539.72--dc23

For more information regarding Larsen and Keller Education and its products, please visit the publisher's website www.larsen-keller.com

Table of Contents

Preface

Particle physics is a branch of physics that is involved in the study of the elementary constituents of matter and radiation, and the interactions between them. It is also referred to as high energy physics because many subatomic particles are absent in nature under normal circumstances but can be created and detected during energetic collisions of other particles. In particle physics, large particle accelerators are used for fundamental research. Particle accelerator is a machine that uses an electric field to speed-up and increase the energy of a beam of particles and the magnetic fields steer and focus them. They are used to study and explore the structure of the atomic nucleus and other aspects of particle physics. The modern particle physics research is focused on subatomic particles such as electrons, protons and neutrons. It also studies particles such as photons, neutrinos and muons, which are produced by radioactive and scattering processes. This book contains studies on the theory and applications of particle physics. It attempts to further enlighten the readers about the new concepts in this field.

The information contained in this book is the result of intensive hard work done by researchers in this field. All due efforts have been made to make this book serve as a complete guiding source for students and researchers. The topics in this book have been comprehensively explained to help readers understand the growing trends in the field.

I would like to thank the entire group of writers who made sincere efforts in this book and my family who supported me in my efforts of working on this book. I take this opportunity to thank all those who have been a guiding force throughout my life.

Zachary Walter

1

Particle Physics: An Overview

Herwig Schopper

Since old ages it has been one of the noble aspirations of humankind to understand the world in which we are living. In addition to our immediate environment, planet earth, two more remote frontiers have attracted interest: the infinitely small and the infinitely large. A flood of new experimental and theoretical results obtained during the past decades has provided a completely new picture of the micro- and macrocosm and surprisingly intimate relations have been discovered between the two. It turned out that the understanding of elementary particles and the forces acting between them is extremely relevant for our perception of the cosmological development. Quite often scientific research is supported because it is the basis for technical progress and for the material well-being of humans. The exploration of the microcosm and the universe contributes to this goal only indirectly by the development of better instruments and new techniques. However, it tries to answer some fundamental questions which are essential to understand the origins, the environment and the conditions for the existence of humankind and thus is an essential part of the cultural heritage.

One of the fundamental questions concerns the nature of matter, the substance of which the stars, the planets and living creatures are made, or to put it in another way—can the many phenomena which we observe in nature be explained on the basis of a few elementary building blocks and forces which act between them. The first attempts go back 2000 years when the Greek philosophers speculated about indestructible atoms, like Democritus, or the four elements and the regular bodies of Plato.

H. Schopper (✉)
CERN, Geneva, Switzerland
e-mail: Herwig.Schopper@cern.ch

Since Newton who introduced infinitely hard smooth balls as constituents of matter[1] and who described gravitation as the first force acting between them, the concept of understanding nature in terms of 'eternal' building blocks hold together by forces has not changed during the past 200 years. What has changed was the nature of the elementary building blocks and new forces were discovered. The chemists discovered the atoms of the 92 elements which, however, contrary to their name, were found to be divisible consisting of a nucleus surrounded by an electron cloud. Then it was found that the atomic nuclei contain protons and neutrons. Around 1930 the world appeared simple with everything consisting of these three particles: protons, neutrons and electrons.

Then came the 'annus mirabilis' 1931 with the discovery of the positron as the first representative of antimatter and the mysterious neutrino in nuclear beta-decay indicating a new force, the weak interaction. In the following decades the 'particle zoo' with all its newly discovered mesons, pions and 'strange' particles was leading to great confusion. Simplicity was restored when all these hundreds of 'elementary 'particles could be understood in terms of a new kind of elementary particles, the quarks and their antiquarks. The systematics of these particles is mainly determined by the strong nuclear force, well described today by the quantum chromodynamics QCD. Whether quarks and gluons (the binding particles of the strong interaction) exist only inside the atomic nuclei or whether a phase transition into a quark-gluon plasma is possible, is one the intriguing questions which still needs an answer.

Impressive progress was made in another domain, in the understanding of the weak nuclear force responsible for radioactive beta-decay and the energy production in the sun. Three kinds of neutrinos (with their associated antiparticles) were found and recently it could be shown that the neutrinos are not massless as had been originally assumed. The mechanism of the weak interaction could be clarified to a large extent by the discovery of its carriers, the W- and Z-particles. All the experimental results obtained so far will be summarized in this volume and the beautiful theoretical developments will be presented. The climax is the establishment of the 'Standard Model of Particle Physics' SM which has been shown to be a renormalizable gauge theory mainly by the LEP precision experiments. The LEP experiments have also shown that there are only three families of quarks and leptons (electron, muon, tau-particle and associated neutrinos), a fact not yet understood.

All the attempts to find experimental deviations from the SM have failed so far. However, the SM cannot be the final theory for the understanding of the microcosm. Its main fault is that it has too many arbitrary parameters (e.g. masses of the particles, values of the coupling constants of the forces, number of quark and lepton families) which have to be determined empirically by experiment. An underlying theory based on first principles is still missing and possible ways into the future will be discussed below.

[1]Isaac Newton, Optics, Query 31, London 1718.

Returning to the 'naïve' point of ultimate building blocks one might ask whether the quarks and leptons are fundamental indivisible particles or whether they have a substructure. Here we are running into a dilemma which was recognised already by the philosopher Immanuel Kant.[2] Either ultimate building blocks are mathematical points and cannot be divided, but then it is difficult to understand how they can have a mass and carry charges and spin. Alternatively, the building blocks might have spatial extension, but then it is hard to understand why they could not be divided into smaller parts. Whenever one meets such a paradox in science it is usually resolved by recognising that a wrong question was asked.

Indeed the recent developments of particle physics indicate that the naïve concept of ultimate building blocks of matter has to be abandoned. The smaller the 'building blocks' are, the higher energies are necessary to break them up. This is simply a consequence of the Heisenberg uncertainty principle of quantum mechanics. In the case of quarks their binding energies become so strong that any energy applied to break them apart is used to produce new quark-antiquark pairs.[3] The existence of antimatter implies also that matter does not have an 'eternal' existence. When matter meets antimatter the two annihilate by being converted into 'pure' energy and in the reverse mode matter can be produced[4] from energy in the form of particle-antiparticle pairs.

One of the most exciting development of physics or in science in general is a change of paradigms. Instead of using building blocks and forces acting between them, it was progressively recognised that symmetry principles are at the basis of our understanding of nature. It seems obvious that laws of nature should be invariant against certain transformations since 'nature does not know' how we observe it. When we make experiments we have to choose the origin of the coordinate system, its orientation in space and the moment in time when we start the observation. These choices are arbitrary and the laws deduced from the observations should not depend on them. It is known since a long time that the invariance of laws of nature against the continuous transformations, i.e. translations and rotations in space and time, give rise to the conservation of momentum, angular momentum and energy, the most basic laws of classical physics.[5] The mirror transitions (i.e. spatial reflection, particle-antiparticle exchange and time reversal) lead to the conservation of parity P, charge parity C and detailed balance rules in reactions, all of which are essential ingredients of quantum mechanics.

The detection of complete parity violation in weak interactions in 1957 was one of the most surprising observations. Many eminent physicists, including Wolfgang

---

[2]Immanuel Kant, Kritik der reinen Vernunft, 1781, see, e.g., Meiner Verlag, Hamburg 1998, or translation by N.K. Smith, London, MacMillan 1929.

[3]The binding energies are comparable to mc^2, where m is the rest mass of a quark and c is the velocity of light.

[4]When Pope Paul John II visited CERN and I explained to him that we can 'create' matter his response was: you can 'produce' matter, but 'creation' is my business.

[5]Emmy Noether, Nachr. d. königl. Gesellschaft d. Wissenschaften zu Göttingen, 1918, page 235.

Pauli, thought that this symmetry could not be violated. Such a believe indeed goes back to Emanuel Kant[2] who claimed that certain 'a priori' concepts have to be valid so that we would be able to explore nature. Since it seemed obvious that nature does not know whether we observe it directly or through a mirror a violation of mirror symmetry seemed unacceptable. This phenomenon is still not understood, although the fact that also C conservation is completely violated and the combined symmetry PC seemed to hold has reduced somewhat the original surprise. The whole situation has become more complicated by the detection that PC is also violated, although very little. A deep understanding of the violation of these 'classical' symmetries is still missing. So far experiments show that the combined symmetry PCT still holds as is required by a very general theorem.

In field theories another class of more abstract symmetries has become important—the gauge symmetries. As is well known from Maxwell's equations the electrodynamic fields are fully determined by the condition that gauge symmetry holds, which means that the electric and magnetic fields are independent against gauge transformations of their potentials. It was discovered that analogous gauge symmetries determine the fields of the strong and weak interactions in which case the (spontaneous) breaking of the symmetries plays a crucial role.

In summary, we have abandoned the description of nature in terms of hard indestructible spheres in favour of abstract ideas—the symmetries and there breaking. From a philosophical point of view one might, in an over-simplistic way, characterize the development as moving away from Democritus to Plato.

Finally, it should be mentioned that in particle physics progress was only possible by an intimate cooperation between theory and experiments. The field has become so complex that by chance discoveries are extremely rare. The guidance by theory is necessary to be able to put reasonable questions to nature. This does not exclude great surprises since many theoretical predictions turned out to be wrong. Indeed most progress could be made by verifying or disproving theories.

Although the Standard Model of Particle Physics SM (with some extensions, e.g. allowing for masses of neutrinos) has achieved a certain maturity by being able to reproduce all experimental results obtained so far, it leaves open many fundamental questions. One particular problem one has gotten accustomed to, concerns P and C violations which are put into the SM 'by hand'. And as has been mentioned above the SM leaves open many other questions which indicate that it cannot be a final theory.

In 2008 I wrote the concluding paragraph of this introduction as "Many arguments indicate that a breakthrough in the understanding of the microcosm will happen when the results of LHC at CERN will become available. LHC will start operation in 2008, but it will probably take several years before the experiments will have sufficient data and one will be able to analyse the complicated events before a major change of our picture will occur, although surprises are not excluded. Hence it seems to be an appropriate time to review the present situation of our understanding of the microcosm". Meanwhile, more than 10 years later, and with the Higgs boson discovered in 2014 at the LHC, the extended SM has been confirmed with unprecedented precision yet the outstanding questions, in particular which path to follow beyond the SM, have remained with us.

2

The Architecture of the Standard Model

Guido Altarelli and Stefano Forte

2.1 Introduction to Chaps. 2, 3 and 4

Stefano Forte

The presentation of the Standard Model in Chap. 2, Chaps. 3 and 4 was originally written by Guido Altarelli in 2007. In this introduction we provide a brief update (with references), and a discussion of the main developments which have taken place since the time of the writing.

Chapter 2 presents the architecture of the Standard Model, the way symmetries are realized and the way this can is described at the quantum level. The structure of the Standard Model is now well-established since half a century or so. The presentation in this chapter highlights the experimental (and thus, to a certain extent, historical Chap. 2) origin of the main structural aspects of the theory. The only aspects of the presentation which require (minimal) updating are the numerical values given for parameters, such as the Fermi coupling constant G_F, see Eq. (2.3). All of these parameters have been known quite accurately since the early 2000s (with the exception of neutrino masses, see Sect. 3.7 of Chap. 3), and thus their values are quite stable. The numbers given below are taken from the then-current edition of the Particle Data Book (PDG) [7]. At any given time, in order to have the most recent and accurate values, the reader should consult the most recent edition

---

The author "G. Altarelli" is deceased at the time of publication.

G. Altarelli
University of Rome 3, Rome, Italy

S. Forte (✉)
Dipartimento di Fisica, Università di Milano, Milano, Italy

of the PDG [25], preferably using the web-based version [26], which is constantly updated.

Chapter 3 presents the Electroweak sector of the Standard Model, which was established as a successful theory by extensive experimentation at the LEP electron-positron collider of CERN in the last decade of the past century, including some aspects of the theory, such as the CKM mechanism for mass mixing (see Sect. 3.6) which were originally often considered to be only approximate. The discovery, at the turn of the century, of neutrino mixing, and thus non-vanishing neutrino masses (see Sect. 3.7) has been the only significant addition to the minimal version of the electroweak theory as formulated in the sixties and seventies of the past century. The general understanding of electroweak interactions was thus essentially settled at the time of the writing of this chapter.

From the experimental point of view, the main development since then is the successful completion of the first two runs of the LHC, which have provided further confirmation of the standard Electroweak theory (see Ref. [27] for a recent review). From a theoretical point of view, the main surprise (from the LHC, but also a number of other experiments) is that there have been no surprises.

First and foremost, the Higgs sector of the Standard Model: after discovery of the Higgs boson in 2012 [28, 29] the Higgs sector has turned out so far to be in agreement with the minimal one-doublet structure presented in Sect. 3.5. The discussion presented there, as well as the phenomenology of the Standard Model Higgs of Sect. 3.13, remain thus essentially unchanged by the Higgs discovery. A theoretical introduction with more specific reference to the LHC can be found in Ref. [30], while the current experimental status of Higgs properties can be found in the continually updated pages of the CERN Higgs cross-section working group [31]. Perhaps, the only real surprise in the Higgs sector of the Standard Model is the extreme closeness of the measured Higgs mass to the critical value required for vacuum stability (see Sect. 3.13.1 below)—a fact with interesting cosmological implications [32]. The discovery of the Higgs has changed somewhat the nature of global fits of Standard Model parameters discussed in Sect. 3.12: with the value of the Higgs mass known, the fit is over-constrained—though the conclusion of global consistency remains unchanged. An updated discussion is given in Ref. [27], as well as in the review on the Electroweak Model by Erler and Freitas in the PDG [26].

Besides Higgs discovery, the general trend of the last several years has been that of the gradual disappearance of all anomalies—instances of discrepancy between Standard Model predictions and the data—either due to more accurate theory calculations (or even the correction of errors: see Sect. 3.9), or to more precise measurements. A case in point is that of the measurements of the electroweak mixing angle, discussed in Sect. 3.12: the tensions or signals of disagreement which are discussed there have all but disappeared, mostly thanks to more accurate theoretical calculations. Another case in which the agreement between Standard Model and experiment is improving (albeit perhaps more slowly) is that of lepton anomalous magnetic moments, discussed in Sect. 3.9. In both cases, updates on the current situation can again been found in Ref. [27], and in the aforementioned PDG review by Erler and Freitas.

Finally, there is a number of cases in which data from LHC experiments (as well as other experiments, specifically in the fields of flavor physics and neutrino physics) have brought more accuracy and more stringent tests, without changing the overall picture. These include gauge boson couplings, discussed in Sects. 3.3–3.4, for which we refer to Ref. [27]; the CKM matrix and flavor physics, discussed in Sect. 3.6, for which we refer to the review by Ceccucci, Ligeti and Sakai in the PDG [26]; neutrino masses and mixings, discussed in Sect. 3.7, for which we refer to the PDG review by Gonzalez-Garcia and Yokohama [26].

This perhaps unexpected success of the Standard Model, and the failure to find any evidence so far of new physics (and in particular supersymmetry) at the LHC has somewhat modified the perspective on the limitations of the Standard Model discussed in Sect. 3.14. Specifically, the significance of the hierarchy problem—the so-called "naturalness" issue—must be questioned, given that it entails new physics which has not be found: a suggestive discussion of this shift in perspective is in Ref. [33]. Yet, the classification of possible new physics scenarios of Sect. 3.14 remains essentially valid: recent updates are in Ref. [34] for supersymmetric models, and in Ref. [35] for non-supersymmetric ones. Consequently, looking for new physics has now become a precision exercise, and this has provided a formidable stimulus to the study of Electroweak radiative corrections, which has been the subject of very intense activity beyond the classic results discussed in Sect. 3.10: a recent detailed review is in Ref. [36].

Chapter 4 is devoted to the theory of strong interactions, Quantum Chromodynamics (QCD). This theory has not changed since its original formulation in the second half of the past century. Specifically, its application to hard processes, which allows for the use of perturbative methods, is firmly rooted in the set of classic results and techniques discussed in Sect. 4.5 below. What did slowly change over the years is the experimental status of QCD. What used to be, in the past century, a theory established qualitatively, has gradually turned into a theory firmly established experimentally—though, at the time this chapter was written, not quite tested to the same precision as the electroweak theory (see Sect. 4.7). Now, after the first two runs of the LHC, it can be stated that the whole of the Standard Model, QCD and the Electroweak theory, are tested to the same very high level of accuracy and precision, typically at the percent or sub-percent level.

Turning QCD into a precision theory has been a pre-requisite for successful physics at the LHC, a hadron collider in which every physical process necessarily involves the strong interaction, since the colliding objects are protons (or nuclei). This has grown into a pressing need as the lack of discovery of new particles or major deviations from Standard Model predictions has turned the search for new physics signals into a precision exercise: it has turned the LHC from an "energy frontier" to a "rarity/accuracy frontier" machine—something that was deemed inconceivable just before the start of its operation [37].

This rapid progress has happened thanks to an ever-increasing set of computational techniques, which, building upon the classic results presented in this chapter, has allowed for an enormous expansion of the set of perturbative computations of processes at colliders which are introduced in Sect. 4.5.4, and discussed in more detail in the context of LHC (and specifically Higgs) physics in Ref. [30].

To begin with, basic quantities such as the running of the coupling, discussed in Sect. 4.4, and $R_{e^+e^-}$, discussed in Sect. 4.5.1 are now know to one extra perturbative order (see the QCD review of the PDG [26] for the current state of the art and full references). These are five-loop perturbative calculations, now made possible thanks to the availability of powerful computing resources. Furthermore, the set of processes discussed in Sect. 4.5.4 has now been extended to include essentially all relevant hadron collider processes, which have been routinely computed to third perturbative order, while the first fourth-order calculations have just started appearing. Again, the QCD review of the PDG [26] provides a useful status update, including comparison between computation and experiment, which refer to cross-sections which span about ten orders of magnitude in size.

This progress has been happening thanks to the development of a vast new set of computational techniques, which, rooted in perturbative QCD, have now spawned a dedicated research field: that of amplitudes [38], which relates phenomenology, quantum field theory, and mathematics. The classic set of methods for "resummation"—the sum of infinite classes of perturbative contributions, discussed specifically in Sect. 4.5.3.1 for deep-inelastic scattering, has been extended well beyond the processes and accuracy discussed in Sect. 4.5.4—an up-to-date list is in the QCD review of the PDG [26]. Moreover, an entirely new set of resummation techniques has been developed, using the methodology of effective field theories: the so-called soft-collinear effective theory (SCET) which provides an extra tool in the resummation box [39]. One remarkable consequence of all these developments is that it is now possible to understand in detail the structure of pure strong interaction events, in which jets of hadrons are produced in the final state, by looking inside these events and tracing their structure in terms of the fundamental fields of QCD—quarks and gluons [40].

One topic in which things have changed rather less is the determination of the strong coupling, discussed in Sect. 4.7. Whereas the agreement between predicted and observed scaling violations discussed in Sect. 4.6.3 is ever more impressive (see the review on structure functions of the PDG [26]) the accuracy on the determination of the strong coupling itself has not improved much. Updated discussions can be found in the QCD review of the PDG, as well as in Ref. [41]. Progress is likely to come from future, more accurate LHC data, as well as from non-perturbative calculations [42] (not discussed here) soon expected to become competitive.

All in all, the dozen or so years since the original writing of these chapter have seen a full vindication of the Standard Model as a correct and accurate theory, and have stimulated a vast number of highly sophisticated experimental and theoretical results which build upon the treatment presented below.

2.2 Introduction

The ultimate goal of fundamental physics is to reduce all natural phenomena to a set of basic laws and theories that, at least in principle, can quantitatively reproduce and predict the experimental observations. At microscopic level all the phenomenology of matter and radiation, including molecular, atomic, nuclear and subnuclear physics, can be understood in terms of three classes of fundamental interactions: strong, electromagnetic and weak interactions. In atoms the electrons are bound to nuclei by electromagnetic forces and the properties of electron clouds explain the complex phenomenology of atoms and molecules. Light is a particular vibration of electric and magnetic fields (an electromagnetic wave). Strong interactions bind the protons and neutrons together in nuclei, being so intensively attractive at short distances that they prevail over the electric repulsion due to the equal sign charges of protons. Protons and neutrons, in turn, are composites of three quarks held together by strong interactions to which quarks and gluons are subject (hence these particles are called "hadrons" from the Greek word for "strong"). To the weak interactions are due the beta radioactivity that makes some nuclei unstable as well as the nuclear reactions that produce the enormous energy radiated by the stars and by our Sun in particular. The weak interactions also cause the disintegration of the neutron, the charged pions, the lightest hadronic particles with strangeness, charm, and beauty (which are "flavour" quantum numbers) as well as the decay of the quark top and of the heavy charged leptons (the muon μ^- and the tau τ^-). In addition all observed neutrino interactions are due to weak forces.

All these interactions are described within the framework of quantum mechanics and relativity, more precisely by a local relativistic quantum field theory. To each particle, described as pointlike, is associated a field with suitable (depending on the particle spin) transformation properties under the Lorentz group (the relativistic space-time coordinate transformations). It is remarkable that the description of all these particle interactions is based on a common principle: "gauge" invariance. A "gauge" symmetry is invariance under transformations that rotate the basic internal degrees of freedom but with rotation angles that depend on the space-time point. At the classical level gauge invariance is a property of the Maxwell equations of electrodynamics and it is in this context that the notion and the name of gauge invariance were introduced. The prototype of all quantum gauge field theories, with a single gauged charge, is QED, Quantum Electro-Dynamics, developed in the years from 1926 until about 1950, which indeed is the quantum version of Maxwell theory. Theories with gauge symmetry, at the renormalizable level, are completely determined given the symmetry group and the representations of the interacting fields. The whole set of strong, electromagnetic and weak interactions is described by a gauge theory, with 12 gauged non-commuting charges, which is called "the Standard Model" of particle interactions (SM). Actually only a subgroup of the SM symmetry is directly reflected in the spectrum of physical states. A part of the electroweak symmetry is hidden by the Higgs mechanism for the spontaneous symmetry breaking of a gauge symmetry.

For all material bodies on the Earth and in all geological, astrophysical and cosmological phenomena a fourth interaction, the gravitational force, plays a dominant role, while it is instead negligible in atomic and nuclear physics. The theory of general relativity is a classic (in the sense of non quantum mechanical) description of gravitation that goes beyond the static approximation described by Newton law and includes dynamical phenomena like, for example, gravitational waves. The problem of the formulation of a quantum theory of gravitational interactions is one of the central problems of contemporary theoretical physics. But quantum effects in gravity become only important for energy concentrations in space-time which are not in practice accessible to experimentation in the laboratory. Thus the search for the correct theory can only be done by a purely speculative approach. All attempts at a description of quantum gravity in terms of a well defined and computable local field theory along similar lines as for the SM have so far failed to lead to a satisfactory framework. Rather, at present the most complete and plausible description of quantum gravity is a theory formulated in terms of non pointlike basic objects, the so called "strings", extended over distances much shorter than those experimentally accessible, that live in a space-time with 10 or 11 dimensions. The additional dimensions beyond the familiar 4 are, typically, compactified which means that they are curled up with a curvature radius of the order of the string dimensions. Present string theory is an all-comprehensive framework that suggests a unified description of all interactions together with gravity of which the SM would be only a low energy or large distance approximation.

A fundamental principle of quantum mechanics, the Heisenberg indetermination principle, implies that, for studying particles with spatial dimensions of order Δx or interactions taking place at distances of order Δx, one needs as a probe a beam of particles (typically produced by an accelerator) with impulse $p \gtrsim \hbar/\Delta x$, where $\hbar$ is the reduced Planck constant ($\hbar = h/2\pi$). Accelerators presently in operation or available in the near future, like the Large Hadron Collider at CERN near Geneva, allow to study collisions between two particles with total center of mass energy up to $2E \sim 2pc \lesssim 14$ TeV. These machines, in principle, can allow to study physics down to distances $\Delta x \gtrsim 10^{-18}$ cm. Thus, on the basis of results from experiments at existing accelerators, we can confirm that, down to distances of that order of magnitude, indeed electrons, quarks and all the fundamental SM particles do not show an appreciable internal structure and look elementary and pointlike. We expect that quantum effects in gravity will certainly become important at distances $\Delta x \gtrsim 10^{-33}$ cm corresponding to energies up to $E \sim M_{Pl}c^2 \sim 10^{19}$ GeV, where M_{Pl} is the Planck mass, related to Newton constant by $G_N = \hbar c/M_{Pl}^2$. At such short distances the particles that so far appeared as pointlike could well reveal an extended structure, like for strings, and be described by a more detailed theoretical framework of which the local quantum field theory description of the SM would be just a low energy/large distance limit.

From the first few moments of the Universe, after the Big Bang, the temperature of the cosmic background went down gradually, starting from $kT \sim M_{Pl}c^2$, where $k = 8.617\ldots 10^{-5}$ eV K^{-1} is the Boltzmann constant, down to the present situation where $T \sim 2.725$ K. Then all stages of high energy physics from string theory, which is a purely speculative framework, down to the SM phenomenology, which is directly accessible to experiment and well tested, are essential for the reconstruction of the evolution of the Universe starting from the Big Bang. This is the basis for the ever increasing relation between high energy physics and cosmology.

2.3 Overview of the Standard Model

The SM is a gauge field theory based on the symmetry group $SU(3) \otimes SU(2) \otimes U(1)$. The transformations of the group act on the basic fields. This group has 8+3+1= 12 generators with a non trivial commutator algebra (if all generators commute the gauge theory is said to be "abelian", while the SM is a "non abelian" gauge theory). $SU(3)$ is the "colour" group of the theory of strong interactions (QCD: Quantum Chromo-Dynamics [1–3]). $SU(2) \otimes U(1)$ describes the electroweak (EW) interactions [4–6] and the electric charge Q, the generator of the QED gauge group $U(1)_Q$, is the sum of T_3, one of the $SU(2)$ generators and of $Y/2$, where Y is the $U(1)$ generator: $Q = T_3 + Y/2$.

In a gauge theory to each generator T is associated a vector boson (also said gauge boson) with the same quantum numbers as T, and, if the gauge symmetry is unbroken, this boson is of vanishing mass. These vector (i.e. of spin 1) bosons act as mediators of the corresponding interactions. For example, in QED the vector boson associated to the generator Q is the photon γ. The interaction between two charged particles in QED, for example two electrons, is mediated by the exchange of one (or seldom more than one) photon emitted by one electron and reabsorbed by the other one. Similarly in the SM there are 8 massless gluons associated to the $SU(3)$ colour generators, while for $SU(2) \otimes U(1)$ there are 4 gauge bosons W^+, W^-, Z^0 and γ. Of these, only the photon γ is massless because the symmetry induced by the other 3 generators is actually spontaneously broken. The masses of W^+, W^- and Z^0 are quite large indeed on the scale of elementary particles: $m_W \sim 80.4$ GeV, $m_Z \sim 91.2$ GeV are as heavy as atoms of intermediate size like rubidium and molibdenum, respectively. In the electroweak theory the breaking of the symmetry is of a particular type, denoted as spontaneous symmetry breaking. In this case charges and currents are as dictated by the symmetry but the fundamental state of minimum energy, the vacuum, is not unique and there is a continuum of degenerate states that all together respect the symmetry (in the sense that the whole vacuum orbit is spanned by applying the symmetry transformations). The symmetry breaking is due to the fact that the system (with infinite volume and infinite number of degrees of freedom) is found in one particular vacuum state, and this choice, which for the SM occurred in the first instants of the Universe life, makes the symmetry violated in the spectrum of states. In a gauge theory like the SM the spontaneous symmetry breaking is realized by the Higgs mechanism (described in detail in Sect. (2.7)): there are a number of scalar (i.e. of zero spin) Higgs bosons with a potential that produces an orbit of degenerate vacuum states. One or more of these scalar Higgs particles must necessarily be present in the spectrum of physical states with masses very close to the range so far explored. It is expected that the Higgs particle(s) will be found at the LHC thus completing the experimental verification of the SM.

The fermionic (all of spin 1/2) matter fields of the SM are quarks and leptons. Each type of quark is a colour triplet (i.e. each quark flavour comes in three colours) and also carries electroweak charges, in particular electric charges +2/3 for up-type quarks and $-1/3$ for down-type quarks. So quarks are subject to all SM interactions.

Leptons are colourless and thus do not interact strongly (they are not hadrons) but have electroweak charges, in particular electric charges -1 for charged leptons (e^-, μ^- and τ^-) while it is 0 for neutrinos (ν_e, ν_μ and ν_τ). Quarks and leptons are grouped in 3 "families" or "generations" with equal quantum numbers but different masses. At present we do not have an explanation for this triple repetition of fermion families:

$$\begin{bmatrix} u\ u\ u\ \nu_e \\ d\ d\ d\ e \end{bmatrix}, \quad \begin{bmatrix} c\ c\ c\ \nu_\mu \\ s\ s\ s\ \mu \end{bmatrix}, \quad \begin{bmatrix} t\ t\ t\ \nu_\tau \\ b\ b\ b\ \tau \end{bmatrix}. \tag{2.1}$$

The QCD sector of the SM has a simple structure but a very rich dynamical content, including the observed complex spectroscopy with a large number of hadrons. The most prominent properties of QCD are asymptotic freedom and confinement. In field theory the effective coupling of a given interaction vertex is modified by the interaction. As a result, the measured intensity of the force depends on the transferred (four)momentum squared, Q^2, among the participants. In QCD the relevant coupling parameter that appears in physical processes is $\alpha_s = e_s^2/4\pi$ where e_s is the coupling constant of the basic interaction vertices of quark and gluons: qqg or ggg (see Eq. (2.30)). Asymptotic freedom means that the effective coupling becomes a function of Q^2: $\alpha_s(Q^2)$ decreases for increasing Q^2 and vanishes asymptotically. Thus, the QCD interaction becomes very weak in processes with large Q^2, called hard processes or deep inelastic processes (i.e. with a final state distribution of momenta and a particle content very different than those in the initial state). One can prove that in 4 space-time dimensions all pure-gauge theories based on a non commuting group of symmetry are asymptotically free and conversely. The effective coupling decreases very slowly at large momenta with the inverse logarithm of Q^2: $\alpha_s(Q^2) = 1/b \log Q^2/\Lambda^2$ where b is a known constant and Λ is an energy of order a few hundred MeV. Since in quantum mechanics large momenta imply short wavelengths, the result is that at short distances the potential between two colour charges is similar to the Coulomb potential, i.e. proportional to $\alpha_s(r)/r$, with an effective colour charge which is small at short distances. On the contrary the interaction strength becomes large at large distances or small transferred momenta, of order $Q \lesssim \Lambda$. In fact all observed hadrons are tightly bound composite states of quarks (baryons are made of qqq and mesons of $q\bar{q}$), with compensating colour charges so that they are overall neutral in colour. In fact, the property of confinement is the impossibility of separating colour charges, like individual quarks and gluons or any other coloured state. This is because in QCD the interaction potential between colour charges increases at long distances linearly in r. When we try to separate the quark and the antiquark that form a colour neutral meson the interaction energy grows until pairs of quarks and antiquarks are created from the vacuum and new neutral mesons are coalesced and observed in the final state instead of free quarks. For example, consider the process $e^+e^- \to q\bar{q}$ at large center of mass energies. The final state quark and antiquark have large energies, so they separate in opposite directions very fast. But the colour confinement forces create new pairs in between them. What is observed is two back-to-back jets of colourless hadrons with a number of slow pions that make the exact separation of the two jets impossible. In some cases a third well separated jet of hadrons is also observed: these events correspond to the radiation of an energetic gluon from the parent quark-antiquark pair.

In the EW sector the SM inherits the phenomenological successes of the old $(V-A)\otimes(V-A)$ four-fermion low-energy description of weak interactions, and provides a well-defined and consistent theoretical framework including weak interactions and quantum electrodynamics in a unified picture. The weak interactions derive their name from their intensity. At low energy the strength of the effective four-fermion interaction of charged currents is determined by the Fermi coupling constant G_F. For example, the effective interaction for muon decay is given by

$$\mathcal{L}_{\text{eff}} = (G_F/\sqrt{2})\left[\bar{\nu}_\mu\gamma_\alpha(1-\gamma_5)\mu\right]\left[\bar{e}\gamma^\alpha(1-\gamma_5)\nu_e\right] , \tag{2.2}$$

with [7]

$$G_F = 1.16639(1)\times 10^{-5}\ \text{GeV}^{-2} . \tag{2.3}$$

In natural units $\hbar = c = 1$, G_F has dimensions of $(\text{mass})^{-2}$. As a result, the intensity of weak interactions at low energy is characterized by $G_F E^2$, where E is the energy scale for a given process ($E \approx m_\mu$ for muon decay). Since

$$G_F E^2 = G_F m_p^2 (E/m_p)^2 \simeq 10^{-5}(E/m_p)^2 , \tag{2.4}$$

where m_p is the proton mass, the weak interactions are indeed weak at low energies (up to energies of order a few ten's of GeV). Effective four fermion couplings for neutral current interactions have comparable intensity and energy behaviour. The quadratic increase with energy cannot continue for ever, because it would lead to a violation of unitarity. In fact, at large energies the propagator effects can no longer be neglected, and the current–current interaction is resolved into current–W gauge boson vertices connected by a W propagator. The strength of the weak interactions at high energies is then measured by g_W, the $W--\mu$–ν_μ coupling, or, even better, by $\alpha_W = g_W^2/4\pi$ analogous to the fine-structure constant α of QED (in Chap. 3, g_W is simply denoted by g or g_2). In the standard EW theory, we have

$$\alpha_W = \sqrt{2}\, G_F\, m_W^2/\pi \cong 1/30 . \tag{2.5}$$

That is, at high energies the weak interactions are no longer so weak.

The range r_W of weak interactions is very short: it is only with the experimental discovery of the W and Z gauge bosons that it could be demonstrated that r_W is non-vanishing. Now we know that

$$r_W = \frac{\hbar}{m_W c} \simeq 2.5\times 10^{-16}\ \text{cm}, \tag{2.6}$$

corresponding to $m_W \simeq 80.4$ GeV. This very large value for the W (or the Z) mass makes a drastic difference, compared with the massless photon and the infinite range of the QED force. The direct experimental limit on the photon mass is [7] $m_\gamma < 6\ 10^{-17}$ eV. Thus, on the one hand, there is very good evidence that the photon is massless. On the other hand, the weak bosons are very heavy. A unified theory of EW interactions has to face this striking difference.

Another apparent obstacle in the way of EW unification is the chiral structure of weak interactions: in the massless limit for fermions, only left-handed quarks and leptons (and right-handed antiquarks and antileptons) are coupled to W's. This clearly implies parity and charge-conjugation violation in weak interactions.

The universality of weak interactions and the algebraic properties of the electromagnetic and weak currents [the conservation of vector currents (CVC), the partial conservation of axial currents (PCAC), the algebra of currents, etc.] have been crucial in pointing to a symmetric role of electromagnetism and weak interactions at a more fundamental level. The old Cabibbo universality [8] for the weak charged current:

$$\begin{aligned} J_\alpha^{\mathrm{weak}} &= \bar{\nu}_\mu \gamma_\alpha (1-\gamma_5)\mu + \bar{\nu}_e \gamma_\alpha (1-\gamma_5) e + \cos\theta_c\, \bar{u}\gamma_\alpha(1-\gamma_5) d + \\ &+ \sin\theta_c\, \bar{u}\gamma_\alpha (1-\gamma_5) s + \dots\,, \end{aligned} \tag{2.7}$$

suitably extended, is naturally implied by the standard EW theory. In this theory the weak gauge bosons couple to all particles with couplings that are proportional to their weak charges, in the same way as the photon couples to all particles in proportion to their electric charges [in Eq. (2.7), $d' = \cos\theta_c\ d + \sin\theta_c\ s$ is the weak-isospin partner of u in a doublet. The (u, d') doublet has the same couplings as the (ν_e, ℓ) and (ν_μ, μ) doublets].

Another crucial feature is that the charged weak interactions are the only known interactions that can change flavour: charged leptons into neutrinos or up-type quarks into down-type quarks. On the contrary, there are no flavour-changing neutral currents at tree level. This is a remarkable property of the weak neutral current, which is explained by the introduction of the Glashow-Iliopoulos-Maiani (GIM) mechanism [9] and has led to the successful prediction of charm.

The natural suppression of flavour-changing neutral currents, the separate conservation of e, μ and τ leptonic flavours that is only broken by the small neutrino masses, the mechanism of CP violation through the phase in the quark-mixing matrix [10], are all crucial features of the SM. Many examples of new physics tend to break the selection rules of the standard theory. Thus the experimental study of rare flavour-changing transitions is an important window on possible new physics.

The SM is a renormalizable field theory which means that the ultra-violet divergences that appear in loop diagrams can be eliminated by a suitable redefinition of the parameters already appearing in the bare lagrangian: masses, couplings and field normalizations. As it will be discussed later, a necessary condition for a theory to be renormalizable is that only operator vertices of dimension not larger than 4 (that is m^4 where m is some mass scale) appear in the lagrangian density $\mathcal{L}$ (itself of dimension 4, because the action S is given by the integral of $\mathcal{L}$ over d^4x and is dimensionless in natural units: $\hbar = c = 1$). Once this condition is added to the specification of a gauge group and of the matter field content the gauge theory lagrangian density is completely specified. We shall see the precise rules to write down the lagrangian of a gauge theory in the next Section.

2.4 The Formalism of Gauge Theories

In this Section we summarize the definition and the structure of a gauge Yang–Mills theory [11]. We will list here the general rules for constructing such a theory. Then these results will be applied to the SM.

Consider a lagrangian density $\mathcal{L}[\phi, \partial_\mu\phi]$ which is invariant under a D dimensional continuous group of transformations:

$$\phi'(x) = U(\theta^A)\phi(x) \qquad (A = 1, 2, \ldots, D)\,. \tag{2.8}$$

with:

$$U(\theta^A) = \exp\,[ig\sum_A \theta^A T^A] \sim 1 + ig\sum_A \theta^A T^A + \ldots, \tag{2.9}$$

The quantities θ^A are numerical parameters, like angles in the particular case of a rotation group in some internal space. The approximate expression on the right is valid for θ^A infinitesimal. Then, g is the coupling constant and T^A are the generators of the group Γ of transformations (2.8) in the (in general reducible) representation of the fields ϕ. Here we restrict ourselves to the case of internal symmetries, so that T^A are matrices that are independent of the space-time coordinates and the arguments of the fields ϕ and ϕ' in Eq. (2.8) is the same. If U is unitary, then the generators T^A are Hermitian, but this need not be the case in general (though it is true for the SM). Similarly if U is a group of matrices with unit determinant, then the traces of T^A vanish: $\mathrm{tr}(T^A) = 0$. The generators T^A are normalized in such a way that for the lowest dimensional non-trivial representation of the group Γ (we use t^A to denote the generators in this particular representation) we have

$$\mathrm{tr}(t^A t^B) = \frac{1}{2}\delta^{AB}\,. \tag{2.10}$$

The generators satisfy the commutation relations

$$[T^A, T^B] = iC_{ABC}T^C\,. \tag{2.11}$$

For $A, B, C \ldots.$ up or down indices make no difference: $T^A = T_A$ etc. The structure constants C_{ABC} are completely antisymmetric in their indices, as can be easily seen. In the following, for each quantity f^A we define

$$\mathbf{f} = \sum_A T^A f^A \,. \tag{2.12}$$

For example, we can rewrite Eq. (2.9) in the form:

$$U(\theta^A) = \exp[ig\boldsymbol{\theta}] \sim 1 + ig\boldsymbol{\theta} + \ldots, \tag{2.13}$$

If we now make the parameters θ^A depend on the space–time coordinates $\theta^A = \theta^A(x_\mu)$, $\mathcal{L}[\phi, \partial_\mu\phi]$ is in general no longer invariant under the gauge transformations $U[\theta^A(x_\mu)]$, because of the derivative terms: indeed $\partial_\mu\phi' = \partial_\mu(U\phi) \neq U\partial_\mu\phi$. Gauge invariance is recovered if the ordinary derivative is replaced by the covariant derivative:

$$D_\mu = \partial_\mu + ig\mathbf{V}_\mu \,, \tag{2.14}$$

where V_μ^A are a set of D gauge vector fields (in one-to-one correspondence with the group generators) with the transformation law

$$\mathbf{V}'_\mu = U\mathbf{V}_\mu U^{-1} - (1/ig)(\partial_\mu U)U^{-1} \,. \tag{2.15}$$

For constant θ^A, $\mathbf{V}$ reduces to a tensor of the adjoint (or regular) representation of the group:

$$\mathbf{V}'_\mu = U\mathbf{V}_\mu U^{-1} \simeq \mathbf{V}_\mu + ig[\boldsymbol{\theta}, \mathbf{V}_\mu] \ \ldots, \tag{2.16}$$

which implies that

$$V'^C_\mu = V^C_\mu - gC_{ABC}\theta^A V^B_\mu \ \ldots, \tag{2.17}$$

where repeated indices are summed up.

As a consequence of Eqs. (2.14) and (2.15), $D_\mu\phi$ has the same transformation properties as ϕ:

$$(D_\mu\phi)' = U(D_\mu\phi) \,. \tag{2.18}$$

In fact

$$\begin{aligned}(D_\mu\phi)' &= (\partial_\mu + ig\mathbf{V}'_\mu)\phi' = (\partial_\mu U)\phi + U\partial_\mu\phi + igU\mathbf{V}_\mu\phi - (\partial_\mu U) \\ \phi &= U(D_\mu\phi) \,. \end{aligned} \tag{2.19}$$

Thus $\mathcal{L}[\phi, D_\mu\phi]$ is indeed invariant under gauge transformations. But, at this stage, the gauge fields V_μ^A appear as external fields that do not propagate. In order to construct a gauge-invariant kinetic energy term for the gauge fields V_μ^A, we consider

$$[D_\mu, D_\nu]\phi = ig\{\partial_\mu \mathbf{V}_\nu - \partial_\nu \mathbf{V}_\mu + ig[\mathbf{V}_\mu, \mathbf{V}_\nu]\}\phi \equiv ig\mathbf{F}_{\mu\nu}\phi \, , \tag{2.20}$$

which is equivalent to

$$F_{\mu\nu}^A = \partial_\mu V_\nu^A - \partial_\nu V_\mu^A - gC_{ABC} V_\mu^B V_\nu^C \, . \tag{2.21}$$

From Eqs. (2.8), (2.18) and (2.20) it follows that the transformation properties of $F_{\mu\nu}^A$ are those of a tensor of the adjoint representation

$$\mathbf{F}'_{\mu\nu} = U\mathbf{F}_{\mu\nu}U^{-1} \, . \tag{2.22}$$

The complete Yang–Mills lagrangian, which is invariant under gauge transformations, can be written in the form

$$\mathcal{L}_{\mathrm{YM}} = -\frac{1}{2}Tr\mathbf{F}_{\mu\nu}\mathbf{F}^{\mu\nu} + \mathcal{L}[\phi, D_\mu\phi] = -\frac{1}{4}\sum_A F_{\mu\nu}^A F^{A\mu\nu} + \mathcal{L}[\phi, D_\mu\phi] \, . \tag{2.23}$$

Note that the kinetic energy term is an operator of dimension 4. Thus if $\mathcal{L}$ is renormalizable, also $\mathcal{L}_{\mathrm{YM}}$ is renormalizable. In fact it is the most general gauge invariant and renormalizable lagrangian density. If we give up renormalizability then more gauge invariant higher dimension terms could be added. It is already clear at this stage that no mass term for gauge bosons of the form $m^2 V_\mu V^\mu$ is allowed by gauge invariance.

For an abelian theory, as for example QED, the gauge transformation reduces to $U[\theta(x)] = \exp[ieQ\theta(x)]$, where Q is the charge generator. The associated gauge field (the photon), according to Eq. (2.15), transforms as

$$V'_\mu = V_\mu - \partial_\mu\theta(x) \, . \tag{2.24}$$

and the familiar gauge transformation by addition of a 4-gradient of a scalar function is recovered. The QED lagrangian density is given by:

$$\mathcal{L} \; = \; -\frac{1}{4}F^{\mu\nu}F_{\mu\nu} \; + \; \sum_\psi \bar{\psi}(i\not{D} - m_\psi)\psi \; . \tag{2.25}$$

Here $\not{D} = D_\mu\gamma^\mu$, where γ^μ are the Dirac matrices and the covariant derivative is given in terms of the photon field A_μ and the charge operator Q by:

$$D_\mu = \partial_\mu + ieA_\mu Q \tag{2.26}$$

and

$$F_{\mu\nu} = \partial_\mu A_\nu - \partial_\nu A_\mu \tag{2.27}$$

Note that in QED one usually takes the e^- to be the particle, so that $Q = -1$ and the covariant derivative is $D_\mu = \partial_\mu - ieA_\mu$ when acting on the electron field. In this case, the $F_{\mu\nu}$ tensor is linear in the gauge field V_μ so that in the absence of matter fields the theory is free. On the other hand, in the non abelian case the $F^A_{\mu\nu}$ tensor contains both linear and quadratic terms in V^A_μ, so that the theory is non-trivial even in the absence of matter fields.

2.5 Application to QCD

According to the formalism of the previous section, the statement that QCD is a renormalizable gauge theory based on the group $SU(3)$ with colour triplet quark matter fields fixes the QCD lagrangian density to be

$$\mathcal{L} = -\frac{1}{4}\sum_{A=1}^{8} F^{A\mu\nu}F^A_{\mu\nu} + \sum_{j=1}^{n_f} \bar{q}_j(i\not{D} - m_j)q_j \tag{2.28}$$

Here q_j are the quark fields (of n_f different flavours) with mass m_j and D_μ is the covariant derivative:

$$D_\mu = \partial_\mu + ie_s\mathbf{g}_\mu; \tag{2.29}$$

e_s is the gauge coupling and later we will mostly use, in analogy with QED

$$\alpha_s = \frac{e_s^2}{4\pi}. \tag{2.30}$$

Also, $\mathbf{g}_\mu = \sum_A t^A g^A_\mu$ where g^A_μ, $A = 1, 8$, are the gluon fields and t^A are the $SU(3)$ group generators in the triplet representation of quarks (i.e. t_A are 3×3 matrices acting on q); the generators obey the commutation relations $[t^A, t^B] = iC_{ABC}t^C$ where C_{ABC} are the complete antisymmetric structure constants of $SU(3)$ (the normalisation of C_{ABC} and of e_s is specified by $Tr[t^A t^B] = \delta^{AB}/2$);

$$F^A_{\mu\nu} = \partial_\mu g^A_\nu - \partial_\nu g^A_\mu - e_s C_{ABC} g^B_\mu g^C_\nu \tag{2.31}$$

Chapter 4 is devoted to a detailed description of the QCD as the theory of strong interactions. The physical vertices in QCD include the gluon-quark-antiquark vertex, analogous to the QED photon-fermion-antifermion coupling, but also the 3-gluon and 4-gluon vertices, of order e_s and e_s^2 respectively, which have no analogue in an abelian theory like QED. In QED the photon is coupled to all electrically charged particles but itself is neutral. In QCD the gluons are coloured hence self-coupled. This is reflected in the fact that in QED $F_{\mu\nu}$ is linear in the gauge field, so that the term $F^2_{\mu\nu}$ in the lagrangian is a pure kinetic term, while in QCD $F^A_{\mu\nu}$ is quadratic in the gauge field so that in $F^{A2}_{\mu\nu}$ we find cubic and quartic vertices beyond the kinetic term. Also instructive is to consider the case of scalar QED:

$$\mathcal{L} = -\frac{1}{4}F^{\mu\nu}F_{\mu\nu} + (D_\mu\phi)^\dagger(D^\mu\phi) - m^2(\phi^\dagger\phi) \quad (2.32)$$

For $Q = 1$ we have:

$$(D_\mu\phi)^\dagger(D^\mu\phi) = (\partial_\mu\phi)^\dagger(\partial^\mu\phi) + ieA_\mu[(\partial^\mu\phi)^\dagger\phi - \phi^\dagger(\partial^\mu\phi)] + e^2A_\mu A^\mu\phi^\dagger\phi \quad (2.33)$$

We see that for a charged boson in QED, given that the kinetic term for bosons is quadratic in the derivative, there is a two-gauge vertex of order e^2. Thus in QCD the 3-gluon vertex is there because the gluon is coloured and the 4-gluon vertex because the gluon is a boson.

2.6 Quantization of a Gauge Theory

The lagrangian density $\mathcal{L}_{YM}$ in Eq. (2.23) fully describes the theory at the classical level. The formulation of the theory at the quantum level requires that a procedure of quantization, of regularization and, finally, of renormalization is also specified. To start with, the formulation of Feynman rules is not straightforward. A first problem, common to all gauge theories, including the abelian case of QED, can be realized by observing that the free equation of motion for V^A_μ, as obtained from Eqs. ((2.21), (2.23)), is given by

$$[\partial^2 g_{\mu\nu} - \partial_\mu\partial_\nu]V^{A\nu} = 0 \quad (2.34)$$

Normally the propagator of the gauge field should be determined by the inverse of the operator $[\partial^2 g_{\mu\nu} - \partial_\mu\partial_\nu]$ which, however, has no inverse, being a projector over the transverse gauge vector states. This difficulty is removed by fixing a particular gauge. If one chooses a covariant gauge condition $\partial^\mu V^A_\mu = 0$ then a gauge fixing term of the form

$$\Delta\mathcal{L}_{GF} = -\frac{1}{2\lambda}\sum_A |\partial^\mu V^A_\mu|^2 \quad (2.35)$$

has to be added to the lagrangian ($1/\lambda$ acts as a lagrangian multiplier). The free equations of motion are now modified as follows:

$$[\partial^2 g_{\mu\nu} - (1 - 1/\lambda)\partial_\mu \partial_\nu]V^{A\nu} = 0. \tag{2.36}$$

This operator now has an inverse whose Fourier transform is given by:

$$D^{AB}_{\mu\nu}(q) = \frac{i}{q^2 + i\epsilon}\,[-g_{\mu\nu} + (1 - \lambda)\frac{q_\mu q_\nu}{q^2 + i\epsilon}]\,\delta^{AB} \tag{2.37}$$

which is the propagator in this class of gauges. The parameter λ can take any value and it disappears from the final expression of any gauge invariant, physical quantity. Commonly used particular cases are $\lambda = 1$ (Feynman gauge) and $\lambda = 0$ (Landau gauge).

While in an abelian theory the gauge fixing term is all that is needed for a correct quantization, in a non abelian theory the formulation of complete Feynman rules involves a further subtlety. This is formally taken into account by introducing a set of D fictitious ghost fields that must be included as internal lines in closed loops (Faddeev-Popov ghosts [12]). Given that gauge fields connected by a gauge transformation describe the same physics, clearly there are less physical degrees of freedom than gauge field components. Ghosts appear, in the form of a transformation Jacobian in the functional integral, in the process of elimination of the redundant variables associated with fields on the same gauge orbit [13]. The correct ghost contributions can be obtained from an additional term in the lagrangian density. For each choice of the gauge fixing term the ghost langrangian is obtained by considering the effect of an infinitesimal gauge transformation $V'^C_\mu = V^C_\mu - gC_{ABC}\theta^A V^B_\mu - \partial_\mu \theta^C$ on the gauge fixing condition. For $\partial^\mu V^C_\mu = 0$ one obtains:

$$\partial^\mu V'^C_\mu = \partial^\mu V^C_\mu - gC_{ABC}\partial^\mu(\theta^A V^B_\mu) - \partial^2\theta^C \; = \; -[\partial^2\delta_{AC} + gC_{ABC}V^B_\mu \partial^\mu]\theta^A \tag{2.38}$$

where the gauge condition $\partial^\mu V^C_\mu = 0$ has been taken into account in the last step. The ghost lagrangian is then given by:

$$\Delta\mathcal{L}_{Ghost} = \bar{\eta}^C[\partial^2\delta_{AC} + gC_{ABC}V^B_\mu \partial^\mu]\eta^A \tag{2.39}$$

where η^A is the ghost field (one for each index A) which has to be treated as a scalar field except that a factor (-1) for each closed loop has to be included as for fermion fields.

Starting from non covariant gauges one can construct ghost-free gauges. An example, also important in other respects, is provided by the set of "axial" gauges: $n^\mu V^A_\mu = 0$ where n_μ is a fixed reference 4-vector (actually for n_μ spacelike one has an axial gauge proper, for $n^2 = 0$ one speaks of a light-like gauge and for n_μ

timelike one has a Coulomb or temporal gauge). The gauge fixing term is of the form:

$$\Delta\mathcal{L}_{GF} = -\frac{1}{2\lambda}\sum_A |n^\mu V_\mu^A|^2 \tag{2.40}$$

With a procedure that can be found in QED textbooks [14] the corresponding propagator, in Fourier space, is found to be:

$$D_{\mu\nu}^{AB}(q) = \frac{i}{q^2 + i\epsilon}\,[-g_{\mu\nu} + \frac{n_\mu q + n_\nu q_\mu}{(nq)} - \frac{n^2 q_\mu q_\nu}{(nq)^2}]\,\delta^{AB} \tag{2.41}$$

In this case there are no ghost interactions because $n^\mu V_\mu^{\prime A}$, obtained by a gauge transformation from $n^\mu V_\mu^A$, contains no gauge fields, once the gauge condition $n^\mu V_\mu^A = 0$ has been taken into account. Thus the ghosts are decoupled and can be ignored.

The introduction of a suitable regularization method that preserves gauge invariance is essential for the definition and the calculation of loop diagrams and for the renormalization programme of the theory. The method that is by now currently adopted is dimensional regularization [15] which consists in the formulation of the theory in n dimensions. All loop integrals have an analytic expression that is actually valid also for non integer values of n. Writing the results for $n = 4 - \epsilon$ the loops are ultraviolet finite for $\epsilon > 0$ and the divergences reappear in the form of poles at $\epsilon = 0$.

2.7 Spontaneous Symmetry Breaking in Gauge Theories

The gauge symmetry of the SM was difficult to discover because it is well hidden in nature. The only observed gauge boson that is massless is the photon. The gluons are presumed massless but cannot be directly observed because of confinement, and the W and Z weak bosons carry a heavy mass. Indeed a major difficulty in unifying the weak and electromagnetic interactions was the fact that e.m. interactions have infinite range ($m_\gamma = 0$), whilst the weak forces have a very short range, owing to $m_{W,Z} \neq 0$.

The solution of this problem is in the concept of spontaneous symmetry breaking, which was borrowed from statistical mechanics.

Consider a ferromagnet at zero magnetic field in the Landau–Ginzburg approximation. The free energy in terms of the temperature T and the magnetization $\mathbf{M}$ can be written as

$$F(\mathbf{M}, T) \simeq F_0(T) + 1/2\,\mu^2(T)\mathbf{M}^2 + 1/4\,\lambda(T)(\mathbf{M}^2)^2 + \dots . \tag{2.42}$$

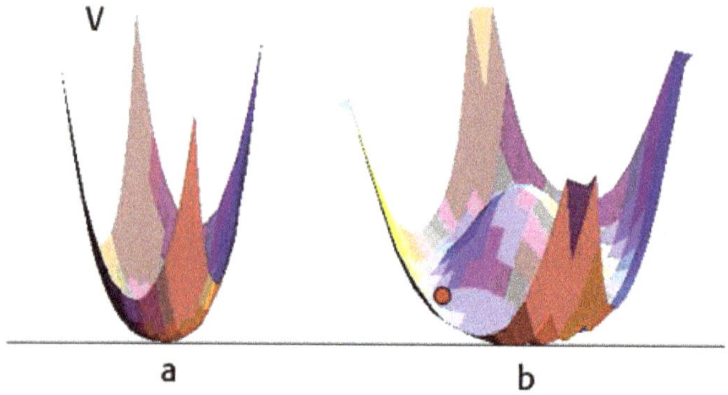

Fig. 2.1 The potential $V = 1/2\ \mu^2\mathbf{M}^2 + 1/4\ \lambda(\mathbf{M}^2)^2$ for positive (**a**) or negative μ^2 (**b**) (for simplicity, **M** is a 2-dimensional vector). The small sphere indicates a possible choice for the direction of **M**

This is an expansion which is valid at small magnetization. The neglect of terms of higher order in $\vec{M}^2$ is the analogue in this context of the renormalizability criterion. Also, $\lambda(T) > 0$ is assumed for stability; F is invariant under rotations, i.e. all directions of **M** in space are equivalent. The minimum condition for F reads

$$\partial F/\partial M_i = 0, \quad [\mu^2(T) + \lambda(T)\mathbf{M}^2]\mathbf{M} = 0\,. \tag{2.43}$$

There are two cases, shown in Fig. 2.1. If $\mu^2 \gtrsim 0$, then the only solution is $\mathbf{M} = 0$, there is no magnetization, and the rotation symmetry is respected. In this case the lowest energy state (in a quantum theory the vacuum) is unique and invariant under rotations. If $\mu^2 < 0$, then another solution appears, which is

$$|\mathbf{M}_0|^2 = -\mu^2/\lambda\,. \tag{2.44}$$

In this case there is a continuous orbit of lowest energy states, all with the same value of $|\mathbf{M}|$ but different orientations. A particular direction chosen by the vector $\mathbf{M}_0$ leads to a breaking of the rotation symmetry.

For a piece of iron we can imagine to bring it to high temperature and let it melt in an external magnetic field **B**. The presence of **B** is an explicit breaking of the rotational symmetry and it induces a non zero magnetization **M** along its direction. Now we lower the temperature while keeping **B** fixed. The critical temperature T_{crit} (Curie temperature) is where $\mu^2(T)$ changes sign: $\mu^2(T_{\text{crit}}) = 0$. For pure iron T_{crit} is below the melting temperature. So at $T = T_{\text{crit}}$ iron is a solid. Below T_{crit} we remove the magnetic field. In a solid the mobility of the magnetic domains is limited and a non vanishing M_0 remains. The form of the free energy becomes rotationally invariant as in Eq. (2.43). But now the system allows a minimum energy state with non vanishing **M** in the direction where **B** was. As a consequence the symmetry is broken by this choice of one particular vacuum state out of a continuum of them.

We now prove the Goldstone theorem [16]. It states that when spontaneous symmetry breaking takes place, there is always a zero-mass mode in the spectrum.

In a classical context this can be proven as follows. Consider a lagrangian

$$\mathcal{L} = \frac{1}{2}|\partial_\mu \phi|^2 - V(\phi). \tag{2.45}$$

The potential $V(\phi)$ can be kept generic at this stage but, in the following, we will be mostly interested in a renormalizable potential of the form (with no more than quartic terms):

$$V(\phi) = -\frac{1}{2}\mu^2 \phi^2 + \frac{1}{4}\lambda \phi^4. \tag{2.46}$$

Here by ϕ we mean a column vector with real components ϕ_i (1=1,2...N) (complex fields can always be decomposed into a pair of real fields), so that, for example, $\phi^2 = \sum_i \phi_i^2$. This particular potential is symmetric under a NxN orthogonal matrix rotation $\phi' = O\phi$, where O is a SO(N) transformation. For simplicity, we have omitted odd powers of ϕ, which means that we assumed an extra discrete symmetry under $\phi \leftrightarrow -\phi$. Note that, for positive μ^2, the mass term in the potential has the "wrong" sign: according to the previous discussion this is the condition for the existence of a non unique lowest energy state. More in general, we only assume here that the potential is symmetric under the infinitesimal transformations

$$\phi \to \phi' = \phi + \delta\phi, \quad \delta\phi_i = i\delta\theta^A t^A_{ij}\phi_j\,. \tag{2.47}$$

where $\delta\theta^A$ are infinitesimal parameters and t^A_{ij} are the matrices that represent the symmetry group on the representation of the fields ϕ_i (a sum over A is understood). The minimum condition on V that identifies the equilibrium position (or the vacuum state in quantum field theory language) is

$$(\partial V/\partial\phi_i)(\phi_i = \phi_i^0) = 0\,. \tag{2.48}$$

The symmetry of V implies that

$$\delta V = (\partial V/\partial\phi_i)\delta\phi_i = i\delta\theta^A(\partial V/\partial\phi_i)t^A_{ij}\phi_j = 0\,. \tag{2.49}$$

By taking a second derivative at the minimum $\phi_i = \phi_i^0$, given by the previous equation, we obtain that, for each A:

$$\frac{\partial^2 V}{\partial\phi_k\partial\phi_i}(\phi_i = \phi_i^0)t^A_{ij}\phi_j^0 + \frac{\partial V}{\partial\phi_i}(\phi_i = \phi_i^0)t^A_{ik} = 0\,. \tag{2.50}$$

The second term vanishes owing to the minimum condition, Eq. (2.48). We then find

$$\frac{\partial^2 V}{\partial\phi_k\partial\phi_i}(\phi_i = \phi_i^0)t^A_{ij}\phi_j^0 = 0\,. \tag{2.51}$$

The second derivatives $M^2_{ki} = (\partial^2 V/\partial\phi_k\partial\phi_i)(\phi_i = \phi^0_i)$ define the squared mass matrix. Thus the above equation in matrix notation can be written as

$$M^2 t^A \phi^0 = 0 \,. \tag{2.52}$$

In the case of no spontaneous symmetry breaking the ground state is unique, all symmetry transformations leave it invariant, so that, for all A, $t^A\phi^0 = 0$. On the contrary, if, for some values of A, the vectors $(t^A\phi^0)$ are non-vanishing, i.e. there is some generator that shifts the ground state into some other state with the same energy (hence the vacuum is not unique), then each $t^A\phi^0 \neq 0$ is an eigenstate of the squared mass matrix with zero eigenvalue. Therefore, a massless mode is associated with each broken generator. The charges of the massless modes (their quantum numbers in quantum language) differ from those of the vacuum (usually taken as all zero) by the values of the t^A charges: one says that the massless modes have the same quantum numbers of the broken generators, i.e. those that do not annihilate the vacuum.

The previous proof of the Goldstone theorem has been given in the classical case. In the quantum case the classical potential corresponds to tree level approximation of the quantum potential. Higher order diagrams with loops introduce quantum corrections. The functional integral formulation of quantum field theory [13, 17] is the most appropriate framework to define and compute, in a loop expansion, the quantum potential which specifies, exactly as described above, the vacuum properties of the quantum theory. If the theory is weakly coupled, e.g. if λ is small, the tree level expression for the potential is not too far from the truth, and the classical situation is a good approximation. We shall see that this is the situation that occurs in the electroweak theory if the Higgs is moderately light (see Chap. 3, Sect. 3.13.1).

We note that for a quantum system with a finite number of degrees of freedom, for example one described by the Schrödinger equation, there are no degenerate vacua: the vacuum is always unique. For example, in the one dimensional Schrödinger problem with a potential:

$$V(x) = -\mu^2/2\, x^2 + \lambda\, x^4/4 \,, \tag{2.53}$$

there are two degenerate minima at $x = \pm x_0 = \sqrt{(\mu^2/\lambda)}$ which we denote by $|+\rangle$ and $|-\rangle$. But the potential is not diagonal in this basis: the off diagonal matrix elements:

$$\langle +|V|-\rangle = \langle -|V|+\rangle \sim \exp(-khd) = \delta \tag{2.54}$$

are different from zero due to the non vanishing amplitude for a tunnel effect between the two vacua, proportional to the exponential of the product of the distance d between the vacua and the height h of the barrier with k a constant (see Fig. 2.2). After diagonalization the eigenvectors are $(|+\rangle + |-\rangle)/\sqrt{2}$ and $(|+\rangle - |-\rangle)/\sqrt{2}$,

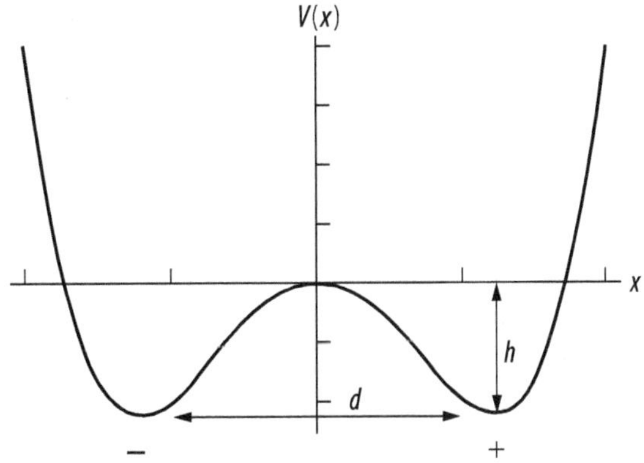

Fig. 2.2 A Schrödinger potential $V(x)$ analogous to the Higgs potential

with different energies (the difference being proportional to δ). Suppose now that you have a sum of n equal terms in the potential, $V = \sum_i V(x_i)$. Then the transition amplitude would be proportional to δ^n and would vanish for infinite n: the probability that all degrees of freedom together jump over the barrier vanishes. In this example there is a discrete number of minimum points. The case of a continuum of minima is obtained, always in the Schrödinger context, if we take

$$V = 1/2\,\mu^2\mathbf{r}^2 + 1/4\,\lambda(\mathbf{r}^2)^2\,, \tag{2.55}$$

with $\mathbf{r} = (x, y, z)$. Also in this case the ground state is unique: it is given by a state with total orbital angular momentum zero, an s-wave state, whose wave function only depends on $|\mathbf{r}|$, independent of all angles. This is a superposition of all directions with the same weight, analogous to what happened in the discrete case. But again, if we replace a single vector $\mathbf{r}$, with a vector field $\mathbf{M}(x)$, that is a different vector at each point in space, the amplitude to go from a minimum state in one direction to another in a different direction goes to zero in the limit of infinite volume. In simple words, the vectors at all points in space have a vanishing small amplitude to make a common rotation, all together at the same time. In the infinite volume limit all vacua along each direction have the same energy and spontaneous symmetry breaking can occur.

The massless Goldstone bosons correspond to a long range force. Unless the massless particles are confined, as for the gluons in QCD, these long range forces would be easily detectable. Thus, in the construction of the EW theory we cannot accept massless physical scalar bosons. Fortunately, when spontaneous symmetry breaking takes place in a gauge theory, the massless Goldstone modes exist, but they are unphysical and disappear from the spectrum. Each of them becomes, in fact, the third helicity state of a gauge boson that takes mass. This is the Higgs mechanism (it should be called Englert-Brout-Higgs mechanism [18], because an equal merit should be credited to the simultaneous paper by Englert and Brout). Consider, for example, the simplest Higgs model described by the lagrangian

$$\mathcal{L} = -\frac{1}{4}\,F^2_{\mu\nu} + |(\partial_\mu + ieA_\mu Q)\phi|^2 + \mu^2\phi^*\phi - \frac{\lambda}{2}(\phi^*\phi)^2\,. \tag{2.56}$$

Note the 'wrong' sign in front of the mass term for the scalar field ϕ, which is necessary for the spontaneous symmetry breaking to take place. The above lagrangian is invariant under the $U(1)$ gauge symmetry

$$A_\mu \to A'_\mu = A_\mu - \partial_\mu \theta(x), \quad \phi \to \phi' = \exp[ieQ\theta(x)]\,\phi. \tag{2.57}$$

For the U(1) charge Q we take $Q\phi = -\phi$, like in QED, where the particle is e^-. Let $\phi^0 = v \neq 0$, with v real, be the ground state that minimizes the potential and induces the spontaneous symmetry breaking. In our case v is given by $v^2 = \mu^2/\lambda$. Making use of gauge invariance, we can do the change of variables

$$\begin{aligned} \phi(x) &\to [v + h(x)/\sqrt{2}]\exp[-i\zeta(x)/v\sqrt{2}]\,, \\ A_\mu(x) &\to A_\mu - \partial_\mu \zeta(x)/ev\sqrt{2}. \end{aligned} \tag{2.58}$$

Then $h = 0$ is the position of the minimum, and the lagrangian becomes

$$\mathcal{L} = -\frac{1}{4}F_{\mu\nu}^2 + e^2v^2A_\mu^2 + \frac{1}{2}e^2h^2A_\mu^2 + \sqrt{2}e^2hvA_\mu^2 + \mathcal{L}(h)\,. \tag{2.59}$$

The field $\zeta(x)$ is the would-be Goldstone boson, as can be seen by considering only the ϕ terms in the lagrangian, i.e. setting $A_\mu = 0$ in Eq. (2.56). In fact in this limit the kinetic term $\partial_\mu\zeta\partial^\mu\zeta$ remains but with no ζ^2 mass term. Instead, in the gauge case of Eq. (2.56), after changing variables in the lagrangian, the field $\zeta(x)$ completely disappears (not even the kinetic term remains), whilst the mass term $e^2v^2A_\mu^2$ for A_μ is now present: the gauge boson mass is $M = \sqrt{2}ev$. The field h describes the massive Higgs particle. Leaving a constant term aside, the last term in Eq. (2.59) is given by:

$$\mathcal{L}(h) = \frac{1}{2}\partial_\mu h\partial^\mu h - h^2\mu^2 + \ldots. \tag{2.60}$$

where the dots stand for cubic and quartic terms in h. We see that the h mass term has the "right" sign, due to the combination of the quadratic terms in h that, after the shift, arise from the quadratic and quartic terms in ϕ. The h mass is given by $m_h^2 = 2\mu^2$.

The Higgs mechanism is realized in well-known physical situations. It was actually discovered in condensed matter physics by Anderson [19]. For a superconductor in the Landau–Ginzburg approximation the free energy can be written as

$$F = F_0 + \frac{1}{2}\mathbf{B}^2 + |(\nabla - 2ie\mathbf{A})\phi|^2/4m - \alpha|\phi|^2 + \beta|\phi|^4\,. \tag{2.61}$$

Here $\mathbf{B}$ is the magnetic field, $|\phi|^2$ is the Cooper pair (e^-e^-) density, $2e$ and $2m$ are the charge and mass of the Cooper pair. The 'wrong' sign of α leads to $\phi \neq 0$ at the minimum. This is precisely the non-relativistic analogue of the Higgs model of the previous example. The Higgs mechanism implies the absence of propagation of massless phonons (states with dispersion relation $\omega = kv$ with constant v). Also the mass term for $\mathbf{A}$ is manifested by the exponential decrease of $\mathbf{B}$ inside the superconductor (Meissner effect).

2.8 Quantization of Spontaneously Broken Gauge Theories: R_ξ Gauges

We have discussed in Sect. 2.6 the problems arising in the quantization of a gauge theory and in the formulation of the correct Feynman rules (gauge fixing terms, ghosts etc.). Here we give a concise account of the corresponding results for spontaneously broken gauge theories. In particular we describe the R_ξ gauge formalism [13, 17, 20]: in this formalism the interplay of transverse and longitudinal gauge boson degrees of freedom is made explicit and their combination leads to the cancellation from physical quantities of the gauge parameter ξ. We work out in detail an abelian example that later will be easy to generalize to the non abelian case.

We restart from the abelian model of Eq. (2.56) (with $Q = -1$). In the treatment presented there the would be Goldstone boson $\zeta(x)$ was completely eliminated from the lagrangian by a non linear field transformation formally identical to a gauge transformation corresponding to the $U(1)$ symmetry of the lagrangian. In that description, in the new variables we eventually obtain a theory with only physical fields: a massive gauge boson A_μ with mass $M = \sqrt{2}ev$ and a Higgs particle h with mass $m_h = \sqrt{2}\mu$. This is called a "unitary" gauge, because only physical fields appear. But for a massive gauge boson the propagator:

$$iD_{\mu\nu}(k) = -i\frac{g_{\mu\nu} - k_\mu k_\nu/M^2}{k^2 - M^2 + i\epsilon}\,, \tag{2.62}$$

has a bad ultraviolet behaviour due to the second term in the numerator. This choice does not prove to be the most convenient for a discussion of the ultraviolet behaviour of the theory. Alternatively one can go to an alternative formulation where the would be Goldstone boson remains in the lagrangian but the complication of keeping spurious degrees of freedom is compensated by having all propagators with good ultraviolet behaviour ("renormalizable" gauges). To this end we replace the non linear transformation for ϕ in Eq. (2.58) with its linear equivalent (after all perturbation theory deals with the small oscillations around the minimum):

$$\phi(x) \to [v + h(x)/\sqrt{2}]\exp[-i\zeta(x)/v\sqrt{2}] \sim [v + h(x)/\sqrt{2} - i\zeta(x)/\sqrt{2}]\,. \tag{2.63}$$

Here we have only applied a shift by the amount v and separated the real and imaginary components of the resulting field with vanishing vacuum expectation value. If we leave A_μ as it is and simply replace the linearized expression for ϕ, we obtain the following quadratic terms (those important for propagators):

$$\mathcal{L}_{\text{quad}} = -\frac{1}{4}\sum_A F^A_{\mu\nu}F^{A\mu\nu} + \frac{1}{2}M^2 A_\mu A^\mu + \\ + \frac{1}{2}(\partial_\mu\zeta)^2 + MA_\mu\partial^\mu\zeta + \frac{1}{2}(\partial_\mu h)^2 - h^2\mu^2 \tag{2.64}$$

The mixing term between A_μ and $\partial_\mu\zeta$ does not allow to directly write diagonal mass matrices. But this mixing term can be eliminated by an appropriate modification of the covariant gauge fixing term given in Eq. (2.35) for the unbroken theory. We now take:

$$\Delta\mathcal{L}_{GF} = -\frac{1}{2\xi}(\partial^\mu A_\mu - \xi M\zeta)^2 \,. \tag{2.65}$$

By adding $\Delta\mathcal{L}_{GF}$ to the quadratic terms in Eq. (2.64) the mixing term cancels (apart from a total derivative that can be omitted) and we have:

$$\mathcal{L}_{\text{quad}} = -\frac{1}{4}\sum_A F^A_{\mu\nu}F^{A\mu\nu} + \frac{1}{2}M^2 A_\mu A^\mu - \frac{1}{2\xi}(\partial^\mu A_\mu)^2 + \\ + \frac{1}{2}(\partial_\mu\zeta)^2 - \frac{\xi}{2}M^2\zeta^2 + \frac{1}{2}(\partial_\mu h)^2 - h^2\mu^2 \tag{2.66}$$

We see that the ζ field appears with a mass $\sqrt{\xi}M$ and its propagator is:

$$iD_\zeta = \frac{i}{k^2 - \xi M^2 + i\epsilon}. \tag{2.67}$$

The propagators of the Higgs field h and of gauge field A_μ are:

$$iD_h = \frac{i}{k^2 - 2\mu^2 + i\epsilon}\,, \tag{2.68}$$

$$iD_{\mu\nu}(k) = \frac{-i}{k^2 - M^2 + i\epsilon}(g_{\mu\nu} - (1-\xi)\frac{k_\mu k_\nu}{k^2 - \xi M^2}) \,. \tag{2.69}$$

As anticipated, all propagators have a good behaviour at large k^2. This class of gauges are called "R_ξ gauges" [20]. Note that for $\xi = 1$ we have a sort of generalization of the Feynman gauge with a Goldstone of mass M and a gauge propagator:

$$iD_{\mu\nu}(k) = \frac{-ig_{\mu\nu}}{k^2 - M^2 + i\epsilon} \,. \tag{2.70}$$

Also for $\xi \to \infty$ the unitary gauge description is recovered in that the Goldstone propagator vanishes and the gauge propagator reproduces that of the unitary gauge in Eq. (2.62). All ξ dependence, including the unphysical singularities of the ζ and A_μ propagators at $k^2 = \xi M^2$, present in individual Feynman diagrams, must cancel in the sum of all contributions to any physical quantity.

An additional complication is that a Faddeev-Popov ghost is also present in R_ξ gauges (while it is absent in an unbroken abelian gauge theory). In fact under an infinitesimal gauge transformation with parameter $\theta(x)$:

$$
\begin{aligned}
A_\mu &\to A_\mu - \partial_\mu \theta \\
\phi &\to (1 - ie\theta)[v + h(x)/\sqrt{2} - i\zeta(x)/\sqrt{2}] ,
\end{aligned}
\tag{2.71}
$$

so that:

$$
\delta A_\mu = -\partial_\mu \theta, \quad \delta h = -e\zeta\theta, \quad \delta\zeta = e\theta\sqrt{2}(v + h/\sqrt{2}) . \tag{2.72}
$$

The gauge fixing condition $\partial_\mu A^\mu - \xi M \zeta = 0$ undergoes the variation:

$$
\partial_\mu A^\mu - \xi M\zeta \to \partial_\mu A^\mu - \xi M\zeta - [\partial^2 + \xi M^2(1 + h/v\sqrt{2})]\theta , \tag{2.73}
$$

where we used $M = \sqrt{2}ev$. From this, recalling the discussion in Sect. 2.6, we see that the ghost is not coupled to the gauge boson (as usual for an abelian gauge theory) but has a coupling to the Higgs field h. The ghost lagrangian is:

$$
\Delta \mathcal{L}_{Ghost} = \bar{\eta}[\partial^2 + \xi M^2(1 + h/v\sqrt{2})]\eta . \tag{2.74}
$$

The ghost mass is seen to be $m_{gh} = \sqrt{\xi} M$ and its propagator is:

$$
iD_{gh} = \frac{i}{k^2 - \xi M^2 + i\epsilon} . \tag{2.75}
$$

The detailed Feynman rules follow from all the basic vertices involving the gauge boson, the Higgs, the would be Goldstone boson and the ghost and can be easily derived, with some algebra, from the total lagrangian including the gauge fixing and ghost additions. The generalization to the non abelian case is in principle straightforward, with some formal complications involving the projectors over the space of the would be Goldstone bosons and over the orthogonal space of the Higgs particles. But for each gauge boson that takes mass M_a we still have a corresponding would be Goldstone boson and a ghost with mass $\sqrt{\xi} M_a$. The Feynman diagrams, both for the abelian and the non abelian case, are listed explicitly, for example, in the Cheng and Li textbook in ref.[17].

We conclude that the renormalizability of non abelian gauge theories, also in presence of spontaneous symmetry breaking, was proven in the fundamental works of t'Hooft and Veltman [21] and discussed in detail in [22].

References

1. H. Fritzsch, M. Gell-Mann and H. Leutwyler, Phys. Lett. B **47**, 365 (1973)
2. D. Gross and F. Wilczek, Phys. Rev. Lett. **30**, 1343 (1973), Phys. Rev. D **8**, 3633 (1973); H.D. Politzer, Phys. Rev. Lett. **30**, 1346 (1973)
3. S. Weinberg, Phys. Rev. Lett. **31**, 494 (1973)
4. S.L. Glashow, Nucl. Phys. **22**, 579 (1961)
5. S. Weinberg, Phys. Rev. Lett. **19**, 1264 (1967)
6. A. Salam, in *Elementary Particle Theory*, ed. N. Svartholm (Almquist and Wiksells, Stockholm, 1969), p. 367
7. Particle Data Group, J. Phys. G **33**, 1 (2006)
8. N. Cabibbo, Phys. Rev. Lett. **10**, 531 (1963)
9. S.L. Glashow, J. Iliopoulos and L. Maiani, Phys. Rev. **96**, 1285 (1970)
10. M. Kobayashi and T. Maskawa, Prog. Theor. Phys. **49**, 652 (1973)
11. C.N. Yang and R. Mills, Phys. Rev. **96**, 191 (1954)
12. R. Feynman, Acta Phys. Pol. **24**, 697 (1963); B. De Witt, Phys. Rev. **162**, 1195, 1239 (1967); L.D. Faddeev and V.N. Popov, Phys. Lett. B **25**, 29 (1967)
13. E.S. Abers and B.W. Lee, Phys. Rep. **9**, 1 (1973)
14. J.D. Bjorken and S. Drell, *Relativistic Quantum Mechanics/Fields*, Vols. I, II, McGraw-Hill, New York, (1965)
15. G.'t Hooft and M. Veltman, Nucl. Phys. B **44**, 189 (1972); C.G. Bollini and J.J. Giambiagi, Phys. Lett. B **40**, 566 (1972); J.F. Ashmore, Nuovo Cim. Lett. **4**, 289 (1972); G.M. Cicuta and E. Montaldi, Nuovo Cim. Lett. **4**, 329 (1972)
16. J. Goldstone, Nuovo Cim. **19**, 154 (1961)
17. C. Itzykson and J. Zuber, *Introduction to Quantum Field Theory*, McGraw-Hill, New York, (1980); T.P. Cheng and L.F. Li, *Gauge Theory of Elementary Particle Physics*, Oxford Univ. Press, New York (1984); M.E. Peskin and D.V. Schroeder, *An Introduction to Quantum Field Theory*, Perseus Books, Cambridge, Mass. (1995); S. Weinberg, *The Quantum Theory of Fields*, Vols. I, II, Cambridge Univ. Press, Cambridge, Mass. (1996); A. Zee, *Quantum Field Theory in a Nutshell*, Princeton Univ. Press, Princeton, N.J. (2003); C.M. Becchi and G. Ridolfi, *An Introduction to Relativistic Processes and the Standard Model of Electroweak Interactions*, Springer, (2006)
18. F. Englert, R. Brout, Phys. Rev. Lett. **13**, 321 (1964); P.W. Higgs, Phys. Lett. **12**, 132 (1964)
19. P.W. Anderson, Phys. Rev. **112**, 1900 (1958); Phys. Rev. **130**, 439 (1963)
20. K. Fujikawa, B.W. Lee and A. Sanda, Phys. Rev. D **6**, 2923 (1972); Y.P. Yao, Phys. Rev. D **7**, 1647 (1973)
21. M. Veltman, Nucl. Phys. B **21**, 288 (1970); G.'t Hooft, Nucl. Phys. B **33**, 173 (1971); **35**, 167 (1971)
22. B.W. Lee and J. Zinn-Justin, Phys. Rev. D **5**, 3121; 3137 (1972); **7**, 1049 (1973)
23. I. J. R. Aitchison and A. J. G. Hey, "Gauge theories in particle physics: A practical introduction. Vol. 1: From relativistic quantum mechanics to QED,"
24. M. Maggiore, "A Modern introduction to quantum field theory," Oxford University Press, 2005. (Oxford Series in Physics, 12. ISBN 0 19 852073 5)
25. M. Tanabashi *et al.* [Particle Data Group], Phys. Rev. D **98** (2018) no.3, 030001.
26. http://pdg.web.cern.ch/pdg/
27. G. Hamel de Monchenault, "Electroweak measurements at the LHC," Ann. Rev. Nucl. Part. Sci. **67** (2017) 19.
28. G. Aad *et al.* [ATLAS Collaboration], Phys. Lett. B **716** (2012) 1 [arXiv:1207.7214 [hep-ex]].
29. S. Chatrchyan *et al.* [CMS Collaboration], Phys. Lett. B **716** (2012) 30 [arXiv:1207.7235 [hep-ex]].
30. T. Plehn, "Lectures on LHC Physics," Lect. Notes Phys. **886** (2015).
31. https://twiki.cern.ch/twiki/bin/view/LHCPhysics/LHCHXSWG

32. J. R. Espinosa, "Cosmological implications of Higgs near-criticality," Phil. Trans. Roy. Soc. Lond. A **376** (2018) no.2114, 20170118.
33. G. F. Giudice, "The Dawn of the Post-Naturalness Era," arXiv:1710.07663 [physics.hist-ph].
34. H. E. Haber and L. Stephenson Haskins, "Supersymmetric Theory and Models," arXiv:1712.05926 [hep-ph].
35. Csáki, Csaba, S. Lombardo and O. Telem, "TASI Lectures on Non-supersymmetric BSM Models," arXiv:1811.04279 [hep-ph].
36. Ansgar Denner and Stefan Dittmaier, "Electroweak Radiative Corrections for Collider Physics" arXiv:1912.06823[hep-ph].
37. Freeman Dyson, "Leaping into the Grand Unknown", New York Rev. Books **56** (2009)
38. J. M. Henn and J. C. Plefka, "Scattering Amplitudes in Gauge Theories," Lect. Notes Phys. **883** (2014)
39. T. Becher, A. Broggio and A. Ferroglia, "Introduction to Soft-Collinear Effective Theory," Lect. Notes Phys. **896** (2015) [arXiv:1410.1892 [hep-ph]].
40. S. Marzani, G. Soyez and M. Spannowsky, "Looking inside jets: an introduction to jet substructure and boosted-object phenomenology," Lect. Notes Phys. **958** (2019) [arXiv:1901.10342 [hep-ph]].
41. G. P. Salam, "The strong coupling: a theoretical perspective," arXiv:1712.05165 [hep-ph].
42. S. Aoki *et al.* [Flavour Lattice Averaging Group], "FLAG Review 2019," arXiv:1902.08191 [hep-lat].

3

Standard Electroweak Theory: Structure and Precision Tests

Guido Altarelli and Stefano Forte

3.1 Introduction

In this chapter,[1] we summarize the structure of the standard EW theory [1] and specify the couplings of the intermediate vector bosons $W^{\pm}$, Z and of the Higgs particle with the fermions and among themselves, as dictated by the gauge symmetry plus the observed matter content and the requirement of renormalizability. We discuss the realization of spontaneous symmetry breaking and of the Higgs mechanism [2]. We then review the phenomenological implications of the EW theory for collider physics (that is we leave aside the classic low energy processes that are well described by the "old" weak interaction theory (see, for example, [3])). Moreover, a detailed description of experiments for precision tests of the EW theory is presented in Chap. 6.

For this discussion we split the lagrangian into two parts by separating the terms with the Higgs field:

$$\mathcal{L} = \mathcal{L}_{\text{gauge}} + \mathcal{L}_{\text{Higgs}} \,. \tag{3.1}$$

Both terms are written down as prescribed by the $SU(2) \otimes U(1)$ gauge symmetry and renormalizability, but the Higgs vacuum expectation value (VEV) induces the

The author "G. Altarelli" is deceased at the time of publication.

[1] See Chap. 2 for a general introduction to Chap. 2–4 with updated references.

G. Altarelli
University of Rome 3, Rome, Italy

S. Forte (✉)
Dipartimento di Fisica, Università di Milano, Milano, Italy

spontaneous symmetry breaking responsible for the non vanishing vector boson and fermion masses.

3.2 The Gauge Sector

We start by specifying $\mathcal{L}_{\rm gauge}$, which involves only gauge bosons and fermions, according to the general formalism of gauge theories discussed in Chap. 2:

$$\mathcal{L}_{\rm gauge} = -\frac{1}{4}\sum_{A=1}^{3} F_{\mu\nu}^{A} F^{A\mu\nu} - \frac{1}{4} B_{\mu\nu} B^{\mu\nu} + \bar{\psi}_L i\gamma^\mu D_\mu \psi_L + \bar{\psi}_R i\gamma^\mu D_\mu \psi_R \,. \tag{3.2}$$

This is the Yang–Mills lagrangian for the gauge group $SU(2) \otimes U(1)$ with fermion matter fields. Here

$$B_{\mu\nu} = \partial_\mu B_\nu - \partial_\nu B_\mu \quad \text{and} \quad F_{\mu\nu}^{A} = \partial_\mu W_\nu^A - \partial_\nu W_\mu^A - g\epsilon_{ABC}\, W_\mu^B W_\nu^C \tag{3.3}$$

are the gauge antisymmetric tensors constructed out of the gauge field B_μ associated with $U(1)$, and W_μ^A corresponding to the three $SU(2)$ generators; ϵ_{ABC} are the group structure constants (see Eqs. (3.8, 3.9)) which, for $SU(2)$, coincide with the totally antisymmetric Levi-Civita tensor (recall the familiar angular momentum commutators). The normalization of the $SU(2)$ gauge coupling g is therefore specified by Eq. (3.3).

The fermion fields are described through their left-hand and right-hand components:

$$\psi_{L,R} = [(1 \mp \gamma_5)/2]\psi, \quad \bar{\psi}_{L,R} = \bar{\psi}[(1 \pm \gamma_5)/2]\,, \tag{3.4}$$

with γ_5 and other Dirac matrices defined as in the book by Bjorken–Drell [4]. In particular, $\gamma_5^2 = 1$, $\gamma_5^\dagger = \gamma_5$. Note that, as given in Eq. (3.4),

$$\bar{\psi}_L = \psi_L^\dagger \gamma_0 = \psi^\dagger[(1-\gamma_5)/2]\gamma_0 = \bar{\psi}\gamma_0[(1-\gamma_5)/2]\gamma_0 = \bar{\psi}[(1+\gamma_5)/2]\,.$$

The matrices $P_\pm = (1 \pm \gamma_5)/2$ are projectors. They satisfy the relations $P_\pm P_\pm = P_\pm$, $P_\pm P_\mp = 0$, $P_+ + P_- = 1$.

The sixteen linearly independent Dirac matrices can be divided into γ_5-even and γ_5-odd according to whether they commute or anticommute with γ_5. For the γ_5-even, we have

$$\bar{\psi}\Gamma_E\psi = \bar{\psi}_L\Gamma_E\psi_R + \bar{\psi}_R\Gamma_E\psi_L \qquad (\Gamma_E \equiv 1, i\gamma_5, \sigma_{\mu\nu})\,, \tag{3.5}$$

whilst for the γ_5-odd,

$$\bar{\psi}\Gamma_O\psi = \bar{\psi}_L\Gamma_O\psi_L + \bar{\psi}_R\Gamma_O\psi_R \qquad (\Gamma_O \equiv \gamma_\mu, \gamma_\mu\gamma_5)\,. \tag{3.6}$$

The standard EW theory is a chiral theory, in the sense that ψ_L and ψ_R behave differently under the gauge group (so that parity and charge conjugation non conservation are made possible in principle). Thus, mass terms for fermions (of the form $\bar{\psi}_L\psi_R$ + h.c.) are forbidden in the symmetric limit. In particular, in the Minimal Standard Model (MSM: i.e. the model that only includes all observed particles plus a single Higgs doublet), all ψ_L are $SU(2)$ doublets while all ψ_R are singlets. But for the moment, by $\psi_{L,R}$ we mean column vectors, including all fermion types in the theory that span generic reducible representations of $SU(2) \otimes U(1)$.

In the absence of mass terms, there are only vector and axial vector interactions in the lagrangian and those have the property of not mixing ψ_L and ψ_R. Fermion masses will be introduced, together with $W^\pm$ and Z masses, by the mechanism of symmetry breaking. The covariant derivatives $D_\mu\psi_{L,R}$ are explicitly given by

$$D_\mu\psi_{L,R} = \left[\partial_\mu + ig\sum_{A=1}^{3} t^A_{L,R}W^A_\mu + ig'\frac{1}{2}Y_{L,R}B_\mu\right]\psi_{L,R}\,, \tag{3.7}$$

where $t^A_{L,R}$ and $1/2Y_{L,R}$ are the $SU(2)$ and $U(1)$ generators, respectively, in the reducible representations $\psi_{L,R}$. The commutation relations of the $SU(2)$ generators are given by

$$[t^A_L, t^B_L] = i\,\epsilon_{ABC}t^C_L \quad \text{and} \quad [t^A_R, t^B_R] = i\epsilon_{ABC}t^C_R\,. \tag{3.8}$$

We use the normalization (3.8) [in the fundamental representation of $SU(2)$]. The electric charge generator Q (in units of e, the positron charge) is given by

$$Q = t^3_L + 1/2\,Y_L = t^3_R + 1/2\,Y_R\,. \tag{3.9}$$

Note that the normalization of the $U(1)$ gauge coupling g' in (3.7) is now specified as a consequence of (3.9). Note that $t^i_R\psi_R = 0$, given that, for all known quark and leptons, ψ_R is a singlet. But in the following, we keep $t^i_R\psi_R$ for generality, in case 1 day a non singlet right-handed fermion is discovered.

3.3 Couplings of Gauge Bosons to Fermions

All fermion couplings of the gauge bosons can be derived directly from Eqs. (3.2) and (3.7). The charged W_μ fields are described by $W^{1,2}_\mu$, while the photon A_μ and weak neutral gauge boson Z_μ are obtained from combinations of W^3_μ and B_μ. The charged-current (CC) couplings are the simplest. One starts from the $W_\mu^{1,2}$ terms in Eqs. (3.2) and (3.7) which can be written as:

$$\begin{aligned} g(t^1 W_\mu^1 + t^2 W_\mu^2) &= g\left\{[(t^1 + it^2)/\sqrt{2}](W_\mu^1 - iW_\mu^2)/\sqrt{2}] + \text{h.c.}\right\} \\ &= g\left\{[(t^+ W_\mu^-)/\sqrt{2}] + \text{h.c.}\right\}, \end{aligned} \tag{3.10}$$

where $t^\pm = t^1 \pm it^2$ and $W^\pm = (W^1 \pm iW^2)/\sqrt{2}$. By applying this generic relation to L and R fermions separately, we obtain the vertex

$$V_{\bar{\psi}\psi W} = g\bar{\psi}\gamma_\mu\left[(t_L^+/\sqrt{2})(1-\gamma_5)/2 + (t_R^+/\sqrt{2})(1+\gamma_5)/2\right]\psi W_\mu^- + \text{h.c.} \tag{3.11}$$

Given that $t_R = 0$ for all fermions in the SM, the charged current is pure $V - A$. In the neutral-current (NC) sector, the photon A_μ and the mediator Z_μ of the weak NC are orthogonal and normalized linear combinations of B_μ and W_μ^3:

$$\begin{aligned} A_\mu &= \cos\theta_W B_\mu + \sin\theta_W W_\mu^3\,, \\ Z_\mu &= -\sin\theta_W B_\mu + \cos\theta_W W_\mu^3\,. \end{aligned} \tag{3.12}$$

and conversely:

$$\begin{aligned} W_\mu^3 &= \sin\theta_W A_\mu + \cos\theta_W Z_\mu\,, \\ B_\mu &= \cos\theta_W A_\mu - \sin\theta_W Z_\mu\,. \end{aligned} \tag{3.13}$$

Equations (3.12) define the weak mixing angle θ_W. We can rewrite the W_μ^3 and B_μ terms in Eqs. (3.2) and (3.7) as follows:

$$\begin{aligned} gt^3 W_\mu^3 + g'Y/2B_\mu &= [gt^3\sin\theta_W + g'(Q - t^3)\cos\theta_W]A_\mu\ + \\ &+\ [gt^3\cos\theta_W - g'(Q - t^3)\sin\theta_W]Z_\mu\,, \end{aligned} \tag{3.14}$$

where Eq. (3.9) for the charge matrix Q was also used. The photon is characterized by equal couplings to left and right fermions with a strength equal to the electric charge. Thus we immediately obtain

$$g\ \sin\theta_W = g'\cos\theta_W = e\,, \tag{3.15}$$

or equivalently,

$$\text{tg}\,\theta_W = g'/g \tag{3.16}$$

Once θ_W has been fixed by the photon couplings, it is a simple matter of algebra to derive the Z couplings, with the result

$$V_{\bar{\psi}\psi Z} = \frac{g}{2\cos\theta_W}\bar{\psi}\gamma_\mu[t_L^3(1-\gamma_5) + t_R^3(1+\gamma_5) - 2Q\sin^2\theta_W]\psi Z^\mu\,, \tag{3.17}$$

where $V_{\bar{\psi}\psi Z}$ is a notation for the vertex. Once again, recall that in the MSM, $t_R^3 = 0$ and $t_L^3 = \pm 1/2$.

In order to derive the effective four-fermion interactions that are equivalent, at low energies, to the CC and NC couplings given in Eqs. (3.11) and (3.17), we anticipate that large masses, as experimentally observed, are provided for $W^\pm$ and Z by $\mathcal{L}_{\text{Higgs}}$. For left–left CC couplings, when the momentum transfer squared can be neglected, with respect to m_W^2, in the propagator of Born diagrams with single W exchange (see, for example, the diagram for μ decay in Fig. 3.1, from Eq. (3.11) we can write

$$\mathcal{L}_{\text{eff}}^{\text{CC}} \simeq \frac{g^2}{8m_W^2}[\bar{\psi}\gamma_\mu(1-\gamma_5)t_L^+\psi][\bar{\psi}\gamma^\mu(1-\gamma_5)t_L^-\psi]\,. \tag{3.18}$$

By specializing further in the case of doublet fields such as $\nu_e - e^-$ or $\nu_\mu - \mu^-$, we obtain the tree-level relation of g with the Fermi coupling constant G_F precisely measured from μ decay (see Chap. 2, Eqs. (2), (3)):

$$G_F/\sqrt{2} = g^2/8m_W^2\,. \tag{3.19}$$

By recalling that $g\ \sin\theta_W = e$, we can also cast this relation in the form

$$m_W = \mu_{\text{Born}}/\sin\theta_W\,, \tag{3.20}$$

with

$$\mu_{\text{Born}} = (\pi\alpha/\sqrt{2}G_F)^{1/2} \simeq 37.2802\,\text{GeV}\,, \tag{3.21}$$

where α is the fine-structure constant of QED ($\alpha \equiv e^2/4\pi = 1/137.036$).

In the same way, for neutral currents we obtain in Born approximation from Eq. (3.17) the effective four-fermion interaction given by

$$\mathcal{L}_{\text{eff}}^{\text{NC}} \simeq \sqrt{2}\,G_F\rho_0\bar{\psi}\gamma_\mu[\ldots]\psi\bar{\psi}\gamma^\mu[\ldots]\psi\,, \tag{3.22}$$

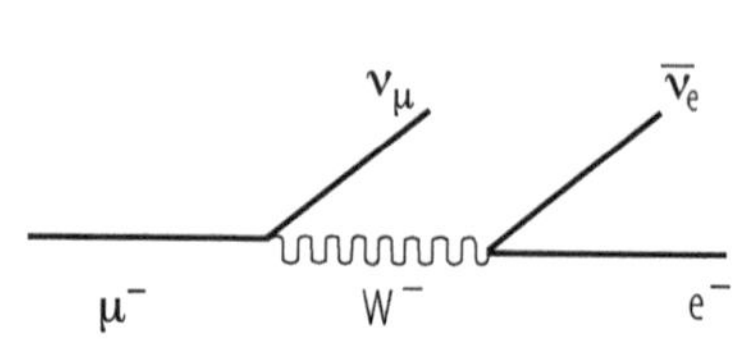

Fig. 3.1 The Born diagram for μ decay

where

$$[\ldots] \equiv t_L^3(1-\gamma_5) + t_R^3(1+\gamma_5) - 2Q\sin^2\theta_W \tag{3.23}$$

and

$$\rho_0 = \frac{m_W^2}{m_Z^2\cos^2\theta_W}. \tag{3.24}$$

All couplings given in this section are obtained at tree level and are modified in higher orders of perturbation theory. In particular, the relations between m_W and $\sin\theta_W$ (Eqs. (3.20) and (3.21)) and the observed values of ρ ($\rho = \rho_0$ at tree level) in different NC processes, are altered by computable EW radiative corrections, as discussed in Sect. (3.11).

The partial width $\Gamma(W \to \bar{f}f')$ is given in Born approximation by the simplest diagram in Fig. 3.2 and one readily obtains from Eq. (3.11) with $t_R = 0$, in the limit of neglecting the fermion masses and summing over all possible f' for a given f:

$$\Gamma(W \to \bar{f}f') = N_C\frac{G_F m_W^3}{6\pi\sqrt{2}} = N_C\frac{\alpha m_W}{12\sin^2\theta_W}, \tag{3.25}$$

where $N_C = 3$ or 1 is the number of colours for quarks or leptons, respectively, and the relations Eqs. (3.15, 3.19) have been used. Here and in the following expressions for the Z widths the one loop QCD corrections for the quark channels can be absorbed in a redefinition of N_C: $N_C \to 3[1+\alpha_s(m_Z)/\pi+\ldots]$. Note that the widths are particularly large because the rate already occurs at order g^2 or G_F. The experimental values of the W total width and the leptonic branching ratio (the average of e, μ and τ modes) are [5, 8] (see Chap. 6):

$$\Gamma_W = 2.147 \pm 0.060\ \text{GeV}, \qquad B(W \to l\nu_l) = 10.80 \pm 0.09. \tag{3.26}$$

The branching ratio B is in very good agreement with the simple approximate formula, derived from Eq. (3.25):

$$B(W \to l\nu_l) \sim \frac{1}{2\cdot 3\cdot(1+\alpha_s(m_Z^2)/\pi)+3} \sim 10.8\%. \tag{3.27}$$

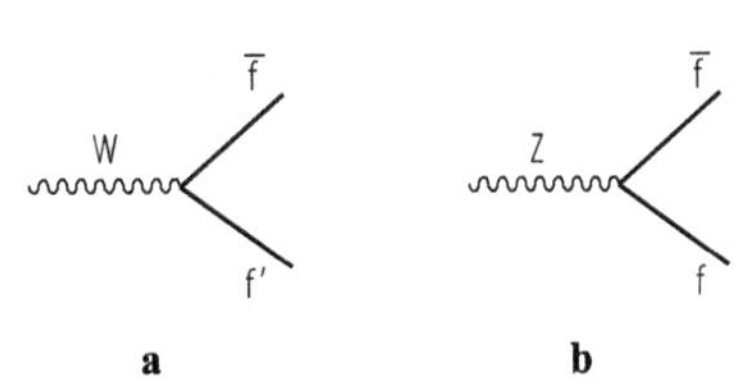

Fig. 3.2 Diagrams for (**a**) the W and (**b**) the Z widths in Born approximation

The denominator corresponds to the sum of the final states $d'\bar{u}$, $s'\bar{c}$, $e^-\bar{\nu}_e$, $\mu^-\bar{\nu}_\mu$, $\tau^-\bar{\nu}_\tau$ (for the definition of d' and s' see Eq. (3.66)).

For $t_R = 0$ the Z coupling to fermions in Eq. (3.17) can be cast into the form:

$$V_{\bar{\psi}_f \psi_f Z} = \frac{g}{2\cos\theta_W} \bar{\psi}_f \gamma_\mu [g_V^f - g_A^f \gamma_5] \psi_f Z^\mu , \tag{3.28}$$

with:

$$g_A^f = t_L^{3f} \quad , \quad g_V^f / g_A^f = 1 - 4|Q_f| \sin^2\theta_W . \tag{3.29}$$

and $t_L^{3f} = \pm 1/2$ for up-type or down-type fermions. In terms of $g_{A,V}$ given in Eqs. (3.29) (the widths are proportional to $(g_V^2 + g_A^2)$), the partial width $\Gamma(Z \to \bar{f} f)$ in Born approximation (see the diagram in Fig. 3.2), for negligible fermion masses, is given by:

$$\begin{aligned}\Gamma(Z \to \bar{f} f) &= N_C \frac{\alpha m_Z}{12 \sin^2 2\theta_W} [1 + (1 - 4|Q_f| \sin^2\theta_W)^2] \\ &= N_C \rho_0 \frac{G_F m_Z^3}{24\pi\sqrt{2}} [1 + (1 - 4|Q_f| \sin^2\theta_W)^2].\end{aligned} \tag{3.30}$$

where $\rho_0 = m_W^2 / m_Z^2 \cos^2\theta_W$ is given in Eq. (3.55). The experimental values of the Z total width and of the partial rates into charged leptons (average of e, μ and τ), into hadrons and into invisible channels are [5, 8] (see Chap. 6):

$$\begin{aligned}\Gamma_Z &= 2.4952 \pm 0.0023 \text{ GeV}, \\ \Gamma_{l^+l^-} &= 83.985 \pm 0.086 \text{ MeV}, \\ \Gamma_h &= 1744.4 \pm 2.0 \text{ MeV}, \\ \Gamma_{inv} &= 499.0 \pm 1.5 \text{ MeV}.\end{aligned} \tag{3.31}$$

The measured value of the Z invisible width, taking radiative corrections into account, leads to the determination of the number of light active neutrinos (see Chap. 6):

$$N_\nu = 2.9841 \pm 0.0083, \tag{3.32}$$

well compatible with the three known neutrinos ν_e, ν_μ and ν_τ; hence there exist only the three known sequential generations of fermions (with light neutrinos), a result with important consequences also in astrophysics and cosmology.

At the Z peak, besides total cross sections, various types of asymmetries have been measured. The results of all asymmetry measurements are quoted in terms of the asymmetry parameter A_f, defined in terms of the effective coupling constants,

g_V^f and g_A^f, as:

$$A_f = 2\frac{g_V^f g_A^f}{g_V^{f2} + g_A^{f2}} = 2\frac{g_V^f/g_A^f}{1+(g_V^f/g_A^f)^2}, \qquad A_{FB}^f = \frac{3}{4}A_e A_f. \quad (3.33)$$

The measurements are: the forward-backward asymmetry ($A_{FB}^f = (3/4)A_e A_f$), the tau polarization (A_τ) and its forward backward asymmetry (A_e) measured at LEP, as well as the left-right and left-right forward-backward asymmetry measured at SLC (A_e and A_f, respectively). Hence the set of partial width and asymmetry results allows the extraction of the effective coupling constants: widths measure $(g_V^2 + g_A^2)$ and asymmetries measure g_V/g_A.

The top quark is heavy enough that it can decay into a real bW pair, which is by far its dominant decay channel. The next mode, $t \to sW$, is suppressed in rate by a factor $|V_{ts}|^2 \sim 1.7 \cdot 10^{-3}$, see Eqs. (3.71–3.73). The associated width, neglecting m_b effects but including 1-loop QCD corrections in the limit $m_W = 0$, is given by (we have omitted a factor $|V_{tb}|^2$ that we set equal to 1):

$$\Gamma(t \to bW^+) = \frac{G_F m_t^3}{8\pi\sqrt{2}}(1 - \frac{m_W^2}{m_t^2})^2(1 + 2\frac{m_W^2}{m_t^2})[1 - \frac{\alpha_s(m_Z)}{3\pi}(\frac{2\pi^2}{3} - \frac{5}{2}) + \ldots]. \quad (3.34)$$

The top quark lifetime is so short, about $0.5 \cdot 10^{-24}$ s, that it decays before hadronizing or forming toponium bound states.

3.4 Gauge Boson Self-interactions

The gauge boson self-interactions can be derived from the $F_{\mu\nu}$ term in $\mathcal{L}_{\text{gauge}}$, by using Eq. (3.12) and $W^\pm = (W^1 \pm iW^2)/\sqrt{2}$.

Defining the three-gauge-boson vertex as in Fig. 3.3 (with all incoming lines), we obtain ($V \equiv \gamma, Z$)

$$V_{W^-W^+V} = ig_{W^-W^+V}[g_{\mu\nu}(p-q)_\lambda + g_{\mu\lambda}(r-p)_\nu + g_{\nu\lambda}(q-r)_\mu], \quad (3.35)$$

with

$$g_{W^-W^+\gamma} = g\ \sin\theta_W = e \quad \text{and} \quad g_{W^-W^+Z} = g\ \cos\theta_W\,. \quad (3.36)$$

Note that the photon coupling to the W is fixed by the electric charge, as imposed by QED gauge invariance. The ZWW coupling is larger by a $\tan\theta_W$ factor. This form of the triple gauge vertex is very special: in general, there could be departures from the above SM expression, even restricting us to Lorentz invariant, em gauge

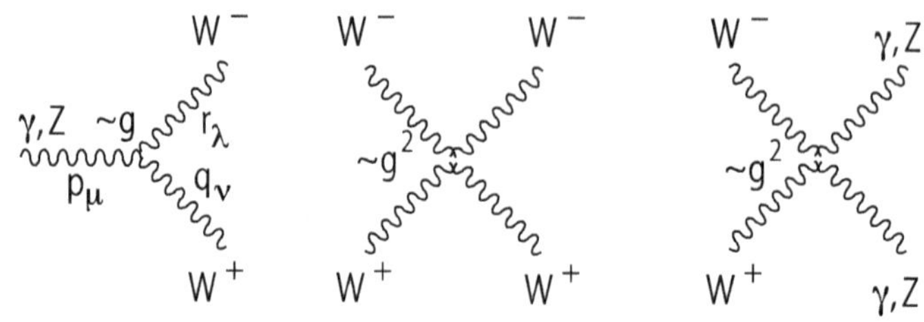

Fig. 3.3 The three- and four-gauge boson vertices. The cubic coupling is of order g, while the quartic one is of order g^2

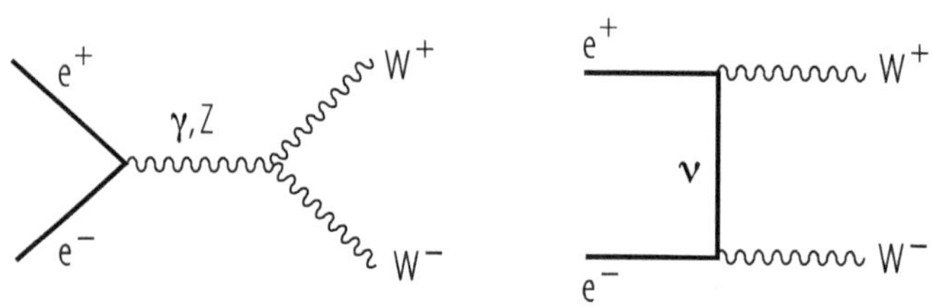

Fig. 3.4 The three- and four-gauge boson vertices. The cubic coupling is of order g, while the quartic one is of order g^2

symmetric and C and P conserving couplings. In fact some small corrections are already induced by the radiative corrections. But, in principle, more important could be the modifications induced by some new physics effect. The experimental testing of the triple gauge vertices has been done mainly at LEP2 and at the Tevatron. At LEP2 the crosssection and angular distributions for the process $e^+e^- \rightarrow W^+W^-$ have been studied (see Chap. 6).

In Born approximation the Feynman diagrams for the LEP2 process are shown in Fig. 3.4 [6]. Besides neutrino exchange which only involves the well established charged current vertex, the triple weak gauge vertices $V_{W^-W^+V}$ appear in the γ and Z exchange diagrams. The Higgs exchange is negligible because the electron mass is very small. The analytic cross section formula in Born approximation can be found, for example, in Ref. [5]. The experimental data are compared with the SM prediction in Chap. 6 [7]. The agreement is very good. Note that the sum of all three exchange amplitudes has a better high energy behaviour. This is due to cancellations among the amplitudes implied by gauge invariance, connected to the fact that the theory is renormalizable (the crosssection can be seen as a contribution to the imaginary part of the $e^+e^- \rightarrow e^+e^-$ amplitude).

The quartic gauge coupling is proportional to $g^2\epsilon_{ABC}W^BW^C\epsilon_{ADE}W^DW^E$. Thus in the term with A $=$ 3 we have four charged W's. For A $=$ 1 or two we have two charged W's and 2 W^3's, each W_3 being a combination of γ and Z according to Eq. (3.13). With a little algebra the quartic vertex can be cast in the form:

$$V_{WWVV} = ig_{WWVV}[2g_{\mu\nu}g_{\lambda\rho} - g_{\mu\lambda}g_{\nu\rho} - g_{\mu\rho}g_{\nu\lambda}]\,, \tag{3.37}$$

where, μ and ν refer to W^+W^+ in the $4W$ vertex and to VV in the $WWVV$ case and:

$$g_{WWWW} = g^2, \quad g_{WW\gamma\gamma} = -e^2, \quad g_{WW\gamma Z} = -eg\cos\theta_W, \quad g_{WWZZ} = -g^2\cos^2\theta_W\,. \tag{3.38}$$

In order to obtain these result for the vertex the reader must duly take into account the factor of $-1/4$ in front of $F^2_{\mu\nu}$ in the lagrangian and the statistical factors which are equal to two for each pair of identical particles (like W^+W^+ or $\gamma\gamma$, for example). The quartic coupling, being quadratic in g, hence small, could not be directly tested so far.

3.5 The Higgs Sector

We now turn to the Higgs sector of the EW lagrangian. The Higgs lagrangian is specified by the gauge principle and the requirement of renormalizability to be

$$\mathcal{L}_{\text{Higgs}} = (D_\mu\phi)^\dagger(D^\mu\phi) - V(\phi^\dagger\phi) - \bar{\psi}_L\Gamma\psi_R\phi - \bar{\psi}_R\Gamma^\dagger\psi_L\phi^\dagger\,, \tag{3.39}$$

where ϕ is a column vector including all Higgs fields; it transforms as a reducible representation of the gauge group. The quantities Γ (which include all coupling constants) are matrices that make the Yukawa couplings invariant under the Lorentz and gauge groups. Without loss of generality, here and in the following, we take Γ to be γ_5-free. The potential $V(\phi^\dagger\phi)$, symmetric under $SU(2)\otimes U(1)$, contains, at most, quartic terms in ϕ so that the theory is renormalizable:

$$V(\phi^\dagger\phi) = -\mu^2\phi^\dagger\phi + \frac{1}{2}\lambda(\phi^\dagger\phi)^2 \tag{3.40}$$

As discussed in Chap. 2, spontaneous symmetry breaking is induced if the minimum of V, which is the classical analogue of the quantum mechanical vacuum state (both are the states of minimum energy), is obtained for non-vanishing ϕ values. Precisely, we denote the vacuum expectation value (VEV) of ϕ, i.e. the position of the minimum, by v (which is a doublet):

$$\langle 0|\phi(x)|0\rangle = v = \begin{pmatrix} 0 \\ v \end{pmatrix} \neq 0\,. \tag{3.41}$$

The reader should be careful that the same symbol is used for the doublet and the only non zero component of the same doublet. The fermion mass matrix is obtained from the Yukawa couplings by replacing $\phi(x)$ by v:

$$M = \bar{\psi}_L\,\mathcal{M}\psi_R + \bar{\psi}_R\mathcal{M}^\dagger\psi_L\,, \tag{3.42}$$

with

$$\mathcal{M} = \Gamma \cdot v\,. \tag{3.43}$$

In the MSM, where all left fermions ψ_L are doublets and all right fermions ψ_R are singlets, only Higgs doublets can contribute to fermion masses. There are enough free couplings in Γ, so that one single complex Higgs doublet is indeed sufficient to generate the most general fermion mass matrix. It is important to observe that by a suitable change of basis we can always make the matrix $\mathcal{M}$ Hermitian and diagonal. In fact, we can make separate unitary transformations on ψ_L and ψ_R according to

$$\psi'_L = U\psi_L\,, \quad \psi'_R = W\psi_R \tag{3.44}$$

and consequently

$$\mathcal{M} \to \mathcal{M}' = U^\dagger \mathcal{M} W\,. \tag{3.45}$$

This transformation does not alter the structure of the fermion couplings in $\mathcal{L}_{\text{symm}}$ (because both the kinetic terms and the couplings to gauge bosons do not mix L and R spinors) except that it leads to the phenomenon of mixing, as we shall see in Sect. (3.6).

If only one Higgs doublet is present, the change of basis that makes $\mathcal{M}$ diagonal will at the same time diagonalize the fermion–Higgs Yukawa couplings. Thus, in this case, no flavour-changing neutral Higgs vertices are present. This is not true, in general, when there are several Higgs doublets. But one Higgs doublet for each electric charge sector i.e. one doublet coupled only to u-type quarks, one doublet to d-type quarks, one doublet to charged leptons (and possibly one for neutrino Dirac masses) would also be all right, because the mass matrices of fermions with different charges are diagonalized separately. For several Higgs doublets in a given charge sector it is also possible to generate CP violation by complex phases in the Higgs couplings. In the presence of six quark flavours, this CP-violation mechanism is not necessary. In fact, at the moment, the simplest model with only one Higgs doublet seems adequate for describing all observed phenomena.

We now consider the gauge-boson masses and their couplings to the Higgs. These effects are induced by the $(D_\mu\phi)^\dagger(D^\mu\phi)$ term in $\mathcal{L}_{\text{Higgs}}$ (Eq. (3.39)), where

$$D_\mu\phi = \left[\partial_\mu + ig\sum_{A=1}^{3} t^A W_\mu^A + ig'(Y/2)B_\mu\right]\phi\,. \tag{3.46}$$

Here t^A and $Y/2$ are the $SU(2)\otimes U(1)$ generators in the reducible representation spanned by ϕ. Not only doublets but all non-singlet Higgs representations can contribute to gauge-boson masses. The condition that the photon remains massless is equivalent to the condition that the vacuum is electrically neutral:

$$Q|v\rangle = (t^3 + \frac{1}{2}Y)|v\rangle = 0 \,. \tag{3.47}$$

We now explicitlly consider the case of a single Higgs doublet:

$$\phi = \begin{pmatrix} \phi^+ \\ \phi^0 \end{pmatrix}, \quad v = \begin{pmatrix} 0 \\ v \end{pmatrix}, \tag{3.48}$$

The charged W mass is given by the quadratic terms in the W field arising from $\mathcal{L}_{\text{Higgs}}$, when $\phi(x)$ is replaced by v in Eq. (3.41). By recalling Eq. (3.10), we obtain

$$m_W^2 W_\mu^+ W^{-\mu} = g^2|(t^+ v/\sqrt{2})|^2 W_\mu^+ W^{-\mu} \,, \tag{3.49}$$

whilst for the Z mass we get [recalling Eqs. (3.12–3.14)]

$$\frac{1}{2} m_Z^2 Z_\mu Z^\mu = |[g \cos\theta_W t^3 - g' \sin\theta_W (Y/2)]v|^2 Z_\mu Z^\mu \,, \tag{3.50}$$

where the factor of 1/2 on the left-hand side is the correct normalization for the definition of the mass of a neutral field. By using Eq. (3.47), relating the action of t^3 and $Y/2$ on the vacuum v, and Eqs. (3.16), we obtain

$$\frac{1}{2} m_Z^2 = (g \cos\theta_W + g' \sin\theta_W)^2 |t^3 v|^2 = (g^2/\cos^2\theta_W)|t^3 v|^2 \,. \tag{3.51}$$

For a Higgs doublet, as in Eq. (3.48), we have

$$|t^+ v|^2 = v^2, \quad |t^3 v|^2 = 1/4 v^2 \,, \tag{3.52}$$

so that

$$m_W^2 = 1/2 g^2 v^2, \quad m_Z^2 = 1/2 g^2 v^2/\cos^2\theta_W \,. \tag{3.53}$$

Note that by using Eq. (3.19) we obtain

$$v = 2^{-3/4} G_F^{-1/2} = 174.1 \text{ GeV} \,. \tag{3.54}$$

It is also evident that for Higgs doublets

$$\rho_0 = \frac{m_W^2}{m_Z^2 \cos^2\theta_W} = 1 \,. \tag{3.55}$$

This relation is typical of one or more Higgs doublets and would be spoiled by the existence of Higgs triplets etc. In general,

$$\rho_0 = \frac{\sum_i ((t_i)^2 - (t_i^3)^2 + t_i)v_i^2}{\sum_i 2(t_i^3)^2 v_i^2} \tag{3.56}$$

for several Higgs bosons with VEVs v_i, weak isospin t_i, and z-component t_i^3. These results are valid at the tree level and are modified by calculable EW radiative corrections, as discussed in Sect. (3.7).

The measured values of the W and Z masses are [5, 8] (see Chap. 6):

$$m_W = 80.398 \pm 0.025\ \text{GeV}, \qquad m_Z = 91.1875 \pm 0.0021\ \text{GeV}. \tag{3.57}$$

In the minimal version of the SM only one Higgs doublet is present. Then the fermion–Higgs couplings are in proportion to the fermion masses. In fact, from the Yukawa couplings $g_{\phi\bar{f}f}(\bar{f}_L\phi f_R + h.c.)$, the mass m_f is obtained by replacing ϕ by v, so that $m_f = g_{\phi\bar{f}f}v$. In the minimal SM three out of the four Hermitian fields are removed from the physical spectrum by the Higgs mechanism and become the longitudinal modes of W^+, W^-, and Z. The fourth neutral Higgs is physical and should be found. If more doublets are present, two more charged and two more neutral Higgs scalars should be around for each additional doublet.

The couplings of the physical Higgs H can be simply obtained from $\mathcal{L}_{\text{Higgs}}$, by the replacement (the remaining three hermitian fields correspond to the would be Goldstone bosons that become the longitudinal modes of $W^\pm$ and Z):

$$\phi(x) = \begin{pmatrix} \phi^+(x) \\ \phi^0(x) \end{pmatrix} \rightarrow \begin{pmatrix} 0 \\ v + (H/\sqrt{2}) \end{pmatrix}, \tag{3.58}$$

[so that $(D_\mu\phi)^\dagger(D^\mu\phi) = 1/2(\partial_\mu H)^2 + \ldots$], with the results

$$\mathcal{L}[H, W, Z] = g^2\frac{v}{\sqrt{2}}W_\mu^+W^{-\mu}H + \frac{g^2}{4}W_\mu^+W^{-\mu}H^2 + \\ + g^2\frac{v}{2\sqrt{2}\cos^2\theta_W}Z_\mu Z^\mu H + \frac{g^2}{8\cos^2\theta_W}Z_\mu Z^\mu H^2 \,. \tag{3.59}$$

Note that the trilinear couplings are nominally of order g^2, but the adimensional coupling constant is actually of order g if we express the couplings in terms of the masses according to Eqs. (3.53):

$$\mathcal{L}[H, W, Z] = gm_W W^+_\mu W^{-\mu} H + \frac{g^2}{4} W^+_\mu W^{-\mu} H^2 + + \frac{gm_Z}{2\cos^2\theta_W} Z_\mu Z^\mu H + \frac{g^2}{8\cos^2\theta_W} Z_\mu Z^\mu H^2 . \tag{3.60}$$

Thus the trilinear couplings of the Higgs to the gauge bosons are also proportional to the masses. The quadrilinear couplings are genuinely of order g^2. Recall that to go from the lagrangian to the Feynman rules for the vertices the statistical factors must be taken into account: for example, the Feynman rule for the $ZZHH$ vertex is $ig_{\mu\nu}g^2/2\cos^2\theta_W$.

The generic coupling of H to a fermion of type f is given by (after diagonalization):

$$\mathcal{L}[H, \bar{\psi}, \psi] = \frac{g_f}{\sqrt{2}} \bar{\psi}\psi H, \tag{3.61}$$

with

$$\frac{g_f}{\sqrt{2}} = \frac{m_f}{\sqrt{2}v} = 2^{1/4} G_F^{1/2} m_f . \tag{3.62}$$

The Higgs self couplings are obtained from the potential in Eq. (3.40) by the replacement in Eq. (3.58). Given that, from the minimum condition:

$$v = \sqrt{\frac{\mu^2}{\lambda}} \tag{3.63}$$

one obtains:

$$V = -\mu^2 (v + \frac{H}{\sqrt{2}})^2 + \frac{\mu^2}{2v^2}(v + \frac{H}{\sqrt{2}})^4 = -\frac{\mu^2 v^2}{2} + \mu^2 H^2 + \frac{\mu^2}{\sqrt{2}v} H^3 + \frac{\mu^2}{8v^2} H^4 \tag{3.64}$$

The constant term can be omitted in our context. We see that the Higgs mass is positive (compare with Eq. (3.40)) and is given by:

$$m_H^2 = 2\mu^2 = 2\lambda v^2 \tag{3.65}$$

We see that for $\sqrt{\lambda} \sim o(1)$ the Higgs mass should be of the order of the weak scale.

The difficulty of the Higgs search is due to the fact that it is heavy and coupled in proportion to mass: it is a heavy particle that must be radiated by another heavy particle. So a lot of phase space and luminosity is needed. At LEP2 the main process for Higgs production was the Higgs-strahlung process $e^+e^- \rightarrow ZH$ shown in

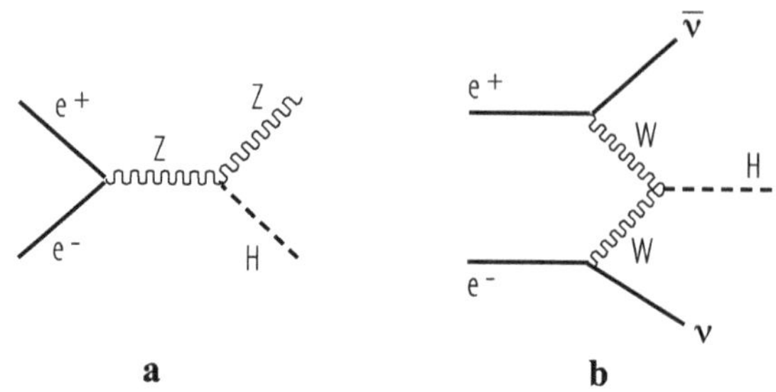

Fig. 3.5 Higgs production diagrams in Born approximation: (**a**) The Higgs-strahlung process $e^+e^- \to ZH$, (**b**) the WW fusion process $e^+e^- \to H\nu\bar{\nu}$

Fig. 3.5 [9]. The alternative process $e^+e^- \to H\nu\bar{\nu}$, via WW fusion, also shown in Fig. 3.5 [10], has a smaller crosssection at LEP2 energies but would become important, even dominant at higher energy e^+e^- colliders, like the ILC or CLIC (the corresponding ZZ fusion process has a much smaller crosssection). The analytic formulae for the crosssections of both processes can be found, for example, in [11]. The direct experimental limit on m_H from LEP2 is $m_H \gtrsim 114$ GeV at 95% c.l. (see Chap. 6).

3.6 The CKM Matrix

Weak charged currents are the only tree level interactions in the SM that change flavour: for example, by emission of a W an up-type quark is turned into a down-type quark, or a ν_l neutrino is turned into a l^- charged lepton (all fermions are letf-handed). If we start from an up quark that is a mass eigenstate, emission of a W turns it into a down-type quark state d' (the weak isospin partner of u) that in general is not a mass eigenstate. The mass eigenstates and the weak eigenstates do not coincide and a unitary transformation connects the two sets:

$$D' = \begin{pmatrix} d' \\ s' \\ b' \end{pmatrix} = V \begin{pmatrix} d \\ s \\ b \end{pmatrix} = VD \tag{3.66}$$

V is the Cabibbo-Kobayashi-Maskawa (CKM) matrix [12] (and similarly we can denote by U the column vector of the three up quark mass eigenstates). Thus in terms of mass eigenstates the charged weak current of quarks is of the form:

$$J^+_\mu \propto \bar{U}\gamma_\mu(1-\gamma_5)t^+ VD \tag{3.67}$$

where

$$V = U_u^\dagger U_d \tag{3.68}$$

Here U_u and U_d are the unitary matrices that operate on left-handed doublets in the diagonalization of the u and d quarks, respectively (see Eq. (3.44)). Since V is unitary (i.e. $VV^\dagger = V^\dagger V = 1$) and commutes with T^2, T_3 and Q (because all d-type quarks have the same isospin and charge), the neutral current couplings are diagonal both in the primed and unprimed basis (if the down-type quark terms in the Z current are written in terms of weak isospin eigenvectors as $\bar{D}'\Gamma D'$, then by changing basis we get $\bar{D}V^\dagger \Gamma V D$ and V and Γ commute because, as seen from Eq. (3.23), Γ is made of Dirac matrices and of T_3 and Q generator matrices). It follows that $\bar{D}'\Gamma D' = \bar{D}\Gamma D$. This is the GIM mechanism [13] that ensures natural flavour conservation of the neutral current couplings at the tree level.

For N generations of quarks, V is a N×N unitary matrix that depends on N^2 real numbers (N^2 complex entries with N^2 unitarity constraints). However, the $2N$ phases of up- and down-type quarks are not observable. Note that an overall phase drops away from the expression of the current in Eq. (3.67), so that only $2N - 1$ phases can affect V. In total, V depends on $N^2 - 2N + 1 = (N-1)^2$ real physical parameters. A similar counting gives $N(N-1)/2$ as the number of independent parameters in an orthogonal N×N matrix. This implies that in V we have $N(N-1)/2$ mixing angles and $(N-1)^2 - N(N-1)/2 = (N-1)(N-2)/2$ phases: for $N = 2$ one mixing angle (the Cabibbo angle θ_C) and no phases, for $N = 3$ three angles (θ_{12}, θ_{13} and θ_{23}) and one phase φ etc.

Given the experimental near diagonal structure of V a convenient parametrisation is the one proposed by Maiani [14]. It can be cast in the form of a product of three independent 2×2 block matrices (s_{ij} and c_{ij} are shorthands for $\sin\theta_{ij}$ and $\cos\theta_{ij}$):

$$V = \begin{pmatrix} 1 & 0 & 0 \\ 0 & c_{23} & s_{23} \\ 0 & -s_{23} & c_{23} \end{pmatrix} \begin{pmatrix} c_{13} & 0 & s_{13}e^{i\varphi} \\ 0 & 1 & 0 \\ -s_{13}e^{-i\varphi} & 0 & c_{13} \end{pmatrix} \begin{pmatrix} c_{12} & s_{12} & 0 \\ -s_{12} & c_{12} & 0 \\ 0 & 0 & 1 \end{pmatrix}. \tag{3.69}$$

The advantage of this parametrization is that the three mixing angles are of different orders of magnitude. In fact, from experiment we know that $s_{12} \equiv \lambda$, $s_{23} \sim o(\lambda^2)$ and $s_{13} \sim o(\lambda^3)$, where $\lambda = \sin\theta_C$ is the sine of the Cabibbo angle, and, as order of magnitude, s_{ij} can be expressed in terms of small powers of λ. More precisely, following Wolfenstein [15] one can set:

$$s_{12} \equiv \lambda, \qquad s_{23} = A\lambda^2, \qquad s_{13}e^{-i\phi} = A\lambda^3(\rho - i\eta) \tag{3.70}$$

As a result, by neglecting terms of higher order in λ one can write down:

$$V = \begin{bmatrix} V_{ud} & V_{us} & V_{ub} \\ V_{cd} & V_{cs} & V_{cb} \\ V_{td} & V_{ts} & V_{tb} \end{bmatrix} \sim \begin{bmatrix} 1 - \frac{\lambda^2}{2} & \lambda & A\lambda^3(\rho - i\eta) \\ -\lambda & 1 - \frac{\lambda^2}{2} & A\lambda^2 \\ A\lambda^3(1 - \rho - i\eta) & -A\lambda^2 & 1 \end{bmatrix} + o(\lambda^4). \tag{3.71}$$

It has become customary to make the replacement $\rho, \eta \rightarrow \bar{\rho}, \bar{\eta}$ with:

$$\rho - i\eta = \frac{\bar{\rho} - i\bar{\eta}}{\sqrt{1-\lambda^2}} \sim (\bar{\rho} - i\bar{\eta})(1 + \lambda^2/2 + \ldots). \tag{3.72}$$

Present values of the CKM parameters as obtained from experiment are [16] [17] (a survey of the current status of the CKM parameters can also be found in Ref. [5]):

$$\begin{aligned}
\lambda &= 0.2258 \pm 0.0014 \\
A &= 0.818 \pm 0.016 \\
\bar{\rho} &= 0.164 \pm 0.029; \quad \bar{\eta} = 0.340 \pm 0.017
\end{aligned} \tag{3.73}$$

A more detailed discussion of the experimental data is given in Chap. 10.

In the SM the non vanishing of the η parameter (related to the phase φ in Eqs. 3.69 and 3.70) is the only source of CP violation. Unitarity of the CKM matrix V implies relations of the form $\sum_a V_{ba} V_{ca}^* = \delta_{bc}$. In most cases these relations do not imply particularly instructive constraints on the Wolfenstein parameters. But when the three terms in the sum are of comparable magnitude we get interesting information. The three numbers which must add to zero form a closed triangle in the complex plane, with sides of comparable length. This is the case for the t-u triangle (unitarity triangle) shown in Fig. 3.6 (or, what is equivalent in first approximation, for the d-b triangle):

$$V_{td} V_{ud}^* + V_{ts} V_{us}^* + V_{tb} V_{ub}^* = 0 \tag{3.74}$$

All terms are of order λ^3. For $\eta = 0$ the triangle would flatten down to vanishing area. In fact the area of the triangle, J of order $J \sim \eta A^2 \lambda^6$, is the Jarlskog invariant [18] (its value is independent of the parametrization). In the SM all CP violating observables must be proportional to J, hence to the area of the triangle or to η. A direct and by now very solid evidence for J non vanishing is obtained from the measurements of ϵ and ϵ' in K decay. Additional direct evidence is being obtained from the experiments on B decays at beauty factories and at the TeVatron where the angles β (the most precisely measured), α and γ have been determined. Together with the available information on the magnitude of the sides all the measurements

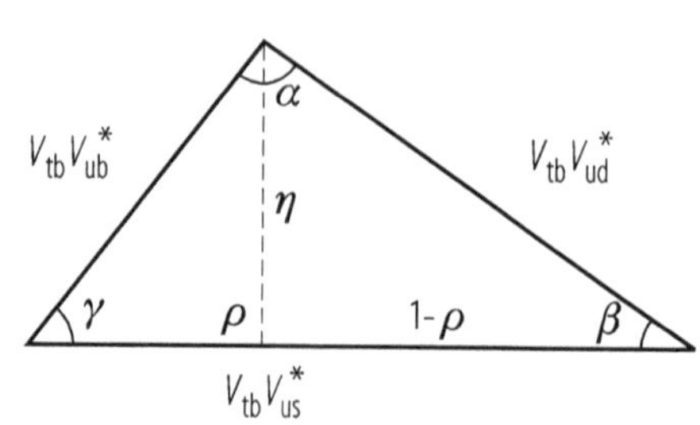

Fig. 3.6 The unitarity triangle corresponding to Eq. (3.74)

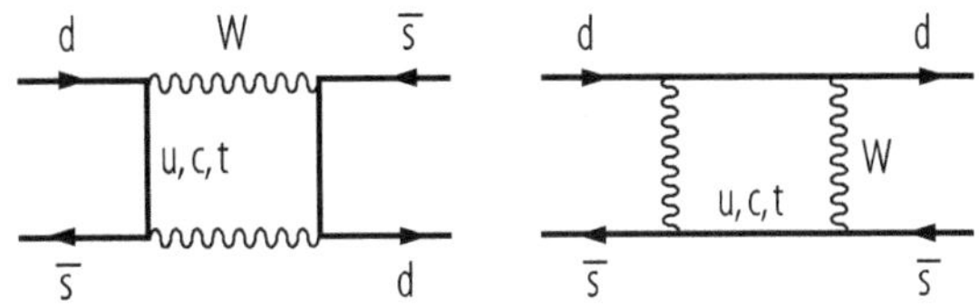

Fig. 3.7 Box diagrams describing $K^0 - \bar{K}^0$ mixing at the quark level at 1-loop

are in good agreement with the predictions from the SM unitary triangle [16, 17] (see Chap. 10).

As we have discussed, due to the GIM mechanism, there are no flavour changing neutral current (FCNC) transitions at tree level in the SM. Transitions with $|\Delta F| = 1, 2$ are induced at one loop level. In particular, meson mixing, i.e. $M \to \bar{M}$ off diagonal $|\Delta F| = 2$ mass matrix elements (with $M = K, D$ or B neutral mesons), are obtained from box diagrams. For example, in the case of $K^0 - \bar{K}^0$ mixing the relevant transition is $\bar{s}d \to s\bar{d}$ (see Fig. 3.7). In the internal quark lines all up-type quarks are exchanged. In the amplitude, two vertices and the connecting propagator (with virtual four momentum p_μ) at one side contribute a factor ($u_i = u, c, t$):

$$F_{GIM} = \sum_i V^*_{u_i s} \frac{1}{\not{p} - m_{ui}} V_{u_i d} , \tag{3.75}$$

which, in the limit of equal m_{ui}, is clearly vanishing due to the unitarity of the CKM matrix V. Thus the result is proportional to mass differences. For $K^0 - \bar{K}^0$ mixing the contribution of virtual u quarks is negligible due to the small value of m_u and the contribution of the t quark is also small due to the mixing factors $V^*_{ts} V_{td} \sim o(A^2\lambda^5)$. The dominant c quark contribution to the real part of the box diagram quark-level amplitude is approximately of the form (see, for example, [19]):

$$ReH_{box} = \frac{G_F^2}{16\pi^2} m_c^2 Re(V^*_{cs} V_{cd})^2 \eta_1 O^{\Delta s=2} , \tag{3.76}$$

where $\eta_1 \sim 0.85$ is a QCD correction factor and $O^{\Delta s=2} = \bar{d}_L \gamma_\mu s_L \; \bar{s}_L \gamma_\mu d_L$ is the 4-quark dimension six relevant operator. To obtain the $K^0 - \bar{K}^0$ mixing its matrix element between meson states must be taken which is parametrized in terms of a "B_K parameter" which is defined in such a way that $B_K = 1$ for vacuum state insertion between the two currents:

$$\langle K^0 | O^{\Delta s=2} | \bar{K}^0 \rangle = \frac{16}{3} f_K m_K^2 B_K , \tag{3.77}$$

where $f_K \sim 113 MeV$ is the kaon pseudoscalar constant. Clearly to the charm contribution in Eq. (3.76) non perturbative additional contributions must be added, some of them of $o(m_K^2/m_c^2)$, because the smallness of m_c makes a completely partonic dominance inadequate. In particular, B_K is best evaluated by QCD lattice simulations. In Eq. (3.76) the factor $o(m_c^2/m_W^2)$ is the "GIM suppression" factor

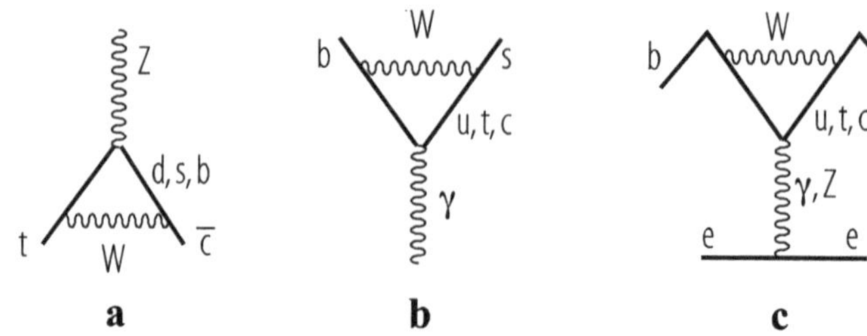

Fig. 3.8 Examples of $|\Delta F| = 1$ transitions at the quark level at 1-loop: (**a**) Diagram for a $Z \to t\,\bar{c}$ vertex, (**b**) $b \to s\,\gamma$, (**c**) a "penguin" diagram for $b \to s\,e^+e^-$

($1/m_W^2$ is hidden in G_F according to Eq. (3.19)). For B mixing the dominant contribution is from the t quark. In this case, the partonic dominance is more realistic and the GIM factor $o(m_t^2/m_W^2)$ is actually larger than one.

All sorts of transitions with $|\Delta F| = 1$ are also induced at loop level. For example, an effective vertex $Z \to t\bar{c}$, which does not exist at tree level, is generated at 1-loop (see Fig. 3.8). Similarly, transitions involving photons or gluons are also possible, like $t \to c\ g$ or $b \to s\ \gamma$ (Fig. 3.8) or $b \to s\ g$. For light fermion exchange in the loop the GIM suppression is also effective in $|\Delta F| = 1$ amplitudes. For example, analogous leptonic transitions like $\mu \to e\ \gamma$ or $\tau \to \mu\ \gamma$ also exist but are extremely small in the SM because the tiny neutrino masses enter in the GIM suppression factor. But new physics effects could well make these rare processes accessible to experiments in the near future. The external Z, photon or gluon can be attached to a pair of light fermions, giving rise to an effective four fermion operator, as in "penguin diagrams" like the one shown in Fig. 3.8 for $b \to s\ l^+l^-$. The inclusive rate $B \to X_s\ \gamma$ with X_s a hadronic state containing a unit of strangeness corresponding to an s-quark, has been precisely measured. The world average result for the branching ratio with $E_\gamma > 1.6$ GeV is [5]:

$$B(B \to X_s\ \gamma)_{exp} = (3.55 \pm 0.26) \cdot 10^{-4}\,. \tag{3.78}$$

The theoretical prediction for this inclusive process is to a large extent free of uncertainties from hadronisation effects and is accessible to perturbation theory as the b-quark is heavy enough. The most complete result at order α_s^2 is at present [20] (and refs. therein):

$$B(B \to X_s\ \gamma)_{th} = (2.98 \pm 0.26) \cdot 10^{-4}\,. \tag{3.79}$$

Note that the theoretical value has recently become smaller than the experimental value. The fair agreement between theory and experiment imposes stringent constraints on possible new physics effects.

3.7 Neutrino Masses

In the minimal version of the SM the right handed neutrinos ν_{iR}, which have no gauge interactions, are not present at all. With no ν_R no Dirac mass is possible for neutrinos. If lepton number conservation is also imposed, then no Majorana mass is allowed either and, as a consequence, all neutrinos are massless. But, at present, from neutrino oscillation experiments (see Chapter 11 of the present work), we know that at least 2 out of the 3 known neutrinos have non vanishing masses: the two mass squared differences measured from solar (Δm^2_{12}) and atmospheric oscillations (Δm^2_{23}) are given by $\Delta m^2_{12} \sim 8\ 10^{-5}\ eV^2$ and $\Delta m^2_{23} \sim 2.5\ 10^{-3}$ [21]. The absolute values of the masses are very small, with an upper limit of a fraction of eV, obtained from laboratory experiments (tritium β decay near the end point: $m_\nu \lesssim 2$ eV [5], absence of visible neutrinoless double β decay : $|m_{ee}| \lesssim 0.3-0.7$ eV (m_{ee} is a combination of neutrino masses; for a review, see, for example [22]) and from cosmological observations: $m_\nu \lesssim 0.1 - 0.7$ eV (depending on the cosmological model assumptions) [23]. If ν_{iR} are added to the minimal model and lepton number is imposed by hand, then neutrino masses would in general appear as Dirac masses, generated by the Higgs mechanism, like for any other fermion. But, for Dirac neutrinos, to explain the extreme smallness of neutrino masses, one should allow for very small Yukawa couplings. However, we stress that, in the SM, baryon B and lepton L number conservation, which are not guaranteed by gauge symmetries (as is the case for the electric charge Q), are understood as "accidental" symmetries, due to the fact that, out of the SM fields, it is not possible to construct gauge invariant operators which are renormalizable (i.e. of operator dimension $d \leq 4$) and violate B and/or L. In fact the SM lagrangian should contain all terms allowed by gauge symmetry and renormalizability. The most general renormalizable lagrangian, built from the SM fields, compatible with the SM gauge symmetry, in absence of ν_{iR}, is automatically B and L conserving. But in presence of ν_{iR}, this is no more true and the right handed Majorana mass term is allowed:

$$M_{RR} = \bar{\nu}^c_{iR} M_{ij} \nu_{jR} = \nu^T_{iR} C M_{ij} \nu_{jR}\ , \tag{3.80}$$

where $\nu^c_{iR} = C\bar{\nu}^T_{iR}$ is the charge conjugated neutrino field and C is the charge conjugation matrix in Dirac spinor space. The Majorana mass term is an operator of dimension $d = 3$ with $\Delta L = 2$. Since the ν_{iR} are gauge singlets the Majorana mass M_{RR} is fully allowed by the gauge symmetry and a coupling with the Higgs is not needed to generate this type of mass. As a consequence, the entries of the mass matrix M_{ij} do not need to be of the order of the EW symmetry breaking scale v and could be much larger. If one starts from the Dirac and RR Majorana mass terms for neutrinos, the resulting mass matrix, in the L, R space, has the form:

$$m_\nu = \begin{bmatrix} 0 & m_D \\ m_D & M \end{bmatrix} \tag{3.81}$$

where m_D and M are the Dirac and Majorana mass matrices (M is the matrix M_{ij} in Eq. (3.80)). The corresponding eigenvalues are three very heavy neutrinos with masses of order M and three light neutrinos with masses

$$m_\nu = -m_D^T M^{-1} m_D \,, \tag{3.82}$$

which are possibly very small if M is large enough. This is the see-saw mechanism for neutrino masses [24]. Note that if no ν_{iR} exist a Majorana mass term could still be built out of ν_{jL}. But ν_{jL} have weak isospin 1/2, being part of the left handed lepton doublet l. Thus, the left handed Majorana mass term has total weak isospin equal to one and needs two Higgs fields to make a gauge invariant term. The resulting mass term:

$$O_5 = \lambda l_i^T \lambda_{ij} l_j HH/M \,, \tag{3.83}$$

with M a large scale (apriori comparable to the scale of M_{RR}) and λ a dimensionless coupling generically of o(1), is a non renormalizable operator of dimension 5. The corresponding mass terms are of the order $m_\nu \sim \lambda v^2/M$, hence of the same generic order of the light neutrino masses from Eq. (3.82).

In conclusion, neutrino masses are believed to be small because neutrinos are Majorana particles with masses inversely proportional to the large scale M of energy where L non conservation is induced. It is interesting that the observed magnitudes of the mass squared splittings of neutrinos are well compatible with a scale M remarkably close to the Grand Unification scale, where in fact L non conservation is naturally expected.

In the previous Section we have discussed flavour mixing for quarks. But, clearly, given that non vanishing neutrino masses have been established, a similar mixing matrix is also introduced in the leptonic sector, but will not be discussed here (see Chapter 11).

3.8 Renormalization of the Electroweak Theory

The Higgs mechanism gives masses to the Z, the $W^\pm$ and to fermions while the lagrangian density is still symmetric. In particular the gauge Ward identities and the symmetric form of the gauge currents are preserved. The validity of these relations is an essential ingredient for renormalizability. In the previous Sections we have specified the Feynman vertices in the "unitary" gauge where only physical particles appear. However, as discussed in Chap. 2, in this gauge the massive gauge boson propagator would have a bad ultraviolet behaviour:

$$W_{\mu\nu} = \frac{-g_{\mu\nu} + \frac{q_\mu q_\nu}{m_W^2}}{q^2 - m_W^2} \,. \tag{3.84}$$

A formulation of the standard EW theory with good apparent ultraviolet behaviour can be obtained by introducing the renormalizable or R_ξ gauges, in analogy with the abelian case discussed in detail in Chap. 2. One parametrizes the Higgs doublet as:

$$\phi = \begin{pmatrix} \phi^+ \\ \phi^0 \end{pmatrix} = \begin{pmatrix} \phi_1 + i\phi_2 \\ \phi_3 + i\phi_4 \end{pmatrix} = \begin{pmatrix} -iw^+ \\ v + \frac{H+iz}{\sqrt{2}} \end{pmatrix}, \tag{3.85}$$

and similarly for $\phi^\dagger$, where w^- appears. The scalar fields $w^\pm$ and z are the pseudo Goldstone bosons associated with the longitudinal modes of the physical vector bosons $W^\pm$ and Z. The R_ξ gauge fixing lagrangian has the form:

$$\Delta\mathcal{L}_{GF} = -\frac{1}{\xi}|\partial^\mu W_\mu - \xi m_W w|^2 - \frac{1}{2\eta}(\partial^\mu Z_\mu - \eta m_Z z)^2 - \frac{1}{2\alpha}(\partial^\mu A_\mu)^2 . \tag{3.86}$$

The $W^\pm$ and Z propagators, as well as those of the scalars $w^\pm$ and z, have exactly the same general forms as for the abelian case in Eqs. (67)–(69) of Chap. 2, with parameters ξ and η, respectively (and the pseudo Goldstone bosons $w^\pm$ and z have masses ξm_W and ηm_Z). In general, a set of associated ghost fields must be added, again in direct analogy with the treatment of R_ξ gauges in the abelian case of Chap. 2. The complete Feynman rules for the standard EW theory can be found in a number of textbooks (see, for example, [25]).

The pseudo Goldstone bosons $w^\pm$ and z are directly related to the longitudinal helicity states of the corresponding massive vector bosons $W^\pm$ and Z. This correspondence materializes in a very interesting "equivalence theorem": at high energies of order E the amplitude for the emission of one or more longitudinal gauge bosons V_L (with $V = W, Z$) becomes equal (apart from terms down by powers of m_V/E) to the amplitude where each longitudinal gauge boson is replaced by the corresponding Goldstone field $w^\pm$ or z [26]. For example, consider top decay with a longitudinal W in the final state: $t \to bW_L^+$. The equivalence theorem asserts that we can compute the dominant contribution to this rate from the simpler $t \to bw^+$ matrix element:

$$\Gamma(t \to bW_L^+) = \Gamma(t \to bw^+)[1 + o(m_W^2/m_t^2)] . \tag{3.87}$$

In fact one finds:

$$\Gamma(t \to bw^+) = \frac{h_t^2}{32\pi} m_t = \frac{G_F m_t^3}{8\pi\sqrt{2}} , \tag{3.88}$$

where $h_t = m_t/v$ is the Yukawa coupling of the top quark (numerically very close to 1), and we used $1/v^2 = 2\sqrt{2}G_F$ (see Eq. (3.54)). If we compare with Eq. (3.34), we see that this expression coincides with the total top width (i.e. including all polarizations for the W in the final state), computed at tree level, apart from terms down by powers of $o(m_W^2/m_t^2)$. In fact, the longitudinal W is dominant in the final state because $h_t >> g^2$. Similarly the equivalence theorem can be applied to find the dominant terms at large $\sqrt{s}$ for the crosssection $e^+e^- \to W_L^+ W_L^-$, or the leading contribution in the limit $m_H >> m_V$ to the width for the decay $\Gamma(H \to VV)$.

The formalism of the R_ξ gauges is also very useful in proving that spontaneously broken gauge theories are renormalizable. In fact, the non singular behaviour of propagators at large momenta is very suggestive of the result. Nevertheless to prove it is by far not a simple matter. The fundamental theorem that in general a gauge theory with spontaneous symmetry breaking and the Higgs mechanism is renormalizable was proven by 't Hooft and Veltman [27, 28].

For a chiral theory like the SM an additional complication arises from the existence of chiral anomalies. But this problem is avoided in the SM because the quantum numbers of the quarks and leptons in each generation imply a remarkable (and, from the point of view of the SM, mysterious) cancellation of the anomaly, as originally observed in Ref. [29]. In quantum field theory one encounters an anomaly when a symmetry of the classical lagrangian is broken by the process of quantization, regularization and renormalization of the theory. Of direct relevance for the EW theory is the Adler-Bell-Jackiw (ABJ) chiral anomaly [30]. The classical lagrangian of a theory with massless fermions is invariant under a U(1) chiral transformations $\psi\prime = e^{i\gamma_5\theta}\psi$. The associated axial Noether current is conserved at the classical level. But, at the quantum level, chiral symmetry is broken due to the ABJ anomaly and the current is not conserved. The chiral breaking is produced by a clash between chiral symmetry, gauge invariance and the regularization procedure.

The anomaly is generated by triangular fermion loops with one axial and two vector vertices (Fig. 3.9). For example, for the Z the axial coupling is proportional to the third component of weak isospin t_3, while the vector coupling is proportional to a linear combination of t_3 and the electric charge Q. Thus in order for the chiral anomaly to vanish all traces of the form $tr\{t_3QQ\}$, $tr\{t_3t_3Q\}$, $tr\{t_3t_3t_3\}$ (and also $tr\{t_+t_-t_3\}$ when charged currents are also included) must vanish, where the trace is extended over all fermions in the theory that can circulate in the loop. Now all these traces happen to vanish for each fermion family separately. For example take $tr\{t_3QQ\}$. In one family there are, with $t_3 = +1/2$, three colours of up quarks with

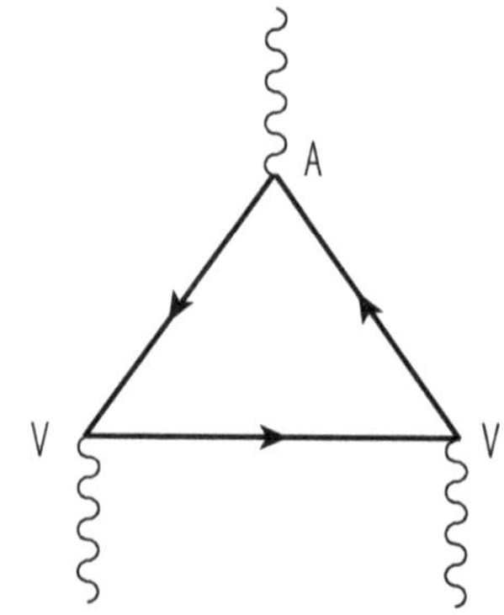

Fig. 3.9 Triangle diagram that generates the ABJ anomaly

charge $Q = +2/3$ and one neutrino with $Q = 0$ and, with $t_3 = -1/2$, three colours of down quarks with charge $Q = -1/3$ and one l^- with $Q = -1$. Thus we obtain $tr\{t_3 QQ\} = 1/2{\cdot}3{\cdot}4/9 - 1/2{\cdot}3{\cdot}1/9 - 1/2{\cdot}1 = 0$. This impressive cancellation suggests an interplay among weak isospin, charge and colour quantum numbers which appears as a miracle from the point of view of the low energy theory but is in fact understandable from the point of view of the high energy theory. For example, in Grand Unified Theories (GUTs) (for reviews, see, for example, [31]) there are similar relations where charge quantization and colour are related: in the five of SU(5) we have the content $(d, d, d, e^+, \bar{\nu})$ and the charge generator has a vanishing trace in each SU(5) representation (the condition of unit determinant, represented by the letter S in the SU(5) group name, translates into zero trace for the generators). Thus the charge of d quarks is $-1/3$ of the positron charge because there are three colours. A whole family fits perfectly in one 16 of SO(10) which is anomaly free. So GUTs can naturally explain the cancellation of the chiral anomaly.

An important implication of chiral anomalies together with the topological properties of the vacuum in non abelian gauge theories is that the conservation of the charges associated to baryon (B) and lepton (L) numbers is broken by the anomaly [32], so that B and L conservation is actually violated in the standard electroweak theory (but B-L remains conserved). B and L are conserved to all orders in the perturbative expansion but the violation occurs via non perturbative instanton effects [33] (the amplitude is proportional to the typical non perturbative factor $\exp -c/g^2$, with c a constant and g the $SU(2)$ gauge coupling). The corresponding effect is totally negligible at zero temperature T, but becomes relevant at temperatures close to the electroweak symmetry breaking scale, precisely at $T \sim o(TeV)$. The non conservation of B+L and the conservation of B−L near the weak scale plays a role in the theory of baryogenesis that quantitatively aims at explaining the observed matter antimatter asymmetry in the Universe (for a recent review, see, for example, [34]; see also Chap. 9).

3.9 QED Tests: Lepton Anomalous Magnetic Moments

The most precise tests of the electroweak theory apply to the QED sector. Here we discuss some recent developments. The anomalous magnetic moments of the electron and of the muon are among the most precise measurements in the whole of physics. The magnetic moment $\vec{\mu}$ and the spin $\vec{S}$ are related by $\vec{\mu} = -ge\vec{S}/2m$, where g is the gyromagnetic ratio ($g = 2$ for a pointlike Dirac particle). The quantity $a = (g-2)/2$ measures the anomalous magnetic moment of the particle. Recently there have been new precise measurements of a_e and a_μ for the electron [35] and the muon [36]:

$$a_e^{exp} = 11596521808.5(7.6) \cdot 10^{-13}, \qquad a_\mu^{exp} = 11659208.0(6.3) \cdot 10^{-10}. \tag{3.89}$$

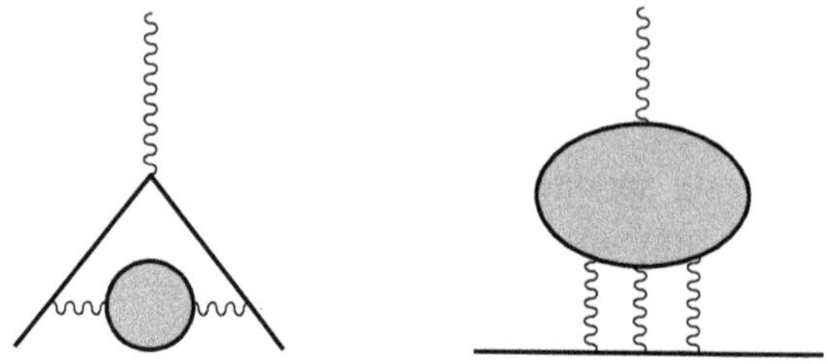

Fig. 3.10 The hadronic contributions to the anomalous magnetic moment: vacuum polarization (left) and light by light scattering (right)

The theoretical calculations in general contain a pure QED part plus the sum of hadronic and weak contribution terms:

$$a = a^{QED} + a^{hadronic} + a^{weak} = \sum_i C_i (\frac{\alpha}{\pi})^i + a^{hadronic} + a^{weak}. \qquad (3.90)$$

The QED part has been computed analytically for $i = 1, 2, 3$, while for $i = 4$ there is a numerical calculation with an error (see, for example, [38] and refs therein). Some terms for $i = 5$ have also been estimated for the muon case. The hadronic contribution is from vacuum polarization insertions and from light by light scattering diagrams (see Fig. 3.10). The weak contribution is from W or Z exchange.

For the electron case the weak contribution is essentially negligible and the hadronic term ($a_e^{hadronic} \sim (16.71 \pm 0.19) \cdot 10^{-13}$) does not introduce an important uncertainty. As a result this measurement can be used to obtain the most precise determination of the fine structure constant [37]:

$$\alpha^{-1} \sim 137.035999710(96)\,, \qquad (3.91)$$

with an uncertainty about 10 times smaller than the previous determination. However, very recently a theoretical error in the α^4 terms was corrected [39]. As a result the value of α^{-1} in Eq. (3.91) is shifted by $-6.41180(73)\ 10^{-7}$ (about 7 σ's). This change has a minor impact in the following discussion of the muon $(g-2)$.

In the muon case the experimental precision is less by about three orders of magnitude, but the sensitivity to new physics effects is typically increased by a factor $(m_\mu/m_e)^2 \sim 4 \cdot 10^4$ (one mass factor arises because the effective operator needs a chirality flip and the second one is because, by definition, one must factor out the Bohr magneton $e/2m$). From the theory side, the QED term (using the value of α from a_e in Eq. (3.91)), and the weak contribution are affected by small errors and are given by (all theory number are taken here from the review [40])

$$a_\mu^{QED} = (116584718.09 \pm 1.6) \cdot 10^{-11}, \qquad a_\mu^{weak} = (154 \pm 2.2) \cdot 10^{-11} \qquad (3.92)$$

The dominant ambiguities arise from the hadronic term. The lowest order (LO) vacuum polarization contribution can be evaluated from the measured cross sections in $e^+e^- \to$ hadrons at low energy via dispersion relations (the largest contribution is from the $\pi\pi$ final state), with the result $a_\mu^{LO} \cdot 10^{-11} = 6909 \pm 44$. The higher order (HO) vacuum polarization contribution (from 2-loop diagrams containing an hadronic insertion) is given by: $a_\mu^{HO} \cdot 10^{-11} = -98 \pm 1$. The contribution of the light by light (LbL) scattering diagrams is estimated to be: $a_\mu^{LbL} \cdot 10^{-11} = 120 \pm 35$. Adding the above contributions up the total hadronic result is reported as:

$$a_\mu^{hadronic} = (6931 \pm 56) \cdot 10^{-11}. \tag{3.93}$$

At face value this would lead to a 3.3σ deviation from the experimental value a_μ^{exp} in Eq. (3.89):

$$a_\mu^{exp} - a_\mu^{th(e^+e^-)} = (275 \pm 84) \cdot 10^{-11}. \tag{3.94}$$

However, the error estimate on the LbL term, mainly a theoretical uncertainty, is not compelling, and it could well be somewhat larger (although probably not by as much as to make the discrepancy to completely disappear). Another puzzle is the fact that, using the conservation of the vector current (CVC) and isospin invariance, which are well established tools at low energy, a_μ^{LO} can also be evaluated from τ decays. But the results on the hadronic contribution from e^+e^- and from τ decay, nominally of comparable accuracy, do not match well, and the discrepancy would be much attenuated if one takes the τ result [41]. Since it is difficult to find a theoretical reason for the e^+e^- vs τ difference, one must conclude that there is something which is not understood either in the data or in the assessment of theoretical errors. The prevailing view is to take the e^+e^- determination as the most directly reliable, which leads to Eq. (3.94), but doubts certainly remain. Finally, we note that, given the great accuracy of the a_μ measurement and the relative importance of the non QED contributions, it is not unreasonable that a first signal of new physics can appear in this quantity.

3.10 Large Radiative Corrections to Electroweak Processes

Since the SM theory is renormalizable higher order perturbative corrections can be reliably computed. Radiative corrections are very important for precision EW tests. The SM inherits all successes of the old V-A theory of charged currents and of QED. Modern tests have focussed on neutral current processes, the W mass and the measurement of triple gauge vertices. For Z physics and the W mass the state of the art computation of radiative corrections include the complete one loop diagrams and selected dominant two loop corrections. In addition some resummation techniques are also implemented, like Dyson resummation of vacuum polarization functions and important renormalization group improvements for large QED and QCD logarithms. We now discuss in more detail sets of large radiative corrections which are particularly significant (for reviews of radiative corrections for LEP1 physics, see, for example: [42]).

Even leaving aside QCD corrections, a set of important quantitative contributions to the radiative corrections arise from large logarithms [e.g. terms of the form $(\alpha/\pi \ln(m_Z/m_{f_{ll}}))^n$ where f_{ll} is a light fermion]. The sequences of leading and close-to-leading logarithms are fixed by well-known and consolidated techniques (β functions, anomalous dimensions, penguin-like diagrams, etc.). For example, large logarithms from pure QED effects dominate the running of α from m_e, the electron mass, up to m_Z. Similarly large logarithms of the form $[\alpha/\pi \ln(m_Z/\mu)]^n$ also enter, for example, in the relation between $\sin^2\theta_W$ at the scales m_Z (LEP, SLC) and μ (e.g. the scale of low-energy neutral-current experiments). Also, large logs from initial state radiation dramatically distort the line shape of the Z resonance as observed at LEP1 and SLC and this effect was accurately taken into account for the measurement of the Z mass and total width. The experimental accuracy on m_Z obtained at LEP1 is $\delta m_Z = \pm 2.1$ MeV (see Chap. 6). Similarly, a measurement of the total width to an accuracy $\delta\Gamma = \pm 2.3$ MeV has been achieved. The prediction of the Z line-shape in the SM to such an accuracy has posed a formidable challenge to theory, which has been successfully met. For the inclusive process $e^+e^- \to f\bar{f}X$, with $f \neq e$ (for a concise discussion, we leave Bhabha scattering aside) and X including γ's and gluons, the physical cross-section can be written in the form of a convolution [42]:

$$\sigma(s) = \int_{z_0}^{1} dz\, \hat{\sigma}(zs) G(z, s)\,, \tag{3.95}$$

where $\hat{\sigma}$ is the reduced cross-section, and $G(z, s)$ is the radiator function that describes the effect of initial-state radiation; $\hat{\sigma}$ includes the purely weak corrections, the effect of final-state radiation (of both γ's and gluons), and also non-factorizable terms (initial- and final-state radiation interferences, boxes, etc.) which, being small, can be treated in lowest order and effectively absorbed in a modified $\hat{\sigma}$. The radiator $G(z, s)$ has an expansion of the form

$$\begin{aligned} G(z, s) = \delta(1-z) + \alpha/\pi(a_{11}L + a_{10}) + (\alpha/\pi)^2(a_{22}L^2 + a_{11}L + a_{20}) + \ \ldots \ + \\ + (\alpha/\pi)^n \sum_{i=0}^{n} a_{ni} L^i\,, \end{aligned} \tag{3.96}$$

where $L = \ln s/m_e^2 \simeq 24.2$ for $\sqrt{s} \simeq m_Z$. All first- and second-order terms are known exactly. The sequence of leading and next-to-leading logs can be exponentiated (closely following the formalism of structure functions in QCD). For $m_Z \approx 91$ GeV, the convolution displaces the peak by $+110$ MeV, and reduces it by a factor of about 0.74. The exponentiation is important in that it amounts to an additional shift of about 14 MeV in the peak position with respect to the one loop radiative correction.

Among the one loop EW radiative corrections, a very remarkable class of contributions are those terms that increase quadratically with the top mass. The sensitivity of radiative corrections to m_t arises from the existence of these terms. The quadratic dependence on m_t (and on other possible widely broken isospin multiplets from new physics) arises because, in spontaneously broken gauge theories, heavy virtual particles do not decouple. On the contrary, in QED or QCD, the running of α and α_s at a scale Q is not affected by heavy quarks with mass $M \gg Q$. According to an intuitive decoupling theorem [43], diagrams with heavy virtual particles of mass M can be ignored at $Q \ll M$ provided that the couplings do not grow with M and that the theory with no heavy particles is still renormalizable. In the spontaneously broken EW gauge theories both requirements are violated. First, one important difference with respect to unbroken gauge theories is in the longitudinal modes of weak gauge bosons. These modes are generated by the Higgs mechanism, and their couplings grow with masses (as is also the case for the physical Higgs couplings). Second the theory without the top quark is no more renormalizable because the gauge symmetry is broken as the (t,b) doublet would not be complete (also the chiral anomaly would not be completely cancelled). With the observed value of m_t the quantitative importance of the terms of order $G_F m_t^2/4\pi^2\sqrt{2}$ is substancial but not dominant (they are enhanced by a factor $m_t^2/m_W^2 \sim 5$ with respect to ordinary terms). Both the large logarithms and the $G_F m_t^2$ terms have a simple structure and are to a large extent universal, i.e. common to a wide class of processes. In particular the $G_F m_t^2$ terms appear in vacuum polarization diagrams which are universal (virtual loops inserted in gauge boson internal lines are independent of the nature of the vertices on each side of the propagator) and in the $Z \to b\bar{b}$ vertex which is not. This vertex is specifically sensitive to the top quark which, being the partner of the b quark in a doublet, runs in the loop. Instead all types of heavy particles could in principle contribute to vacuum polarization diagrams. The study of universal vacuum polarization contributions, also called "oblique" corrections, and of top enhanced terms is important for an understanding of the pattern of radiative corrections. More in general, the important consequence of non decoupling is that precision tests of the electroweak theory may apriori be sensitive to new physics even if the new particles are too heavy for their direct production, but aposteriori no signal of deviation has clearly emerged.

While radiative corrections are quite sensitive to the top mass, they are unfortunately much less dependent on the Higgs mass. If they were sufficiently sensitive by now we would precisely know the mass of the Higgs. But the dependence of one loop diagrams on m_H is only logarithmic: $\sim G_F m_W^2 \log(m_H^2/m_W^2)$. Quadratic terms $\sim G_F^2 m_H^2$ only appear at two loops [44] and are too small to be detectable. The difference with the top case is that the splitting $m_t^2 - m_b^2$ is a direct breaking of the gauge symmetry that already affects the 1- loop corrections, while the Higgs couplings are "custodial" SU(2) symmetric in lowest order.

3.11 Electroweak Precision Tests in the SM and Beyond

For the analysis of electroweak data in the SM one starts from the input parameters: as is the case in any renormalizable theory, masses and couplings have to be specified from outside. One can trade one parameter for another and this freedom is used to select the best measured ones as input parameters. Some of them, α, G_F and m_Z, are very precisely known, as we have seen, some other ones, $m_{f_{light}}$, m_t and $\alpha_s(m_Z)$ are less well determined while m_H is largely unknown. Among the light fermions, the quark masses are badly known, but fortunately, for the calculation of radiative corrections, they can be replaced by $\alpha(m_Z)$, the value of the QED running coupling at the Z mass scale. The value of the hadronic contribution to the running, embodied in the value of $\Delta\alpha^{(5)}_{had}(m_Z^2)$ (see Table 3.1, [8]) is obtained through dispersion relations from the data on $e^+e^- \to$ hadrons at moderate centre-of-mass energies. From the input parameters one computes the radiative corrections to a sufficient precision to match the experimental accuracy. Then one compares the theoretical predictions with the data for the numerous observables which have been measured [45], checks the consistency of the theory and derives constraints on m_t, $\alpha_s(m_Z)$ and m_H. A detailed discussion of all experimental aspects of precision tests of the EW theory is presented in Chap. 6.

The basic tree level relations:

$$\frac{g^2}{8m_W^2} = \frac{G_F}{\sqrt{2}}, \qquad g^2 \sin^2\theta_W = e^2 = 4\pi\alpha \tag{3.97}$$

can be combined into

$$\sin^2\theta_W = \frac{\pi\alpha}{\sqrt{2}G_F m_W^2} \tag{3.98}$$

Always at tree level, a different definition of $\sin^2\theta_W$ is from the gauge boson masses:

$$\frac{m_W^2}{m_Z^2\cos^2\theta_W} = \rho_0 = 1 \quad \Longrightarrow \quad \sin^2\theta_W = 1 - \frac{m_W^2}{m_Z^2} \tag{3.99}$$

where $\rho_0 = 1$ assuming that there are only Higgs doublets. The last two relations can be put into the convenient form

$$(1 - \frac{m_W^2}{m_Z^2})\frac{m_W^2}{m_Z^2} = \frac{\pi\alpha}{\sqrt{2}G_F m_Z^2} \tag{3.100}$$

Beyond tree level, these relations are modified by radiative corrections:

$$(1-\frac{m_W^2}{m_Z^2})\frac{m_W^2}{m_Z^2}=\frac{\pi\alpha(m_Z)}{\sqrt{2}G_F m_Z^2}\frac{1}{1-\Delta r_W}$$

$$\frac{m_W^2}{m_Z^2\cos^2\theta_W}=1+\Delta\rho_m \tag{3.101}$$

The Z and W masses are to be precisely defined in terms of the pole position in the respective propagators. Then, in the first relation the replacement of α with the running coupling at the Z mass $\alpha(m_Z)$ makes Δr_W completely determined at 1-loop by purely weak corrections (G_F is protected from logarithmic running as an indirect consequence of (V-A) current conservation in the massless theory). This relation defines Δr_W unambigously, once the meaning of $\alpha(m_Z)$ is specified (for example, $\bar{M}S$). On the contrary, in the second relation $\Delta\rho_m$ depends on the definition of $\sin^2\theta_W$ beyond the tree level. For LEP physics $\sin^2\theta_W$ is usually defined from the $Z\to\mu^+\mu^-$ effective vertex. At the tree level the vector and axial-vector couplings g_V^μ and g_A^μ are given in Eqs. (3.29). Beyond the tree level a corrected vertex can be written down in terms of modified effective couplings. Then $\sin^2\theta_W\equiv\sin^2\theta_{eff}$ is in general defined through the muon vertex:

$$g_V^\mu/g_A^\mu=1-4\sin^2\theta_{eff}$$

$$\sin^2\theta_{eff}=(1+\Delta k)s_0^2,\qquad s_0^2c_0^2=\frac{\pi\alpha(m_Z)}{\sqrt{2}G_F m_Z^2}$$

$$g_A^{\mu 2}=\frac{1}{4}(1+\Delta\rho) \tag{3.102}$$

We see that s_0^2 and c_0^2 are "improved" Born approximations (by including the running of α) for $\sin^2\theta_{eff}$ and $\cos^2\theta_{eff}$. Actually, since in the SM lepton universality is only broken by masses and is in agreement with experiment within the present accuracy, in practice the muon channel can be replaced with the average over charged leptons.

We can write a symbolic equation that summarizes the status of what has been computed up to now for the radiative corrections (we list some recent work on each item from where older references can be retrieved) Δr_W [46], $\Delta\rho$ [47] and Δk [48]:

$$\Delta r_W,\Delta\rho,\Delta k=g^2\frac{m_t^2}{m_W^2}(1+\alpha_s+\alpha_s^2)+g^2(1+\alpha_s+\sim\alpha_s^2)+g^4\frac{m_t^4}{m_W^4}+g^4\frac{m_t^2}{m_W^2}+\dots \tag{3.103}$$

The meaning of this relation is that the one loop terms of order g^2 are completely known, together with their first order QCD corrections (the second order QCD corrections are only estimated for the g^2 terms not enhanced by m_t^2/m_W^2), and the terms of order g^4 enhanced by the ratios m_t^4/m_W^4 or m_t^2/m_W^2 are also known.

In the SM the quantities Δr_W, $\Delta\rho$, Δk, for sufficiently large m_t, are all dominated by quadratic terms in m_t of order $G_F m_t^2$. The quantity $\Delta\rho_m$ is not independent and can expressed in terms of them. As new physics can more easily be disentangled if not masked by large conventional m_t effects, it is convenient to keep $\Delta\rho$ while trading Δr_W and Δk for two quantities with no contributions of order $G_F m_t^2$. One thus introduces the following linear combinations (epsilon parameters) [49]:

$$
\begin{aligned}
\epsilon_1 &= \Delta\rho, \\
\epsilon_2 &= c_0^2\Delta\rho + \frac{s_0^2\Delta r_W}{c_0^2 - s_0^2} - 2s_0^2\Delta k, \\
\epsilon_3 &= c_0^2\Delta\rho + (c_0^2 - s_0^2)\Delta k.
\end{aligned} \tag{3.104}
$$

The quantities ϵ_2 and ϵ_3 no longer contain terms of order $G_F m_t^2$ but only logarithmic terms in m_t. The leading terms for large Higgs mass, which are logarithmic, are contained in ϵ_1 and ϵ_3. To complete the set of top-enhanced radiative corrections one adds ϵ_b defined from the loop corrections to the $Zb\bar{b}$ vertex. One modifies g_V^b and g_A^b as follows:

$$
\begin{aligned}
g_A^b &= -\frac{1}{2}(1 + \frac{\Delta\rho}{2})(1 + \epsilon_b), \\
\frac{g_V^b}{g_A^b} &= \frac{1 - 4/3\sin^2\theta_{eff} + \epsilon_b}{1 + \epsilon_b}.
\end{aligned} \tag{3.105}
$$

ϵ_b can be measured from $R_b = \Gamma(Z \to b\bar{b})/\Gamma(Z \to \text{hadrons})$ (see Table 3.1). This is clearly not the most general deviation from the SM in the $Z \to b\bar{b}$ vertex but ϵ_b is the quantity where the large m_t corrections are located in the SM. Thus, summarizing, in the SM one has the following "large" asymptotic contributions:

$$
\begin{aligned}
\epsilon_1 &= \frac{3G_F m_t^2}{8\pi^2\sqrt{2}} - \frac{3G_F m_W^2}{4\pi^2\sqrt{2}}\tan^2\theta_W \ln\frac{m_H}{m_Z} + \ldots, \\
\epsilon_2 &= -\frac{G_F m_W^2}{2\pi^2\sqrt{2}}\ln\frac{m_t}{m_Z} + \ldots, \\
\epsilon_3 &= \frac{G_F m_W^2}{12\pi^2\sqrt{2}}\ln\frac{m_H}{m_Z} - \frac{G_F m_W^2}{6\pi^2\sqrt{2}}\ln\frac{m_t}{m_Z} \ldots, \\
\epsilon_b &= -\frac{G_F m_t^2}{4\pi^2\sqrt{2}} + \ldots.
\end{aligned} \tag{3.106}
$$

The ϵ_i parameters vanish in the limit where only tree level SM effects are kept plus pure QED and/or QCD corrections. So they describe the effects of quantum corrections (i.e. loops) from weak interactions. A similar set of parameters are the S, T, U parameters [50]: the shifts induced by new physics on S, T and U are proportional to those induced on ϵ_3, ϵ_1 and ϵ_2, respectively. In principle, with no model dependence, one can measure the four ϵ_i from the basic observables of LEP physics $\Gamma(Z \to \mu^+\mu^-)$, A^{μ}_{FB} and R_b on the Z peak plus m_W. With increasing model dependence, one can include other measurements in the fit for the ϵ_i. For example, use lepton universality to average the μ with the e and τ final states, or include all lepton asymmetries and so on. The present experimental values of the ϵ_i, obtained from a fit of all LEP1-SLD measurements plus m_W, are given by The LEP Electroweak Working Group [8]:

$$\begin{aligned} &\epsilon_1 \cdot 10^3 = 5.4 \pm 1.0, \quad &\epsilon_2 \cdot 10^3 = -8.9 \pm 1.2, \\ &\epsilon_3 \cdot 10^3 = 5.34 \pm 0.94, \quad &\epsilon_b \cdot 10^3 = -5.0 \pm 1.6. \end{aligned} \tag{3.107}$$

Note that the ϵ parameters are of order a few in 10^{-3} and are known with an accuracy in the range 15–30%. As discussed in the next Section, these values are in agreement with the SM with a light Higgs. All models of new physics must be compared with these findings and pass this difficult test.

3.12 Results of the SM Analysis of Precision Tests

The electroweak Z pole measurements, combining the results of all the experiments, are summarised in Table 3.1. The various asymmetries determine the effective electroweak mixing angle for leptons with highest sensitivity. The weighted average of these results, including small correlations, is:

$$\sin^2\theta_{eff} = 0.23153 \pm 0.00016, \tag{3.108}$$

Note, however, that this average has a χ^2 of 11.8 for 5 degrees of freedom, corresponding to a probability of a few %. The χ^2 is pushed up by the two most precise measurements of $\sin^2\theta_{eff}$, namely those derived from the measurements of A_l by SLD, dominated by the left-right asymmetry A^0_{LR}, and of the forward-backward asymmetry measured in $b\bar{b}$ production at LEP, $A^{0,b}_{FB}$, which differ by about 3σs.

We now discuss fitting the data in the SM. One can think of different types of fit, depending on which experimental results are included or which answers one wants to obtain. For example, in Table 3.2 we present in column 1 a fit of all Z pole data plus m_W and Γ_W (this is interesting as it shows the value of m_t obtained indirectly from radiative corrections, to be compared with the value of m_t measured in production experiments), in column 2 a fit of all Z pole data plus

Table 3.1 Summary of electroweak precision measurements at high Q^2 [8]

Observable	Measurement	SM fit
m_Z [GeV]	91.1875 ± 0.0021	91.1875
Γ_Z [GeV]	2.4952 ± 0.0023	2.4957
σ_h^0 [nb]	41.540 ± 0.037	41.477
R_l^0	20.767 ± 0.025	20.744
$AFB^{0,l}$	0.01714 ± 0.00095	0.01645
A_l (SLD)	0.1513 ± 0.0021	0.1481
A_l (P_τ)	0.1465 ± 0.0032	0.1481
R_b^0	0.21629 ± 0.00066	0.21586
R_c^0	0.1721 ± 0.0030	0.1722
$A_{FB}^{0,b}$	0.0992 ± 0.0016	0.1038
$A_{FB}^{0,c}$	0.0707 ± 0.0035	0.0742
A_b	0.923 ± 0.020	0.935
A_c	0.670 ± 0.027	0.668
$\sin^2\theta_{eff}$ (Q_{FB}^{had})	0.2324 ± 0.0012	0.2314
m_W [GeV]	80.398 ± 0.025	80.374
Γ_W [GeV]	2.140 ± 0.060	2.091
m_t [GeV $(p\bar{p})$	170.9 ± 1.8	171.3
$\Delta\alpha_{had}^{(5)}(m_Z^2)$	0.02758 ± 0.00035	0.02768

The first block shows the Z-pole measurements. The second block shows additional results from other experiments: the mass and the width of the W boson measured at the Tevatron and at LEP-2, the mass of the top quark measured at the Tevatron, and the contribution to α of the hadronic vacuum polarization. The SM fit results are derived from the SM analysis of these results

m_t (here it is m_W which is indirectly determined), and, finally, in column 3 a fit of all the data listed in Table 3.1 (which is the most relevant fit for constraining m_H). From the fit in column 1 of Table 3.2 we see that the extracted value of m_t is in good agreement with the direct measurement (see Table 3.1). Similarly we see that the experimental measurement of m_W in Table 3.1 is larger by about one standard deviation with respect to the value from the fit in column 2. We have seen that quantum corrections depend only logarithmically on m_H. In spite of this small sensitivity, the measurements are precise enough that one still obtains a quantitative indication of the mass range. From the fit in column 3 we obtain: $\log_{10} m_H(\text{GeV}) = 1.88 \pm 0.16$ (or $m_H = 76^{+34}_{-24}$ GeV). This result on the Higgs mass is particularly remarkable. The value of $\log_{10} m_H(\text{GeV})$ is compatible with the small window between ~2 and ~3 which is allowed, on the one side, by the direct search limit ($m_H > 114$ GeV from LEP-2 [8]), and, on the other side, by the theoretical upper limit on the Higgs mass in the minimal SM, $m_H \lesssim 600-800$ GeV [51].

Thus the whole picture of a perturbative theory with a fundamental Higgs is well supported by the data on radiative corrections. It is important that there is a clear indication for a particularly light Higgs: at 95% c.l. $m_H \lesssim 182$ GeV (including

Table 3.2 Standard Model fits of electroweak data [8]

Fit	1	2	3
Measurements	m_W	m_t	m_t, m_W
m_t (GeV)	178.9^{+12}_{-9}	170.9 ± 1.8	171.3 ± 1.7
m_H (GeV)	145^{+240}_{-81}	99^{+52}_{-35}	76^{+34}_{-24}
log $[m_H(\text{GeV})]$	$2.16 \pm +0.39$	2.00 ± 0.19	1.88 ± 0.16
$\alpha_s(m_Z)$	0.1190 ± 0.0028	0.1189 ± 0.0027	0.1185 ± 0.0026
m_W (MeV)	80385 ± 19	80360 ± 20	80374 ± 15

All fits use the Z pole results and $\Delta\alpha^{(5)}_{had}(m_Z^2)$ as listed in Table 3.1. In addition, the measurements listed on top of each column are included as well. The fitted W mass is also shown [8] (the directly measured value is $m_W = 80398 \pm 25$ MeV)

the input from the direct search result). This is quite encouraging for the ongoing search for the Higgs particle. More general, if the Higgs couplings are removed from the Lagrangian the resulting theory is non renormalizable. A cutoff Λ must be introduced. In the quantum corrections $\log m_H$ is then replaced by $\log \Lambda$ plus a constant. The precise determination of the associated finite terms would be lost (that is, the value of the mass in the denominator in the argument of the logarithm). A heavy Higgs would need some unfortunate accident: the finite terms, different in the new theory from those of the SM, should by chance compensate for the heavy Higgs in a few key parameters of the radiative corrections (mainly ϵ_1 and ϵ_3, see, for example, [49]). Alternatively, additional new physics, for example in the form of effective contact terms added to the minimal SM lagrangian, should accidentally do the compensation, which again needs some sort of conspiracy.

To the list of precision tests of the SM one should add the results on low energy tests obtained from neutrino and antineutrino deep inelastic scattering (NuTeV [52]), parity violation in Cs atoms (APV [53]) and the recent measurement of the parity-violating asymmetry in Moller scattering [54] (see Chap. 6). When these experimental results are compared with the SM predictions the agreement is good except for the NuTeV result that shows a deviation by three standard deviations. The NuTeV measurement is quoted as a measurement of $\sin^2\theta_W = 1 - m_W^2/m_Z^2$ from the ratio of neutral to charged current deep inelastic cross-sections from ν_μ and $\bar{\nu}_\mu$ using the Fermilab beams. But it has been argued and it is now generally accepted that the NuTeV anomaly probably simply arises from an underestimation of the theoretical uncertainty in the QCD analysis needed to extract $\sin^2\theta_W$. In fact, the lowest order QCD parton formalism on which the analysis has been based is too crude to match the experimental accuracy.

When confronted with these results, on the whole the SM performs rather well, so that it is fair to say that no clear indication for new physics emerges from the data. However, as already mentioned, one problem is that the two most precise measurements of $\sin^2\theta_{\text{eff}}$ from A_{LR} and A^b_{FB} differ by about 3σs. In general, there appears to be a discrepancy between $\sin^2\theta_{\text{eff}}$ measured from leptonic asymmetries ($(\sin^2\theta_{\text{eff}})_l$) and from hadronic asymmetries ($(\sin^2\theta_{\text{eff}})_h$). In fact, the result from

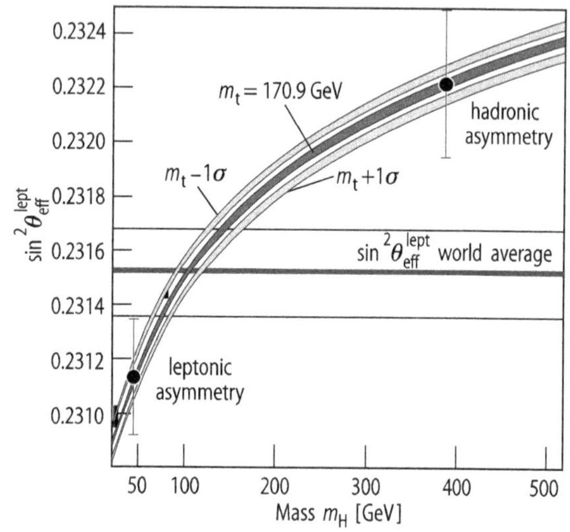

Fig. 3.11 The data for $\sin^2\theta_{\mathrm{eff}}^{\mathrm{lept}}$ are plotted vs m_H. The theoretical prediction for the measured value of m_t is also shown. For presentation purposes the measured points are shown each at the m_H value that would ideally correspond to it given the central value of m_t (updated from [55])

A_{LR} is in good agreement with the leptonic asymmetries measured at LEP, while all hadronic asymmetries, though their errors are large, are better compatible with the result of A_{FB}^b. These two results for $\sin^2\theta_{\mathrm{eff}}$ are shown in Fig. 3.11 [55]. Each of them is plotted at the m_H value that would correspond to it given the central value of m_t. Of course, the value for m_H indicated by each $\sin^2\theta_{\mathrm{eff}}$ has an horizontal ambiguity determined by the measurement error and the width of the $\pm 1\sigma$ band for m_t. Even taking this spread into account it is clear that the implications on m_H are sizably different. One might imagine that some new physics effect could be hidden in the $Zb\bar{b}$ vertex. Like for the top quark mass there could be other non decoupling effects from new heavy states or a mixing of the b quark with some other heavy quark. However, it is well known that this discrepancy is not easily explained in terms of some new physics effect in the $Zb\bar{b}$ vertex. A rather large change with respect to the SM of the b-quark right handed coupling to the Z is needed in order to reproduce the measured discrepancy (precisely a $\sim$30% change in the right-handed coupling), an effect too large to be a loop effect but which could be produced at the tree level, e.g., by mixing of the b quark with a new heavy vectorlike quark [56]), or some mixing of the Z with ad hoc heavy states [57]. But then this effect should normally also appear in the direct measurement of A_b performed at SLD using the left-right polarized b asymmetry, even within the moderate precision of this result. The measurements of neither A_b at SLD nor R_b confirm the need of a new effect. Alternatively, the observed discrepancy could be simply due to a large statistical fluctuation or an unknown experimental problem. As a consequence of this problem, the ambiguity in the measured value of $\sin^2\theta_{\mathrm{eff}}$ is in practice larger than the nominal error, reported in Eq. 3.108, obtained from averaging all the existing determinations, and the interpretation of precision tests is less sharp than it would otherwise be.

We have already observed that the experimental value of m_W (with good agreement between LEP and the Tevatron) is a bit high compared to the SM prediction (see Fig. 3.12). The value of m_H indicated by m_W is on the low side, just in the same interval as for $\sin^2\theta_{\mathrm{eff}}^{\mathrm{lept}}$ measured from leptonic asymmetries.

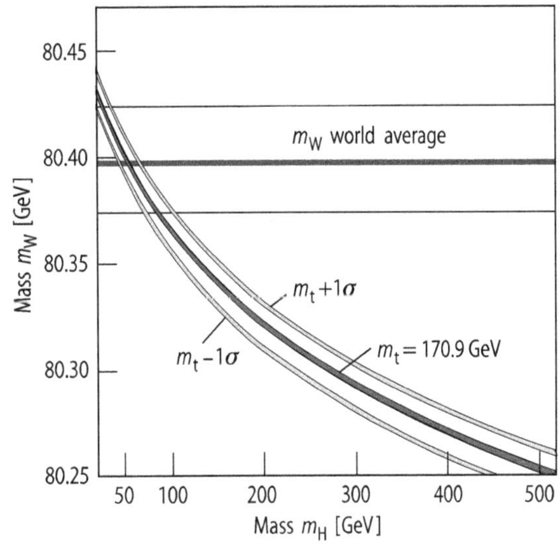

Fig. 3.12 The data for m_W are plotted vs m_H. The theoretical prediction for the measured value of m_t is also shown (updated from [55])

In conclusion, overall the validity of the SM has been confirmed to a level that we can say was unexpected at the beginning. In the present data there is no significant evidence for departures from the SM, no compelling evidence of new physics. The impressive success of the SM poses strong limitations on the possible forms of new physics.

3.13 Phenomenology of the SM Higgs

The Higgs problem is really central in particle physics today. On the one hand, the experimental verification of the Standard Model (SM) cannot be considered complete until the structure of the Higgs sector is not established by experiment. On the other hand, the Higgs is also related to most of the major problems of particle physics, like the flavour problem and the hierarchy problem, the latter strongly suggesting the need for new physics near the weak scale. In turn the discovery of new physics could clarify the dark matter identity. It is clear that the fact that some sort of Higgs mechanism is at work has already been established. The W or the Z with longitudinal polarization that we observe are not present in an unbroken gauge theory (massless spin-1 particles, like the photon, are transversely polarized). The longitudinal degree of freedom for the W or the Z is borrowed from the Higgs sector and is an evidence for it. Also, it has been verified that the gauge symmetry is unbroken in the vertices of the theory: all currents and charges are indeed symmetric. Yet there is obvious evidence that the symmetry is instead badly broken in the masses. Not only the W and the Z have large masses, but the large splitting of, for example, the t-b doublet shows that even a global weak SU(2) is not at all respected by the fermion spectrum. This is a clear signal of spontaneous symmetry breaking and the implementation of spontaneous symmetry breaking in a gauge theory is via the Higgs mechanism. The big remaining questions are about the nature and the properties of the Higgs particle(s). The present experimental information on the Higgs sector, is surprisingly limited and can be summarized in a few lines, as follows. First, the relation $M_W^2 = M_Z^2 \cos^2\theta_W$, Eq. (3.55), modified by small, computable radiative corrections, has been experimentally proven. This relation means that the effective Higgs (be it fundamental or composite) is indeed a weak isospin doublet. The Higgs particle has not been found but, in the SM, its mass can well be larger than the present direct lower limit $m_H \gtrsim 114\,\text{GeV}$ (at 95% c.l.) obtained from searches at LEP-2. The radiative corrections computed in the SM when compared to the data on precision electroweak tests lead to a clear indication for a light Higgs, not too far from the present lower bound. The exact experimental upper limit for m_H in the SM depends on the value of the top quark mass m_t. The CDF and D0 combined value after Run II is at present [8] $m_t = 170.9 \pm 1.8\,\text{GeV}$ (it went down with respect to the value $m_t = 178 \pm 4.3\,\text{GeV}$ from Run I and also the experimental error is now sizably reduced). As a consequence the present limit on m_H is more stringent [8]: $m_H < 182\,\text{GeV}$ (at 95% c.l., after including the information from the 114 GeV direct bound). On the Higgs the LHC will address the following questions : one doublet, more doublets, additional singlets? SM Higgs or SUSY Higgses? Fundamental or composite (of fermions, of WW...)? Pseudo-Goldstone boson of an enlarged symmetry? A manifestation of large extra dimensions (5th component of a gauge boson, an effect of orbifolding or of boundary conditions...)? Or some combination of the above or something so far unthought of? Here in the following we will summarize the main properties of the SM Higgs that provide an essential basis for the planning and the interpretation of the LHC Higgs programme. We start from the mass, then the width and the branching ratios and, finally, the most important production channels.

3.13.1 Theoretical Bounds on the SM Higgs Mass

It is well known [58–60] that in the SM with only one Higgs doublet a lower limit on m_H can be derived from the requirement of vacuum stability (or, in milder form, of a moderate instability, compatible with the lifetime of the Universe [61]). The limit is a function of m_t and of the energy scale Λ where the model breaks down and new physics appears. The Higgs mass enters because it fixes the initial value of the quartic Higgs coupling λ for its running up to the large scale Λ. Similarly an upper bound on m_H (with mild dependence on m_t) is obtained, as described in [62] and refs. therein, from the requirement that for λ no Landau pole appears up to the scale Λ, or in simpler terms, that the perturbative description of the theory remains valid up to Λ. We now briefly recall the derivation of these limits.

The possible instability of the Higgs potential $V[\phi]$ is generated by the quantum loop corrections to the classical expression of $V[\phi]$. At large ϕ the derivative

$V'[\phi]$ could become negative and the potential would become unbound from below. The one-loop corrections to $V[\phi]$ in the SM are well known and change the dominant term at large ϕ according to $\lambda\phi^4 \rightarrow (\lambda + \gamma \log \phi^2/\Lambda^2)\phi^4$. This one-loop approximation is not enough in this case, because it fails at large enough ϕ, when $\gamma \log \phi^2/\Lambda^2$ becomes of order one. The renormalization group improved version of the corrected potential leads to the replacement $\lambda\phi^4 \rightarrow \lambda(\Lambda)\phi'^4(\Lambda)$ where $\lambda(\Lambda)$ is the running coupling and $\phi'(\mu) = \phi \exp \int^t \gamma(t')dt'$, with $\gamma(t)$ being an anomalous dimension function and $t = \log\Lambda/v$ (v is the vacuum expectation value $v = (2\sqrt{2}G_F)^{-1/2}$). As a result, the positivity condition for the potential amounts to the requirement that the running coupling $\lambda(\Lambda)$ never becomes negative. A more precise calculation, which also takes into account the quadratic term in the potential, confirms that the requirements of positive $\lambda(\Lambda)$ leads to the correct bound down to scales Λ as low as $\sim$1 TeV. The running of $\lambda(\Lambda)$ at one loop is given by:

$$\frac{d\lambda}{dt} = \frac{3}{4\pi^2}[\lambda^2 + 3\lambda h_t^2 - 9h_t^4 + \text{small gauge and Yukawa terms}]\,, \tag{3.109}$$

with the normalization such that at $t = 0, \lambda = \lambda_0 = m_H^2/2v^2$ and the top Yukawa coupling $h_t^0 = m_t/v$. We see that, for m_H small and m_t fixed at its measured value, λ decreases with t and can become negative. If one requires that λ remains positive up to $\Lambda = 10^{15}$–10^{19} GeV, then the resulting bound on m_H in the SM with only one Higgs doublet is given by, (also including the effect of the two-loop beta function terms) [60] :

$$m_H(\text{GeV}) > 128.4 + 2.1\,[m_t - 170.9] - 4.5\,\frac{\alpha_s(m_Z) - 0.118}{0.006}\,. \tag{3.110}$$

Note that this limit is evaded in models with more Higgs doublets. In this case the limit applies to some average mass but the lightest Higgs particle can well be below, as it is the case in the minimal SUSY extension of the SM (MSSM).

The upper limit on the Higgs mass in the SM is clearly important for assessing the chances of success of the LHC as an accelerator designed to solve the Higgs problem. The upper limit [62] arises from the requirement that the Landau pole associated with the non asymptotically free behaviour of the $\lambda\phi^4$ theory does not occur below the scale Λ. The initial value of λ at the weak scale increases with m_H and the derivative is positive at large λ (because of the positive λ^2 term—the $\lambda\varphi^4$ theory is not asymptotically free—which overwhelms the negative top-Yukawa term). Thus, if m_H is too large, the point where λ computed from the perturbative beta function becomes infinite (the Landau pole) occurs at too low an energy. Of course in the vicinity of the Landau pole the 2-loop evaluation of the beta function is not reliable. Indeed the limit indicates the frontier of the domain where the theory is well described by the perturbative expansion. Thus the quantitative evaluation of the limit is only indicative, although it has been to some extent supported by simulations of the Higgs sector of the EW theory on the lattice. For the upper limit on m_H one finds [62]

$$m_H \lesssim 180\, GeV \text{ for } \Lambda \sim M_{GUT} - M_{Pl}$$
$$m_H \lesssim 0.5 - 0.8\, TeV \text{ for } \Lambda \sim 1\, TeV. \tag{3.111}$$

In conclusion, for $m_t \sim 171$ GeV, only a small range of values for m_H is allowed, $130 < m_H < \sim 200$ GeV, if the SM holds up to $\Lambda \sim M_{GUT}$ or M_{Pl}.

An additional argument indicating that the solution of the Higgs problem cannot be too far away is the fact that, in the absence of a Higgs particle or of an alternative mechanism, violations of unitarity appear in some scattering amplitudes at energies in the few TeV range [63]. In particular, amplitudes involving longitudinal gauge bosons (those most directly related to the Higgs sector) are affected. For example, at tree level in the absence of Higgs exchange, for $s >> m_Z^2$ one obtains:

$$A(W_L^+ W_L^- \to Z_L Z_L)_{no\ Higgs} \sim i \frac{s}{v^2} \tag{3.112}$$

In the SM this unacceptable large energy behaviour is quenched by the Higgs exchange diagram contribution:

$$A(W_L^+ W_L^- \to Z_L Z_L)_{Higgs} \sim -i \frac{s^2}{v^2(s - m_H^2)} \tag{3.113}$$

Thus the total result in the SM is:

$$A(W_L^+ W_L^- \to Z_L Z_L)_{SM} \sim -i \frac{s m_H^2}{v^2(s - m_H^2)} \tag{3.114}$$

which at large energies saturates at a constant value. To be compatible with unitarity bounds one needs $m_H^2 < 4\pi\sqrt{2}/G_F$ or $m_H < 1.5$ TeV. Both the Landau pole and the unitarity argument show that, if the Higgs is too heavy, the SM becomes a non perturbative theory at energies of o(1 TeV). In conclusion, these arguments imply that the SM Higgs cannot escape detection at the LHC.

3.13.2 SM Higgs Decays

The total width and the branching ratios for the SM Higgs as function of m_H are given in Figs. 3.13 and 3.14, respectively [64].

Since the couplings of the Higgs particle are in proportion to masses, when m_H increases the Higgs becomes strongly coupled. This is reflected in the sharp rise of the total width with m_H. For m_H near its present lower bound of 114 GeV, the width is below 5 MeV, much less than for the W or the Z which have a comparable mass.

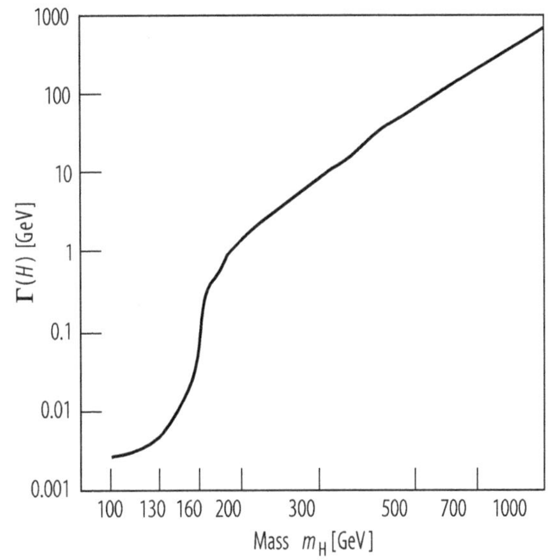

Fig. 3.13 The total width of the SM Higgs boson [64]

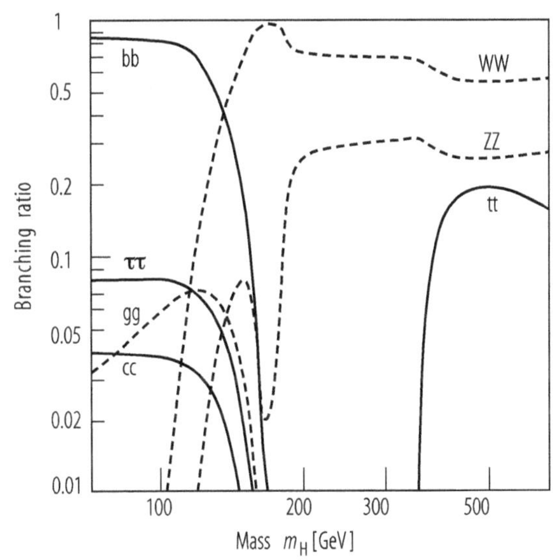

Fig. 3.14 The branching ratios of the SM Higgs boson [65]

The dominant channel for such a Higgs is $H \to b\bar{b}$. In Born approximation the partial width into a fermion pair is given by Djouadi [64] and Haber [66]:

$$\Gamma(H \to f\bar{f}) = N_C \frac{G_F}{4\pi\sqrt{2}} m_H m_f^2 \beta_f^3 \tag{3.115}$$

where $\beta_f = (1 - 4m_f^2/m_H^2)^{1/2}$. The factor of β^3 appears because the fermion pair must be in a p-state of orbital angular momentum for a Higgs with scalar coupling, because of parity (this factor would be β for a pseudoscalar coupling). We see that the width is suppressed by a factor m_f^2/m_H^2 with respect to the natural size $G_F m_H^3$

for the width of a particle of mass m_H decaying through a diagram with only one weak vertex.

A glance to the branching ratios shows that the branching ratio into τ pairs is larger by more than a factor of two with respect to the $c\bar{c}$ channel. This is at first sight surprising because the colour factor N_C favours the quark channels and the masses of τ's and of D mesons are quite similar. This is due to the fact that the QCD corrections replace the charm mass at the scale of charm with the charm mass at the scale m_H, which is lower by about a factor of 2.5. The masses run logarithmically in QCD, similar to the coupling constant. The corresponding logs are already present in the 1-loop QCD correction that amounts to the replacement $m_q^2 \to m_q^2[1 + 2\alpha_s/\pi(\log m_q^2/m_H^2 + 3/2)] \sim m_q^2(m_H^2)$.

The Higgs width sharply increases as the WW threshold is approached. For decay into a real pair of V's, with $V = W, Z$, one obtains in Born approximation [64, 66]:

$$\Gamma(H \to VV) = \frac{G_F m_H^3}{16\pi\sqrt{2}} \delta_V \beta_W (1 - 4x + 12x^2) \tag{3.116}$$

where $\beta_W = \sqrt{1-4x}$ with $x = m_V^2/m_H^2$ and $\delta_W = 2$, $\delta_Z = 1$. Much above threshold the VV channels are dominant and the total width, given approximately by:

$$\Gamma_H \sim 0.5 \text{ TeV}(\frac{m_H}{1 \text{ TeV}})^3 \tag{3.117}$$

becomes very large, signalling that the Higgs sector is becoming strongly interacting (recall the upper limit on the SM Higgs mass in Eq. (3.111)). The VV dominates over the $t\bar{t}$ because of the β threshold factors that disfavour the fermion channel and, at large m_H, by the cubic versus linear behaviour with m_H of the partial widths for VV versus $t\bar{t}$. Below the VV threshold the decays into virtual V particles is important: VV^* and V^*V^*. Note in particular the dip of the ZZ branching ratio just below the ZZ threshold: this is due to the fact that the W is lighter than the Z and the opening of its threshold depletes all other branching ratios. When the ZZ threshold is also passed then the ZZ branching fraction comes back to the ratio of approximately 1:2 with the WW channel (just the number of degrees of freedom: two hermitian fields for the W, one for the Z).

The decay channels into $\gamma\gamma$, $Z\gamma$ and gg proceed through loop diagrams, with the contributions from W (only for $\gamma\gamma$ and $Z\gamma$) and from fermion loops (for all) (Fig. 3.15).

We reproduce here the results for $\Gamma(H \to \gamma\gamma)$ and $\Gamma(H \to gg)$ [64, 66]:

$$\Gamma(H \to \gamma\gamma) = \frac{G_F \alpha^2 m_H^3}{128\pi^3\sqrt{2}} |A_W(\tau_W) + \sum_f N_C Q_f^2 A_f(\tau_f)|^2 \tag{3.118}$$

$$\Gamma(H \to gg) = \frac{G_F \alpha_s^2 m_H^3}{64\pi^3\sqrt{2}} |\sum_{f=Q} A_f(\tau_f)|^2 \tag{3.119}$$

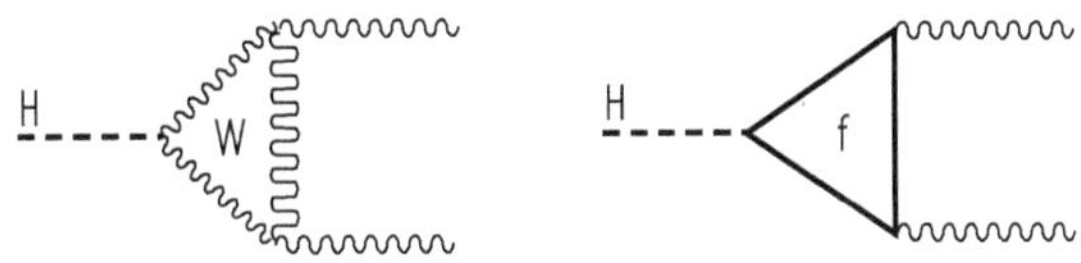

Fig. 3.15 One-loop diagrams for Higgs decay into $\gamma\gamma$, $Z\gamma$ and gg

where $\tau_i = m_H^2/4m_i^2$ and:

$$A_f(\tau) = \frac{2}{\tau^2}[\tau + (\tau - 1)f(\tau)]$$
$$A_W(\tau) = -\frac{1}{\tau^2}[2\tau^2 + 3\tau + 3(2\tau - 1)f(\tau)] \quad (3.120)$$

with:

$$f(\tau) = \arcsin^2\sqrt{\tau} \quad \text{for } \tau \leq 1$$
$$f(\tau) = -\frac{1}{4}[\log\frac{1+\sqrt{1-\tau^{-1}}}{1-\sqrt{1-\tau^{-1}}} - i\pi]^2 \quad \text{for } \tau > 1 \quad (3.121)$$

For $H \to \gamma\gamma$ (as well as for $H \to Z\gamma$) the W loop is the dominant contribution at small and moderate m_H. We recall that the $\gamma\gamma$ mode can be a possible channel for Higgs discovery only for m_H near its lower bound (i.e for $114 < m_H < 150$ GeV). In this domain of m_H we have $\Gamma(H \to \gamma\gamma)$~6–23 KeV. For example, in the limit $m_H << 4m_i^2$, or $\tau \to 0$, we have $A_W(0) = -7$ and $A_f(0) = 4/3$. The two contributions become comparable only for $m_H \sim 650$ GeV where the two amplitudes, still of opposite sign, nearly cancel. The top loop is dominant among fermions (lighter fermions are suppressed by m_f^2/m_H^2 modulo logs) and, as we have seen, it approaches a constant for large m_t. Thus the fermion loop amplitude for the Higgs would be sensitive to effects from very heavy fermions, in particular the $H \to gg$ effective vertex would be sensitive to all possible very heavy coloured quarks. As discussed in the QCD Chapter (Chap. 4) the $gg \to H$ vertex provides one of the main production channels for the Higgs at hadron colliders.

3.14 Limitations of the Standard Model

No signal of new physics has been found neither in electroweak precision tests nor in flavour physics. Given the success of the SM why are we not satisfied with this theory? Why not just find the Higgs particle, for completeness, and declare that particle physics is closed? The reason is that there are both conceptual problems and phenomenological indications for physics beyond the SM. On the conceptual side the most obvious problems are that quantum gravity is not included in the SM and the related hierarchy problem. Among the main phenomenological hints for new physics we can list coupling unification, dark matter, neutrino masses (discussed in Sect. (3.7)), baryogenesis and the cosmological vacuum energy.

The computed evolution with energy of the effective SM gauge couplings clearly points towards the unification of the electro-weak and strong forces (GUT's) at scales of energy $M_{GUT} \sim 10^{15} - 10^{16}$ GeV [31] which are close to the scale of quantum gravity, $M_{Pl} \sim 10^{19}$ GeV. One is led to imagine a unified theory of all interactions also including gravity (at present superstrings provide the best attempt at such a theory). Thus GUT's and the realm of quantum gravity set a very distant energy horizon that modern particle theory cannot ignore. Can the SM without new physics be valid up to such large energies? One can imagine that some obvious problems could be postponed to the more fundamental theory at the Planck mass. For example, the explanation of the three generations of fermions and the understanding of fermion masses and mixing angles can be postponed. But other problems must find their solution in the low energy theory. In particular, the structure of the SM could not naturally explain the relative smallness of the weak scale of mass, set by the Higgs mechanism at $\mu \sim 1/\sqrt{G_F} \sim 250$ GeV with G_F being the Fermi coupling constant. This so-called hierarchy problem is due to the instability of the SM with respect to quantum corrections. This is related to the presence of fundamental scalar fields in the theory with quadratic mass divergences and no protective extra symmetry at $\mu = 0$. For fermion masses, first, the divergences are logarithmic and, second, they are forbidden by the $SU(2) \otimes U(1)$ gauge symmetry plus the fact that at $m = 0$ an additional symmetry, i.e. chiral symmetry, is restored. Here, when talking of divergences, we are not worried of actual infinities. The theory is renormalizable and finite once the dependence on the cut off Λ is absorbed in a redefinition of masses and couplings. Rather the hierarchy problem is one of naturalness. We can look at the cut off as a parameterization of our ignorance on the new physics that will modify the theory at large energy scales. Then it is relevant to look at the dependence of physical quantities on the cut off and to demand that no unexplained enormously accurate cancellations arise.

The hierarchy problem can be put in very practical terms (the "little hierarchy problem"): loop corrections to the Higgs mass squared are quadratic in Λ. The most pressing problem is from the top loop. With $m_h^2 = m_{bare}^2 + \delta m_h^2$ the top loop gives

$$\delta m^2_{h|top} \sim -\frac{3G_F}{2\sqrt{2}\pi^2} m_t^2 \Lambda^2 \sim -(0.2\Lambda)^2 \tag{3.122}$$

If we demand that the correction does not exceed the light Higgs mass indicated by the precision tests, Λ must be close, $\Lambda \sim o(1\,TeV)$. Similar constraints arise from the quadratic Λ dependence of loops with gauge bosons and scalars, which, however, lead to less pressing bounds. So the hierarchy problem demands new physics to be very close (in particular the mechanism that quenches the top loop). Actually, this new physics must be rather special, because it must be very close, yet its effects are not clearly visible neither in precision electroweak tests (the "LEP Paradox" [67]) nor in flavour changing processes and CP violation. Examples of proposed classes of solutions for the hierarchy problem are: (1) Supersymmetry

[68]. In the limit of exact boson-fermion symmetry the quadratic divergences of bosons cancel so that only log divergences remain. However, exact SUSY is clearly unrealistic. For approximate SUSY (with soft breaking terms), which is the basis for all practical models, Λ is replaced by the splitting of SUSY multiplets, $\Lambda^2 \sim m_{SUSY}^2 - m_{ord}^2$. In particular, the top loop is quenched by partial cancellation with s-top exchange, so the s-top cannot be too heavy. (2) Technicolor [69]. The Higgs system is a condensate of new fermions. There is no fundamental scalar Higgs sector, hence no quadratic divergences associated to the μ^2 mass in the scalar potential. This mechanism needs a very strong binding force, $\Lambda_{TC} \sim 10^3 \; \Lambda_{QCD}$. It is difficult to arrange that such nearby strong force is not showing up in precision tests. Hence this class of models has been disfavoured by LEP, although some special class of models have been devised aposteriori, like walking TC, top-color assisted TC etc (for recent reviews, see, for example, [69]). (3) Extra dimensions (for a recent review, see, for example, [70]). The idea is that M_{Pl} appears very large, or equivalently that gravity appears very weak, because we are fooled by hidden extra dimensions so that either the real gravity scale is reduced down to a lower scale, even possibly down to $o(1\ TeV)$ or the intensity of gravity is red shifted away by an exponential warping factor [71]. This possibility is very exciting in itself and it is really remarkable that it is compatible with experiment. It provides a very rich framework with many different scenarios. (4) "Little Higgs" models [72]. In these models the Higgs is a pseudo-Goldstone boson and extra symmetries allow $m_h \neq 0$ only at two-loop level, so that Λ can be as large as $o(10\,\mathrm{TeV})$ with the Higgs within present bounds (the top loop is quenched by exchange of heavy vectorlike new quarks with charge 2/3). The physics beyond the SM will be discussed in Chap. 8.

Acknowledgments I am very grateful to Giuseppe Degrassi, Paolo Gambino, Martin Grunewald, Vittorio Lubicz for their help and advise.

References

1. S.L. Glashow, Nucl. Phys. **22** (1961) 579; S. Weinberg, Phys. Rev. Lett. **19** (1967) 1264; A. Salam, in *Elementary Particle Theory*, ed. N. Svartholm (Almquist and Wiksells, Stockholm, 1969), p. 367.
2. F. Englert, R. Brout, Phys.Rev.Lett.**13**, 321 (1964); P.W. Higgs, Phys.Lett. **12**,132 (1964).
3. E.D.Commins, *Weak interactions*, Mc Graw Hill, 1973; L. V. Okun, *Leptons and Quarks*, North Holland, 1982; D. Bailin, *Weak Interactions*, 2nd e., Hilger, 1982; H. M. Georgi, *Weak interactions and modern particle theory*, Benjamin, 1984.
4. J.D. Bjorken and S. Drell, *Relativistic Quantum Mechanics/Fields*, Vols. I, II, McGraw-Hill, New York, (1965).
5. Particle Data Group, The Journal of Physics G 33(2006)1.
6. G.Altarelli, T.Sjöstrand and F.Zwirner (eds.), "Physics at LEP2", CERN Report 95-03.
7. The ALEPH, DELPHI, L3, OPAL, SLD Collaborations and the LEP Electroweak Working Group, *A Combination of Preliminary Electroweak Measurements and Constraints on the Standard Model*, hep-ex/0312023, and references therein.
8. The LEP Electroweak Working Group, http://lepewwg.web.cern.ch/LEPEWWG/.

9. J. Ellis, J. M. Gaillard and D. V. Nanopoulos, Nucl. Phys. B106 (1976) 292; J. D. Bjorken, SLAC Report 198 (1976).
10. G. Altarelli, B. Mele and F. Pitolli, Nucl. Phys. B287 (1987) 205.
11. G. Altarelli, T. Sjostrand and F. Zwirner, *Physics at LEP2*, CERN Report 96-01, (1996), pag. 361–366.
12. N. Cabibbo, Phys. Rev. Lett., 10 (1963) 531; M. Kobayashi and T. Maskawa, Progr. Th. Phys. 49 (1973) 652.
13. S.L. Glashow, J. Iliopoulos and L. Maiani, Phys. Rev. D2 (1970) 1285.
14. L. Maiani, *Proc. Int. Symp. on Lepton and Photon Interactions at High Energy*, Hamburg,1977.
15. L. Wolfenstein, Phys. Rev. Lett., 51 (1983) 1945.
16. The Unitary Triangle Fit Group, http://www.utfit.org
17. http://ckmfitter.in2p3.fr/
18. C. Jarlskog, Phys. Rev. Lett., 55 (1985) 1039.
19. J. F. Donoghue, E. Golowich and B. Holstein, *Dynamics of the standard model*, Cambridge Univ. Press, 1992.
20. T. Becher and M. Neubert, Phys.Rev.Lett.98:022003,2007, [hep-ph 0610067]; see also M. Misiak et al, Phys.Rev.Lett.98:022002,2007 [hep-ph/0609232].
21. G.L. Fogli et al, *Proceedings of the 40th Rencontres de Moriond on Electroweak Interactions and Unified Theories*, La Thuile, Italy, hep-ph/0506307; M. Maltoni et al, New J. Phys. 6 (2004) 122, [hep-ph/0405172].
22. K. Zuber, Acta Phys.Polon.B37:1905–1921,2006; [nucl-ex/0610007].
23. J. Lesgourgues and S. Pastor, Phys.Rept.429:307–379,2006; [astro-ph/0603494].
24. P. Minkowski, Phys. Lett. B67 (1977) 421; T. Yanagida, in *Proc. of the Workshop on Unified Theory and Baryon Number in the Universe*, KEK (1979); S. L. Glashow, in *Quarks and Leptons'*, Cargèse, ed. M. Lévy et al., Plenum (1980); M. Gell-Mann, P. Ramond and R. Slansky, in *Supergravity*, Stony Brook (1979); R. N. Mohapatra and G. Senjanovic, Phys. Rev. Lett., 44 (1980) 912.
25. Ta-Pei Cheng and Ling-Fong Li, *Gauge theory of elementary particle physics*, Oxford Univ. Press, (1988); D. Bailin and A. Love,*Introduction to gauge field theory (Revised edition)*, (1993); D. Y. Bardin and G. Passarino, *The standard model in the making : precision study of the electroweak interactions*; Oxford Clarendon Press, (1999).
26. J.M. Cornwall, D.N. Levin and G. Tiktopoulos, Phys. Rev. **D10**, 1145 (1974); C.E. Vayonakis, Lett. Nuovo Cim. **17**, 383 (1976); B.W. Lee, C. Quigg and H. Thacker, Phys. Rev. **D16**, 1519 (1977); M. S. Chanowitz and M. K. Gaillard, Nucl. Phys.**B261**, 379 (1985).
27. M. Veltman, Nucl. Phys.**B21**, 288 (1970); G.'t Hooft, Nucl. Phys. **B33**, 173 (1971); **B35**, 167 (1971).
28. B.W. Lee and J. Zinn-Justin, Phys. Rev. **D5**, 3121; 3137 (1972); **D7**, 1049 (1973).
29. C. Bouchiat, J. Iliopoulos and P. Meyer, Phys.Lett.B38:519,1972.
30. S. L. Adler, Phys.Rev.177:2426,1969; S. L. Adler and W. A. Bardeen, Phys.Rev.182:1517,1969; J.S. Bell and R. Jackiw, Nuovo Cim. A60:47,1969; W. A. Bardeen, Phys. Rev. **184** (1969) 1848.
31. G.G.Ross, "Grand Unified Theories", Benjamin, 1985; R.N.Mohapatra, "Unification and Supersymmetry" Springer-Verlag, 1986; S. Raby, hep-ph/0608183. [32] G. 't Hooft, Phys.Rev.Lett.37:8,1976; Phys. Rev. D **14** (1976) 3432 [Erratum-ibid. D **18** (1978) 2199].
32. G. 't Hooft, Phys.Rev.Lett.37:8,1976.
33. A. Belavin, A. Polyakov, A. Shvarts and Y. Tyupkin, Phys. Lett. B **59** (1975) 85.
34. W. Buchmuller, R.D. Peccei and T. Yanagida, Ann.Rev.Nucl.Part.Sci.55:311,2005; [hep-ph 0502169].
35. B. Odom, D. Hanneke, B. D'Urso and G. Gabrielse, Phys. Rev. Lett. 97: 030801 (2006).
36. Muon g-2 Collab., G. W. Bennett et al, Phys. Rev. D73, 072003 (2006).
37. G. Gabrielse et al, Phys.Rev.Lett. 97:030802,2006.
38. T. Kinoshita and N. Nio, Phys. Rev. D 73, 013003(2006).
39. T. Aoyama, M. Hayakawa, T. Kinoshita and M. Nio, Phys. Rev. Lett. **99** (2007) 110406

[arXiv:0706.3496 [hep-ph]]. T. Aoyama, M Hayakawa, T. Kinoshita and N. Nio, hep-ph/0706.3496.
40. M. Passera, Nucl. Phys. Proc. Suppl. **169** (2007) 213 [hep-ph/0702027 [HEP-PH]].
41. M. Davier, Nucl. Phys. Proc. Suppl. **169** (2007) 288 [hep-ph/0701163].
42. G. Altarelli, R. Kleiss and C.Verzegnassi (eds.), Z Physics at LEP1 (CERN 89- 08, Geneva, 1989), Vols. 1–3; Precision Calculations for the Z Resonance, ed. by D.Bardin, W.Hollik and G.Passarino, CERN Rep 95-03 (1995); M.I. Vysotskii, V.A. Novikov, L.B. Okun and A.N. Rozanov, hep-ph/9606253 or Phys.Usp.39:503–538,1996.
43. T. Appelquist and J. Carazzone, Phys.Rev.D11:2856,1975.
44. J. van der Bij and M.J.G. Veltman, Nucl.Phys.B231:205,1984.
45. The LEP and SLD Collaborations, Phys. Rep. **427** (2006) 257, hep-ex/0509008.
46. M. Awramik and M. Czakon, Phys. Rev. Lett. **89** (2002) 241801, [hep-ph/0208113]. M. Awramik, M. Czakon, A. Onishchenko and O. Veretin, Phys. Rev. D **68** (2003) 053004, [hep-ph/0209084]. M. Awramik and M. Czakon, Phys. Lett. B **568** (2003) 48, [hep-ph/0305248]. A. Freitas, W. Hollik, W. Walter and G. Weiglein, Phys. Lett. B **495** (2000) 338, [Erratum-ibid. B **570** (2003) 260], [hep-ph/0007091]. A. Freitas, W. Hollik, W. Walter and G. Weiglein, Nucl. Phys. B **632** (2002) 189, [Erratum-ibid. B **666** (2003) 305], [hep-ph/0202131]. A. Onishchenko and O. Veretin, Phys. Lett. B **551** (2003) 111 [hep-ph/0209010].
47. K. G. Chetyrkin, M. Faisst, J. H. Kuhn, P. Maierhofer and C. Sturm, Phys. Rev. Lett. **97** (2006) 102003, hep-ph/0605201; R. Boughezal and M. Czakon, Nucl. Phys. B **755** (2006) 221, [hep-ph/0606232].
48. M. Awramik, M. Czakon, A. Freitas and G. Weiglein, Phys. Rev. Lett. **93** (2004) 201805, [hep-ph/0407317]; M. Awramik, M. Czakon and A. Freitas, hep-ph/0608099; Phys. Lett. B **642** (2006) 563, [hep-ph/0605339]; W. Hollik, U. Meier and S. Uccirati, hep-ph/0610312.
49. G.Altarelli, R.Barbieri and F.Caravaglios, Int. J. Mod. Phys. A 13(1998)1031 and references therein.
50. M.E. Peskin and T. Takeuchi, Phys. Rev. Lett. 65 (1990) 964; Phys. Rev. D46 (1991) 381.
51. See, for example, M. Lindner, *Z. Phys.* **31**, 295 (1986); T. Hambye and K. Riesselmann, *Phys. Rev.* **D55**, 7255 (1997); hep-ph/9610272.
52. The NuTeV Collaboration, G.P. Zeller et al., Phys. Rev. Lett. **88** (2002) 091802.
53. M. Yu. Kuchiev and V. V. Flambaum, hep-ph/0305053.
54. The SLAC E158 Collaboration, P.L. Anthony et al., hep-ex/0312035, hep-ex/0403010.
55. P. Gambino, Int.J.Mod.Phys.A19:808,2004, [hep-ph/0311257].
56. D. Choudhury, T.M.P. Tait and C.E.M. Wagner, Phys. Rev. D 65 (2002) 053002, hep-ph/0109097.
57. A. Djouadi, G. Moreau and F. Richard, Nucl. Phys. B **773** (2007) 43 [hep-ph/0610173].
58. N. Cabibbo, L. Maiani, G. Parisi and R. Petronzio, *Nucl. Phys.* **B158**,295 (1979).
59. M. Sher, *Phys. Rep.* **179**, 273 (1989); *Phys. Lett.* **B317**, 159 (1993).
60. G. Altarelli and G. Isidori, *Phys. Lett.* **B337**, 141 (1994); J.A. Casas, J.R. Espinosa and M. Quirós, *Phys. Lett.* **B342**, 171 (1995); J.A. Casas et al., *Nucl. Phys.* **B436**, 3 (1995); **EB439**, 466 (1995); M. Carena and C.E.M. Wagner, *Nucl. Phys.* **B452**, 45 (1995).
61. G. Isidori, G. Ridolfi and A. Strumia; *Nucl. Phys.* **B609**,387 (2001).
62. T. Hambye and K.Riesselmann, Phys.Rev.D55:7255,1997.
63. B. W. Lee, C. Quigg and H.B. Thacker, Phys.Rev. D16:1519,1977.
64. A. Djouadi, hep-ph/0503172.
65. E. Accomando et al., Phys. Rep. 299 (1998) 1.
66. H. E. Haber, G. Kane, S. Dawson and J. F. Gunion, *The Higgs Hunter's Guide*, Westview, 1990.
67. R. Barbieri and A. Strumia, hep-ph/0007265.
68. H.P. Nilles, Phys. Rep. C110 (1984) 1; H.E. Haber and G.L. Kane, Phys. Rep. C117 (1985) 75; R. Barbieri, Riv. Nuovo Cim. 11 (1988) 1; S. P. Martin, hep-ph/9709356; M. Drees, R. Godbole and P. Roy, *Theory and Phenomenology of Sparticles*, World Sci. (2004).
69. K. Lane, hep-ph/0202255; R.S. Chivukula, hep-ph/0011264.

70. R. Rattazzi, hep-ph/ 0607055.
71. L. Randall and R. Sundrum, Phys. Rev. Lett. 83 (1999) 3370; 83 (1999) 4690. W.D. Goldberger and M. B. Wise, Phys. Rev. Letters 83 (1999) 4922.
72. M. Schmaltz, hep-ph/0210415;H. C. Cheng and I. Low, JHEP **0408** (2004) 061 [hep-ph/0405243]; J. Hubisz, P. Meade, A. Noble and M. Perelstein, JHEP **0601** (2006) 135 [hep-ph/0506042].

4

Strong Interactions and Quantum Chromo-Dynamics

Guido Altarelli and Stefano Forte

4.1 Introduction

This Chapter[1] is devoted to a concise introduction to Quantum Chromo-Dynamics (QCD), the theory of strong interactions [1–3]. We start with a general introduction where a broad overview of the strong interactions is presented. The basic principles and the main applications of perturbative QCD will be discussed first (for reviews of the subject, see, for example, [4–6]). Then the methods of non perturbative QCD will be introduced, first the analytic approaches and then the simulations of the theory on a discrete space-time lattice.The main emphasis will be on ideas with a minimum of technicalities.

As discussed in Chap. 2 the QCD theory of strong interactions is an unbroken gauge theory based on the group $SU(3)$ of colour. The eight massless gauge bosons are the gluons g^A_μ and matter fields are colour triplets of quarks q^a_i (in different flavours i). Quarks and gluons are the only fundamental fields of the Standard Model (SM) with strong interactions (hadrons). As discussed in Chap. 2, the statement that

The author "G. Altarelli" is deceased at the time of publication.

[1] See Chap. 2 for a general introduction to Chaps. 2–4 with updated references.

G. Altarelli
University of Rome 3, Rome, Italy

S. Forte (✉)
Dipartimento di Fisica, Università di Milano, Milano, Italy

QCD is a renormalisable gauge theory based on the group $SU(3)$ with colour triplet quark matter fields [7] fixes the QCD lagrangian density to be:

$$\mathcal{L} = -\frac{1}{4}\sum_{A=1}^{8} F^{A\mu\nu}F^{A}_{\mu\nu} + \sum_{j=1}^{n_f}\bar{q}_j(i\not{D} - m_j)q_j \tag{4.1}$$

Here: q_j are the quark fields (of n_f different flavours) with mass m_j; $\not{D} = D_\mu\gamma^\mu$, where γ^μ are the Dirac matrices and D_μ is the covariant derivative:

$$D_\mu = \partial_\mu + ie_s\mathbf{g}_\mu; \tag{4.2}$$

e_s is the gauge coupling, later we will mostly use, in analogy with QED

$$\alpha_s = \frac{e_s^2}{4\pi}; \tag{4.3}$$

$\mathbf{g}_\mu = \sum_A t^A g^A_\mu$ where g^A_μ, $A = 1, 8$, are the gluon fields and t^A are the $SU(3)$ group generators in the triplet representation of quarks (i.e. t_A are 3×3 matrices acting on q); the generators obey the commutation relations $[t^A, t^B] = iC_{ABC}t^C$ where C_{ABC} are the complete antisymmetric structure constants of $SU(3)$ (the normalisation of C_{ABC} and of e_s is specified by $Tr[t^At^B] = \delta^{AB}/2$);

$$F^A_{\mu\nu} = \partial_\mu g^A_\nu - \partial_\nu g^A_\mu - e_s C_{ABC} g^B_\mu g^C_\nu \tag{4.4}$$

For quantisation the classical Lagrangian in Eq. (4.1) must be enlarged to contain gauge fixing and ghost terms, as described in Chap. 2. The Feynman rules of QCD are listed in Fig. 4.1. The physical vertices in QCD include the gluon-quark-antiquark vertex, analogous to the QED photon-fermion-antifermion coupling, but also the 3-gluon and 4-gluon vertices, of order e_s and e_s^2 respectively, which have no analogue in an abelian theory like QED.

The QCD lagrangian in Eq. (4.1) has a simple structure but a very rich dynamical content. It gives rise to a complex spectrum of hadrons, it implies the striking properties of confinement and asymptotic freedom, is endowed with an approximate chiral symmetry which is spontaneously broken, has a highly non trivial topological vacuum structure (instantons, $U(1)_A$ symmetry breaking, strong CP violation (which is a problematic item in QCD possibly connected with new physics, like axions), ...), an intriguing phase transition diagram (colour deconfinement, quark-gluon plasma, chiral symmetry restoration, colour superconductivity, ...).

Confinement is the property that no isolated coloured charge can exist but only colour singlet particles. For example, the potential between a quark and an antiquark has been studied on the lattice. It has a Coulomb part at short distances and a linearly rising term at long distances:

$$V_{q\bar{q}} \approx C_F[\frac{\alpha_s(r)}{r} + \ldots. + \sigma r] \tag{4.5}$$

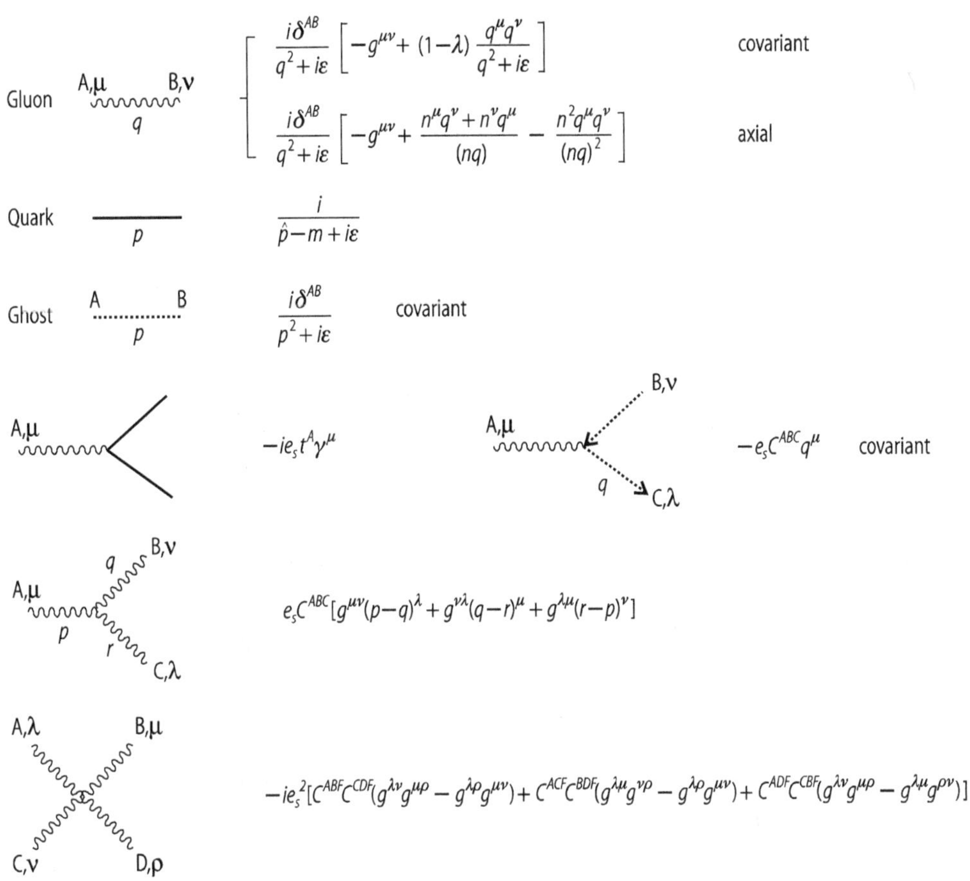

Fig. 4.1 Feynman rules for QCD. The solid lines represent the fermions, the curly lines the gluons, and the dotted lines represent the ghosts (see Chap. 2). The gauge parameter is denoted by λ. The 3-gluon vertex is written as if all gluon lines are outgoing

where

$$C_F = \frac{1}{N_C}\sum_A t^A t^A = \frac{N_C^2-1}{2N_C} \tag{4.6}$$

with N_C the number of colours ($N_C = 3$ in QCD). The scale dependence of α_s (the distance r is Fourier-conjugate to momentum transfer) will be explained in detail in the following. The understanding of the confinement mechanism has much improved thanks to lattice simulations of QCD at finite temperatures and densities. The slope decreases with increasing temperature until it vanishes at a critical temperature T_C. Above T_C the slope remains zero. The phase transitions of colour deconfinement and of chiral restauration appear to happen together on the lattice. A rapid transition is observed in lattice simulations where the energy density $\epsilon(T)$ is seen to sharply increase near the critical temperature for deconfinement and chiral restauration. The critical parameters and the nature of the phase transition depend on the number of quark flavours n_f and on their masses. For example, for $n_f = 2$ or $2 + 1$ (i.e. two light u and d quarks and one heavier s quark), $T_C \sim 175\,\mathrm{MeV}$ and $\epsilon(T_C) \sim 0.5 - 1.0\,\mathrm{GeV/fm^3}$. For realistic values of the masses m_s and $m_{u,d}$ the phase transition appears to be a second order one, while it becomes first order for very small or very large $m_{u,d,s}$. The hadronic phase and the deconfined phase are separated by a crossover line at small densities and by a critical line at high densities. Determining the exact location of the critical point in T and μ_B is an important challenge for theory which is also important for the interpretation of heavy ion collision experiments. At high densities the colour superconducting phase is also present with bosonic diquarks acting as Cooper pairs.

A large investment is being done in experiments of heavy ion collisions with the aim of finding some evidence of the quark gluon plasma phase. Many exciting results have been found at the CERN SPS in the past years and more recently at RHIC. The status of the experimental search for the quark-gluon plasma will be reviewed in Chap. 7.

The linearly rising term in the potential makes it energetically impossible to separate a $q - \bar{q}$ pair. If the pair is created at one space-time point, for example in e^+e^- annihilation, and then the quark and the antiquark start moving away from each other in the center of mass frame, it soon becomes energetically favourable to create additional pairs, smoothly distributed in rapidity between the two leading charges, which neutralise colour and allow the final state to be reorganised into two jets of colourless hadrons, that communicate in the central region by a number of "wee" hadrons with small energy. It is just like the familiar example of the broken magnet: if you try to isolate a magnetic pole by stretching a dipole, the magnet breaks down and two new poles appear at the breaking point.

Confinement is essential to explain why nuclear forces have very short range while massless gluon exchange would be long range. Nucleons are colour singlets and they cannot exchange colour octet gluons but only colourless states. The lightest colour singlet hadronic particles are pions. So the range of nuclear forces is fixed by the pion mass $r \simeq m_\pi^{-1} \simeq 10^{-13}\ cm : V \approx \exp(-m_\pi r)/r$.

Why $SU(N_C = 3)_{colour}$? The selection of $SU(3)$ as colour gauge group is unique in view of a number of constraints. (a) The group must admit complex representations because it must be able to distinguish a quark from an antiquark. In fact there are meson states made up of $q\bar{q}$ but not analogous qq bound states. Among simple groups this restricts the choice to $SU(N)$ with $N \geq 3$, $SO(4N+2)$ with $N \geq 2$ (taking into account that $SO(6)$ has the same algebra as $SU(4)$) and $E(6)$. (b) The group must admit a completely antisymmetric colour singlet baryon made up of 3 quarks: qqq. In fact, from the study of hadron spectroscopy we know that the low lying baryons, completing an octet and a decuplet of (flavour) $SU(3)$ (the approximate symmetry that rotate the three light quarks u, d and s), are made up of three quarks and are colour singlets. The qqq wave function must be completely antisymmetric in colour in order to agree with Fermi statistics. Indeed if we consider, for example, a N^{*++} with spin z-component $+3/2$, this is made up of ($u \Uparrow u \Uparrow u \Uparrow$) in an s-state. Thus its wave function is totally symmetric in space, spin and flavour so that complete antisymmetry in colour is required by Fermi statistics. In QCD this requirement is very simply satisfied by $\epsilon_{abc}q^a q^b q^c$ where a, b, c are $SU(3)_{colour}$ indices. (c) The choice of $SU(N_C = 3)_{colour}$ is confirmed by many processes that directly measure N_C. Some examples are listed here. The total rate for hadronic production in e^+e^- annihilation is linear in N_C. Precisely if we consider $R = \sigma(e^+e^- \to hadrons)/\sigma_{point}(e^+e^- \to \mu^+\mu^-)$ above $b\bar{b}$ threshold and below m_Z and we neglect small computable radiative corrections (that will be discussed later in Sect. 4.5) we have a sum of individual contributions (proportional to Q^2, where Q is the electric charge in units of the proton charge) from $q\bar{q}$ final states with $q = u, c, d, s, b$:

$$R \approx N_C[2 \cdot \frac{4}{9} + 3 \cdot \frac{1}{9}] \approx N_C \frac{11}{9} \tag{4.7}$$

The data neatly indicate $N_C = 3$ as seen from Fig. 4.2 [9]. The slight excess of the data with respect to the value 11/3 is due to the QCD radiative corrections (Sect. 4.5). Similarly we can consider the branching ratio $B(W^- \to e^-\bar{\nu})$, again in Born approximation. The possible fermion-antifermion ($f\bar{f}$) final states are for $f = e^-, \mu^-, \tau^-, d, s$ (there is no $f = b$ because the top quark is too heavy for $b\bar{t}$ to occur). Each channel gives the same contribution, except that for quarks we have N_C colours:

$$B(W^- \to e^-\bar{\nu}) \approx \frac{1}{3 + 2N_C} \tag{4.8}$$

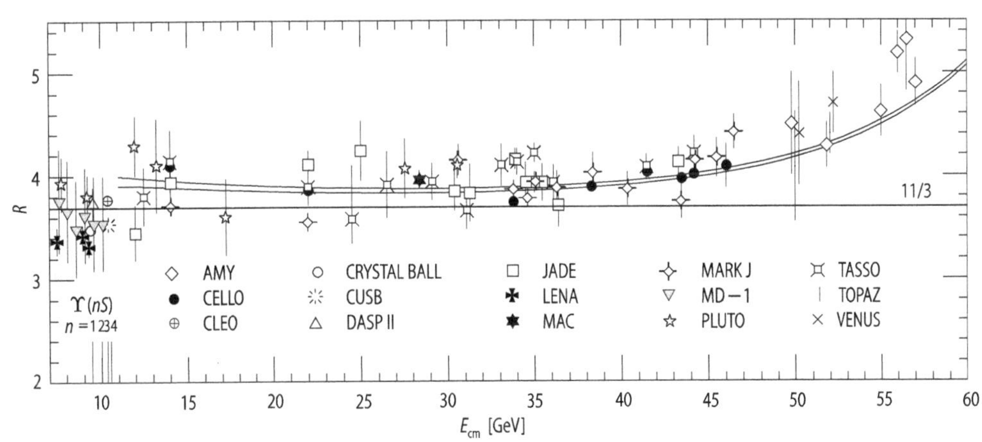

Fig. 4.2 Comparison of the data on $R = \sigma(e^+e^- \to hadrons)/\sigma_{point}(e^+e^- \to \mu^+\mu^-)$ with the QCD prediction [9]. $N_C = 3$ is indicated

For $N_C = 3$ we obtain $B = 11\%$ and the experimental number is $B = 10.7\%$. Another analogous example is the branching ratio $B(\tau^- \to e^- \bar{\nu}_e \nu_\tau)$. From the final state channels with $f = e^-,\ \mu^-,\ d$ we find

$$B(\tau^- \to e^- \bar{\nu}_e \nu_\tau) \approx \frac{1}{2 + N_C} \tag{4.9}$$

For $N_C = 3$ we obtain $B = 20\%$ and the experimental number is $B = 18\%$ (the less accuracy in this case is explained by the larger radiative and phase-space corrections because the mass of τ^- is much smaller than m_W). An important process that is quadratic in N_C is the rate $\Gamma(\pi^0 \to 2\gamma)$. This rate can be reliably calculated from a solid theorem in field theory which has to do with the chiral anomaly:

$$\Gamma(\pi^0 \to 2\gamma) \approx (\frac{N_C}{3})^2 \frac{\alpha^2 m_{\pi^0}^3}{32\pi^3 f_\pi^2} = (7.73 \pm 0.04)(\frac{N_C}{3})^2 \text{ eV} \tag{4.10}$$

where the prediction is obtained for $f_\pi = (130.7 \pm 0.37)$ MeV. The experimental result is $\Gamma = (7.7 \pm 0.5)$ eV in remarkable agreement with $N_C = 3$. There are many more experimental confirmations that $N_C = 3$: for example the rate for Drell-Yan processes (see Sect. 5.4) is inversely proportional to N_C.

How do we get testable predictions from QCD? On the one hand there are non perturbative methods. The most important at present is the technique of lattice simulations: it is based on first principles, it has produced very valuable results on confinement, phase transitions, bound states, hadronic matrix elements and so on, and it is by now an established basic tool. The main limitation is from computing power and therefore there is continuous progress and a lot of good perspectives for the future. Another class of approaches is based on effective lagrangians which provide simpler approximations than the full theory, valid in some definite domain of physical conditions. Chiral lagrangians are based on soft pion theorems and are valid for suitable processes at energies below 1 GeV. Heavy quark effective theories are obtained from expanding in inverse powers of the heavy quark mass and are mainly important for the study of b and, to less accuracy, c decays. The approach of QCD sum rules has led to interesting results but appears to offer not much potential for further development. Similarly specific potential models for quarkonium have a limited range of application. On the other hand, the perturbative approach, based on asymptotic freedom, still remains the main quantitative connection to experiment, due to its wide range of applicability to all sorts of "hard" processes. To perturbative QCD will be devoted the next sections.

4.2 Massless QCD and Scale Invariance

As discussed in Chap. 2, the QCD lagrangian in Eq. (4.1) only specifies the theory at the classical level. The procedure for quantisation of gauge theories involves a number of complications that arise from the fact that not all degrees of freedom of gauge fields are physical because of the constraints from gauge invariance which can be used to eliminate the dependent variables. This is already true for abelian theories and we are familiar with the QED case. One introduces a gauge fixing term (an additional term in the lagrangian density that acts as a Lagrange multiplier in the action extremisation). One can choose to preserve manifest Lorentz invariance. In this case, one adopts a covariant gauge, like the Lorentz gauge, and in QED one proceeds according to the formalism of Gupta-Bleuler. Or one can give up explicit formal covariance and work in a non covariant gauge, like the Coulomb or the axial gauges, and only quantise the physical degrees of freedom (in QED the transverse components of the photon field). While this is all for an abelian gauge theory, in the non-abelian case some additional complications arise, in particular the necessity to introduce ghosts for the formulation of Feynman rules. As we have seen, there are in general as many ghost fields as gauge bosons and they appear in the form of a transformation Jacobian in the Feynman diagram functional integral. Ghosts only propagate in closed loops and their vertices with gluons can be included as additional terms in the lagrangian density which are fixed once the gauge fixing terms and their infinitesimal gauge transformations are specified. Finally the complete Feynman rules in a given gauge can be obtained and they appear in Fig. 4.1.

Once the Feynman rules are derived we have a formal perturbative expansion but loop diagrams generate infinities. First a regularisation must be introduced, compatible with gauge symmetry and Lorentz invariance. This is possible in QCD. In principle one can introduce a cut-off K (with dimensions of energy), for example, a' la Pauli-Villars. But at present the universally adopted regularisation procedure is dimensional regularisation that we will briefly describe later on. After regularisation the next step is renormalisation. In a renormalisable theory (like for all gauge theories in four spacetime dimensions and for QCD in particular) the dependence on the cutoff can be completely reabsorbed in a redefinition of particle masses, of gauge coupling(s) and of wave function normalisations. After renormalisation is achieved the perturbative definition of the quantum theory that corresponds to a classical lagrangian like in Eq. (4.1) is completed. In the QCD Lagrangian of Eq. (4.1) quark masses are the only parameters with physical dimensions (we work in the natural system of units $\hbar = c = 1$). Naively we would expect that massless QCD is scale invariant. This is actually true at the classical level. Scale invariance implies that dimensionless observables should not depend on the absolute scale of energy but only on ratios of energy-dimensional variables. The massless limit should be relevant for the asymptotic large energy limit of processes which are non singular for $m \to 0$.

The naive expectation that massless QCD should be scale invariant is false in the quantum theory. The scale symmetry of the classical theory is unavoidably destroyed by the regularisation and renormalisation procedure which introduce a dimensional parameter in the quantum version of the theory. When a symmetry of the classical theory is necessarily destroyed by quantisation, regularisation and renormalisation one talks of an "anomaly". So, in this sense, scale invariance in massless QCD is anomalous.

While massless QCD is finally not scale invariant, the departures from scaling are asymptotically small, logarithmic and computable. In massive QCD there are additional mass corrections suppressed by powers of m/E, where E is the energy scale (for non singular processes in the limit $m \to 0$). At the parton level (q and g) we can conceive to apply the asymptotics from massless QCD to processes and observables (we use the word "processes" for both) with the following properties ("hard processes"). (a) All relevant energy variables must be large:

$$E_i = z_i Q, \qquad Q >> m_j; \qquad z_i\text{: scaling variables o(1)} \tag{4.11}$$

(b) There should be no infrared singularities (one talks of "infrared safe" processes). (c) The processes concerned must be finite for $m \to 0$ (no mass singularities). To possibly satisfy these criteria processes must be as "inclusive" as possible: one should include all final states with massless gluon emission and add all mass degenerate final states (given that quarks are massless also $q - \bar{q}$ pairs can be massless if "collinear", that is moving together in the same direction at the common speed of light).

In perturbative QCD one computes inclusive rates for partons (the fields in the lagrangian, that is, in QCD, quarks and gluons) and takes them as equal to rates for hadrons. Partons and hadrons are considered as two equivalent sets of complete states. This is called "global duality" and it is rather safe in the rare instance of a totally inclusive final state. It is less so for distributions, like distributions in the invariant mass M ("local duality") where it can be reliable only if smeared over a sufficiently wide bin in M.

Let us discuss more in detail infrared and collinear safety. Consider, for example, a quark virtual line that ends up into a real quark plus a real gluon (Fig. 4.3).

For the propagator we have:

$$\text{propagator} = \frac{1}{(p+k)^2 - m^2} = \frac{1}{2(p \cdot k)} = \frac{1}{2E_k E_p} \cdot \frac{1}{1 - \beta_p \cos\theta} \tag{4.12}$$

Since the gluon is massless, E_k can vanish and this corresponds to an infrared singularity. Remember that we have to take the square of the amplitude and integrate over the final state phase space, or, in this case, all together, dE_k/E_k. Indeed we get $1/E_k^2$ from the squared amplitude and $d^3k/E_k \sim E_k dE_k$ from the phase space.

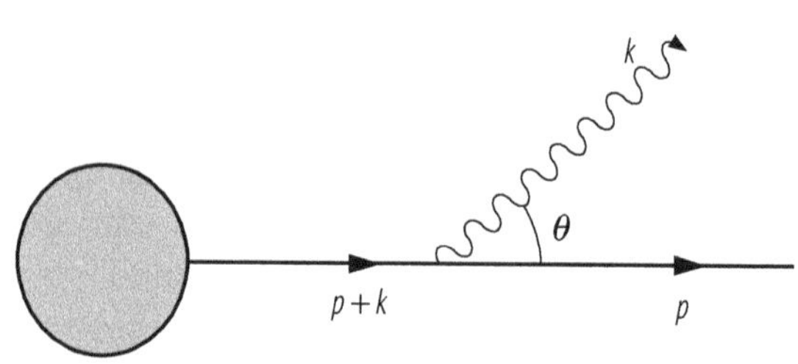

Fig. 4.3 The splitting of a virtual quark into a quark and a gluon

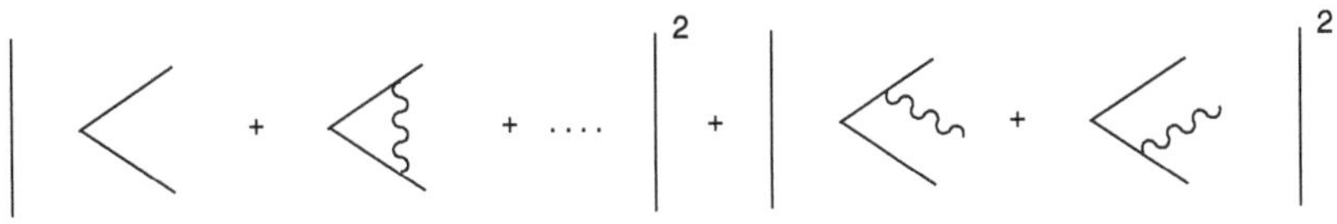

Fig. 4.4 The diagrams contributing to the total cross-section $e^+e^- \to$ hadrons at order α_s. For simplicity, only the final state quarks and (virtual or real) gluons are drawn

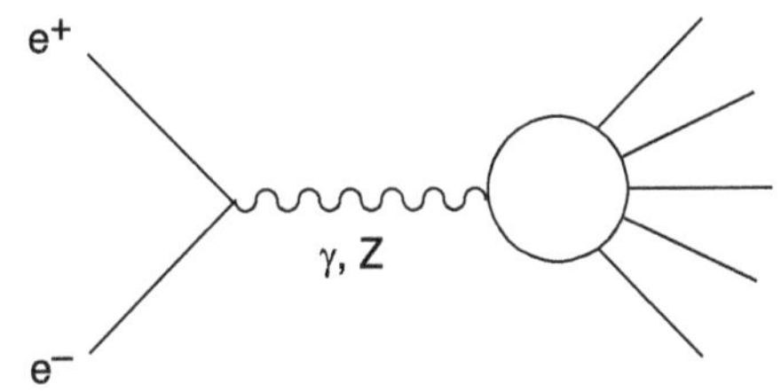

Fig. 4.5 The total cross-section $e^+e^- \to$ hadrons

Also, for $m \to 0$, $\beta_p = \sqrt{1 - m^2/E_p^2} \to 1$ and $(1-\beta_p \cos\theta)$ vanishes at $\cos\theta = 1$. This leads to a collinear mass singularity.

There are two very important theorems on infrared and mass singularities. The first one is the Bloch-Nordsieck theorem [8]: infrared singularities cancel between real and virtual diagrams (see Fig. 4.4) when all resolution indistinguishable final states are added up. For example, for each real detector there is a minimum energy of gluon radiation that can be detected. For the cancellation of infrared divergences, one should add all possible gluon emission with a total energy below the detectable minimum. The second one is the Kinoshita-Lee, Nauenberg theorem [10]: mass singularities connected with an external particle of mass m are canceled if all degenerate states (that is with the same mass) are summed up. That is for a final state particle of mass m we should add all final states that in the limit $m \to 0$ have the same mass, also including gluons and massless pairs. If a completely inclusive final state is taken, only the mass singularities from the initial state particles remain (we shall see that they will be absorbed inside the non perturbative parton densities, which are probability densities of finding the given parton in the initial hadron).

Hard processes to which the massless QCD asymptotics can possibly apply must be infrared and collinear safe, that is they must satisfy the requirements from the Bloch-Nordsieck and the Kinoshita-Lee-Nauenberg theorems. We give now some examples of important hard processes. One of the simplest hard processes is the totally inclusive cross section for hadron production in e^+e^- annihilation, Fig. 4.5, parameterised in terms of the already mentioned dimensionless observable $R = \sigma(e^+e^- \to hadrons)/\sigma_{point}(e^+e^- \to \mu^+\mu^-)$. The pointlike cross section in the denominator is given by $\sigma_{point} = 4\pi\alpha^2/3s$, where $s = Q^2 = 4E^2$ is the squared total center of mass energy and Q is the mass of the exchanged virtual gauge boson. At parton level the final state is $(q\bar{q} + n\ g + n'\ q'\bar{q}')$ and n and n' are limited at each order of perturbation theory. It is assumed that the conversion of partons into

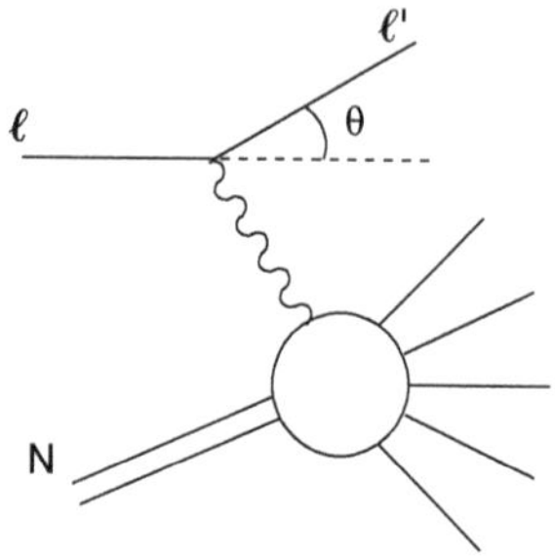

Fig. 4.6 Deep inelastic lepto-production

hadrons does not affect the rate (it happens with probability 1). We have already mentioned that in order for this to be true within a given accuracy an averaging over a sufficiently large bin of Q must be understood. The binning width is larger in the vicinity of thresholds: for example when one goes across the charm $c\bar{c}$ threshold the physical cross-section shows resonance bumps which are absent in the smooth partonic counterpart which however gives an average of the cross-section.

A very important class of hard processes is Deep Inelastic Scattering (DIS)

$$l + N \to l' + X \qquad l = e^{\pm}, \mu^{\pm}, \nu, \bar{\nu} \tag{4.13}$$

which has played and still plays a very important role for our understanding of QCD and of nucleon structure. For the processes in Eq. (4.13), shown in Fig. 4.6, we have, in the lab system where the nucleon of mass m is at rest:

$$Q^2 = -q^2 = -(k-k')^2 = 4EE' \sin^2\theta/2; \qquad m\nu = (p.q); \qquad x = \frac{Q^2}{2m\nu} \tag{4.14}$$

In this case the virtual momentum q of the gauge boson is spacelike. x is the familiar Bjorken variable. The DIS processes in QCD will be extensively discussed in Sect. 4.5

4.3 The Renormalisation Group and Asymptotic Freedom

In this section we aim at providing a reasonably detailed introduction to the renormalisation group formalism and the concept of running coupling which leads to the result that QCD has the property of asymptotic freedom. We start with a summary on how renormalisation works.

In the simplest conceptual situation imagine that we implement regularisation of divergent integrals by introducing a dimensional cut-off K that respects gauge and Lorentz invariance. The dependence of renormalised quantities on K is eliminated by absorbing it into a redefinition of m (the quark mass: for simplicity we assume a single flavour here), the gauge coupling e (can be e in QED or e_s in QCD) and the wave function renormalisation factors $Z_{q,g}^{1/2}$ for q and g, using suitable renormalisation conditions (that is precise definitions of m, g and Z that can be implemented order by order in perturbation theory). For example we can define the renormalised mass m as the position of the pole in the quark propagator and, similarly, the normalisation Z_q as the residue at the pole:

$$\text{Propagator} = \frac{Z_q}{p^2 - m^2} + \text{no} - \text{pole terms} \tag{4.15}$$

The renormalised coupling e can be defined in terms of a renormalised 3-point vertex at some specified values of the external momenta. Precisely, we consider a one particle irreducible vertex (1PI). We recall that a connected Green function is the sum of all connected diagrams, while 1PI Green functions are the sum of all diagrams that cannot be separated into two disconnected parts by cutting only one line.

We now become more specific by concentrating on the case of massless QCD. If we start from a vanishing mass at the classical (or "bare") level, $m_0 = 0$, the mass is not renormalised because it is protected by a symmetry, chiral symmetry. The conserved currents of chiral symmetry are axial currents: $\bar{q}\gamma_\mu\gamma_5 q$. The divergence of the axial current gives, by using the Dirac equation, $\partial^\mu(\bar{q}\gamma_\mu\gamma_5 q) = 2m\bar{q}\gamma_5 q$. So the axial current and the corresponding axial charge are conserved in the massless limit. Since QCD is a vector theory we have not to worry about chiral anomalies in this respect. So one can choose a regularisation that preserves chiral symmetry besides gauge and Lorentz symmetry. Then the renormalised mass remains zero. The renormalised propagator has the form in Eq. (4.15) with $m = 0$.

The renormalised coupling e_s can be defined from the renormalised 1PI 3-gluon vertex at a scale $-\mu^2$ (Fig. 4.7):

$$V_{bare}(p^2, q^2, r^2) = Z V_{ren}(p^2, q^2, r^2), \quad Z = Z_g^{-3/2}, \quad V_{ren}(-\mu^2, -\mu^2, -\mu^2) \to e_s \tag{4.16}$$

We could as well use the quark-gluon vertex or any other vertex which coincides with e_0 in lowest order (even the ghost-gluon vertex, if we want). With a regularisation and renormalisation that preserves gauge invariance we are guaranteed that all these different definitions are equivalent.

Here V_{bare} is what is obtained from computing the Feynman diagrams including, for example, the 1-loop corrections at the lowest non trivial order (V_{bare} is defined

Fig. 4.7 Diagrams contributing to the 1PI 3-gluon vertex at the one-loop approximation level

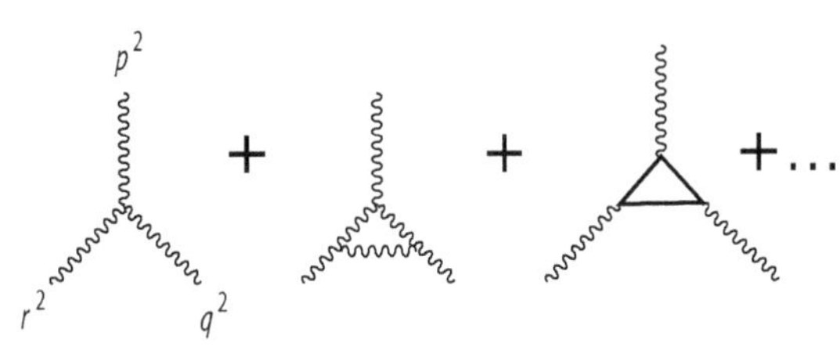

as the scalar function multiplying the vertex tensor, normalised in such a way that it coincides with e_{s0} in lowest order). V_{bare} contains the cut-off K but does not know about μ. Z is a factor that depends both on the cut-off and on μ but not on momenta. Because of infrared singularities the defining scale μ cannot vanish. The negative value $-\mu^2 < 0$ is chosen to stay away from physical cuts (a gluon with negative virtual mass cannot decay). Similarly, in the massless theory, we can define Z_g^{-1} as the inverse gluon propagator (the 1PI 2-point function) at the same scale $-\mu^2$ (the vanishing mass of the gluon is guaranteed by gauge invariance).

After computing all 1-loop diagrams indicated in Fig. 4.7 we have:

$$\begin{aligned} V_{bare}(p^2, p^2, p^2) &= e_{0s}[1 + c\alpha_{0s} \cdot \log \frac{K^2}{p^2} + \ldots] = \\ &= [1 + c\alpha_s \cdot \log \frac{K^2}{-\mu^2} + \ \ldots] e_{0s}[1 + c\alpha_s \cdot \log \frac{-\mu^2}{p^2} + \ldots] \\ &= Z_V^{-1} e_{0s}[1 + c\alpha_s \cdot \log \frac{-\mu^2}{p^2} + \ldots] \\ &= [1 + d\alpha_s \cdot \log \frac{K^2}{-\mu^2} + \ \ldots] e_s[1 + c\alpha_s \cdot \log \frac{-\mu^2}{p^2} + \ldots] \\ &= Z_g^{-3/2} V_{ren} \end{aligned} \tag{4.17}$$

Note the replacement of e_0 with e in the second step, compensated by changing c into d in the first bracket (corresponding to $e_0 = Z_g^{-3/2} Z_V e$). The definition of e_s demands that one precisely specifies what is included in Z. For this, in a given renormalisation scheme, a prescription is fixed to specify the finite terms that go into Z (i.e. the terms of order α_s that accompany $\log K^2$). Then V_{ren} is specified and the renormalised coupling is defined from it according to Eq. (4.16). For example, in the momentum subtraction scheme we define $V_{ren}(p^2, p^2, p^2) = e_s + V_{bare}(p^2, p^2, p^2) - V_{bare}(-\mu^2, -\mu^2, -\mu^2)$, which is equivalent to say, at 1-loop, that all finite terms that do not vanish at $p^2 = -\mu^2$ are included in Z.

A crucial observation is that V_{bare} depends on K but not on μ, which is only introduced when Z, V_{ren} and hence α_s are defined. (From here on, for shorthand, we write α to indicate either the QED coupling or the QCD coupling α_s). More in general for a generic Green function G, we similarly have:

$$G_{bare}(K^2, \alpha_0, p_i^2) = Z_G G_{ren}(\mu^2, \alpha, p_i^2) \tag{4.18}$$

so that we have:

$$\frac{dG_{bare}}{d \log \mu^2} = \frac{d}{d \log \mu^2}[Z_G G_{ren}] = 0 \tag{4.19}$$

or

$$Z_G[\frac{\partial}{\partial \log \mu^2} + \frac{\partial \alpha}{\partial \log \mu^2}\frac{\partial}{\partial \alpha} + \frac{1}{Z_G}\frac{\partial Z_G}{\partial \log \mu^2}]G_{ren} = 0 \tag{4.20}$$

Finally the renormalisation group equation (RGE) can be written as:

$$[\frac{\partial}{\partial \log \mu^2} + \beta(\alpha)\frac{\partial}{\partial \alpha} + \gamma_G(\alpha)]G_{ren} = 0 \tag{4.21}$$

where

$$\beta(\alpha) = \frac{\partial \alpha}{\partial \log \mu^2} \tag{4.22}$$

and

$$\gamma_G(\alpha) = \frac{\partial \log Z_G}{\partial \log \mu^2} \tag{4.23}$$

Note that $\beta(\alpha)$ does not depend on which Green function G we are considering, but it is a property of the theory and the renormalisation scheme adopted, while $\gamma_G(\alpha)$ also depends on G. Strictly speaking the RGE as written above is only valid in the Landau gauge ($\lambda = 0$). In other gauges an additional term that takes the variation of the gauge fixing parameter λ should also be included. We omit this term, for simplicity, as it is not relevant at the 1-loop level.

Assume that we want to apply the RGE to some hard process at a large scale Q, related to a Green function G that we can always take as dimensionless (by multiplication by a suitable power of Q). Since the interesting dependence on Q will be logarithmic we introduce the variable t as :

$$t = \log \frac{Q^2}{\mu^2} \tag{4.24}$$

Then we can write $G_{ren} \equiv F(t, \alpha, x_i)$ where x_i are scaling variables (we often omit to write them in the following). In the naive scaling limit F should be independent of t. To find the actual dependence on t, we want to solve the RGE

$$[-\frac{\partial}{\partial t} + \beta(\alpha)\frac{\partial}{\partial \alpha} + \gamma_G(\alpha)]G_{ren} = 0 \tag{4.25}$$

with a given boundary condition at $t = 0$ (or $Q^2 = \mu^2$): $F(0, \alpha)$.

We first solve the RGE in the simplest case that $\gamma_G(\alpha) = 0$. This is not an unphysical case: for example, it applies to $R_{e^+e^-}$ where the vanishing of γ is related to the non renormalisation of the electric charge in QCD (otherwise the proton and the electron charge would not exactly compensate: this will be better explained in Sect. 4.5). So we consider the equation:

$$[-\frac{\partial}{\partial t} + \beta(\alpha)\frac{\partial}{\partial \alpha}]G_{ren} = 0 \tag{4.26}$$

The solution is simply

$$F(t, \alpha) = F[0, \alpha(t)] \tag{4.27}$$

where the "running coupling" $\alpha(t)$ is defined by:

$$t = \int_{\alpha}^{\alpha(t)} \frac{1}{\beta(\alpha')} d\alpha' \tag{4.28}$$

Note that from this definition it follows that $\alpha(0) = \alpha$, so that the boundary condition is also satisfied. To prove that $F[0, \alpha(t)]$ is indeed the solution, we first take derivatives with respect of t and α (the two independent variables) of both sides of Eq. (4.28). By taking d/dt we obtain

$$1 = \frac{1}{\beta(\alpha(t)} \frac{\partial \alpha(t)}{\partial t} \tag{4.29}$$

We then take $d/d\alpha$ and obtain

$$0 = -\frac{1}{\beta(\alpha)} + \frac{1}{\beta(\alpha(t)} \frac{\partial \alpha(t)}{\partial \alpha} \tag{4.30}$$

These two relations make explicit the dependence of the running coupling on t and α:

$$\frac{\partial \alpha(t)}{\partial t} = \beta(\alpha(t)) \tag{4.31}$$

$$\frac{\partial \alpha(t)}{\partial \alpha} = \frac{\beta(\alpha(t))}{\beta(\alpha)} \tag{4.32}$$

Using these two equations one immediately checks that $F[0, \alpha(t)]$ is indeed the solution.

Similarly, one finds that the solution of the more general equation with $\gamma \neq 0$, Eq. (4.25), is given by:

$$F(t, \alpha) = F[0, \alpha(t)] \exp \int_{\alpha}^{\alpha(t)} \frac{\gamma(\alpha')}{\beta(\alpha')} d\alpha' \tag{4.33}$$

In fact the sum of the two derivatives acting on the factor $F[0, \alpha(t)]$ vanishes and the exponential is by itself a solution of the complete equation. Note that the boundary condition is also satisfied.

The important point is the appearance of the running coupling that determines the asymptotic departures from scaling. The next step is to study the functional form of the running coupling. From Eq. (4.31) we see that the rate of change with t of the running coupling is determined by the β function. In turn $\beta(\alpha)$ is determined by the μ dependence of the renormalised coupling through Eq. (4.22). Clearly there is no dependence on μ of the basic 3-gluon vertex in lowest order (order e). The dependence starts at 1-loop, that is at order e^3 (one extra gluon has to be emitted and reabsorbed). Thus we obtain that in perturbation theory:

$$\frac{\partial e}{\partial \log \mu^2} \propto e^3 \tag{4.34}$$

Recalling that $\alpha = e^2/4\pi$, we have:

$$\frac{\partial \alpha}{\partial \log \mu^2} \propto 2e \frac{\partial e}{\partial \log \mu^2} \propto e^4 \propto \alpha^2 \tag{4.35}$$

Thus the behaviour of $\beta(\alpha)$ in perturbation theory is as follows:

$$\beta(\alpha) = \pm b\alpha^2[1 + b'\alpha + \ldots] \tag{4.36}$$

Since the sign of the leading term is crucial in the following discussion, we stipulate that always $b > 0$ and we make the sign explicit in front.

Let us make the procedure for computing the 1-loop beta function in QCD (or, similarly, in QED) more precise. The result of the 1loop 1PI diagrams for V_{ren} can be written down as (we denote e_s and α_s by e and α, for shorthand):

$$V_{ren} = e[1 + \alpha B_{3g} \log \frac{\mu^2}{-p^2} + \ldots] \tag{4.37}$$

V_{ren} satisfies the RGE:

$$[\frac{\partial}{\partial \log \mu^2} + \beta(\alpha)\frac{\partial e}{\partial \alpha}\frac{\partial}{\partial e} - \frac{3}{2}\gamma_g(\alpha)]V_{ren} = 0 \tag{4.38}$$

With respect to Eq. (4.21) the beta function term has been rewritten taking into account that V_{ren} starts with e and the anomalous dimension term arises from a factor $Z_g^{-1/2}$ for each gluon leg. In general for a n-leg 1PI Green function $V_{n,bare} = Z_g^{-n/2} V_{n,ren}$, if all external legs are gluons. Note that in the particular case of $V = V_3$ that is used to define e other Z factors are absorbed in the replacement

$Z_V^{-1} Z_g^{3/2} e_0 = e$. At 1-loop accuracy we replace $\beta(\alpha) = -b\alpha^2$ and $\gamma_g(\alpha) = \gamma_g^{(1)}\alpha$. All together one obtains:

$$b = 2(B_{3g} - \frac{3}{2}\gamma_g^{(1)}) \tag{4.39}$$

Similarly we can write the diagrammatic expression and the RGE for the 1PI 2-gluon Green function which is the inverse gluon propagator Π (a scalar function after removing the gauge invariant tensor):

$$\Pi_{ren} \;=\; [1 + \alpha B_{2g} \log \frac{\mu^2}{-p^2} + \ldots] \tag{4.40}$$

and

$$[\frac{\partial}{\partial \log \mu^2} \;+\; \beta(\alpha)\frac{\partial}{\partial \alpha} \;-\; \gamma_g(\alpha)]\Pi_{ren} \;=\; 0 \tag{4.41}$$

Notice that the normalisation and the phase of Π are specified by the lowest order term being one. In this case the β function term is negligible being of order α^2 (because Π is a function of e only through α). and we obtain:

$$\gamma_g^{(1)} = B_{2g} \tag{4.42}$$

Thus, finally:

$$b = 2(B_{3g} - \frac{3}{2}B_{2g}) \tag{4.43}$$

By direct calculation at 1-loop one finds:

$$\text{QED}: \quad \beta(\alpha) \;\sim\; +b\alpha^2 + \ldots.. \qquad b \;=\; \sum_i \frac{N_C Q_i^2}{3\pi} \tag{4.44}$$

where $N_C = 3$ for quarks and $N_C = 1$ for leptons and the sum runs over all fermions of charge $Q_i e$ that are coupled. Also, one finds:

$$\text{QCD}: \quad \beta(\alpha) \;\sim\; -\,b\alpha^2 + \ldots.. \qquad b \;=\; \frac{11N_C - 2n_f}{12\pi} \tag{4.45}$$

where, as usual, n_f is the number of coupled flavours of quarks (we assume here that $n_f \;\leq\; 16$ so that $b > 0$ in QCD). If $\alpha(t)$ is small we can compute $\beta(\alpha(t))$ in perturbation theory. The sign in front of b then decides the slope of the coupling: $\alpha(t)$ increases with t (or Q^2) if β is positive at small α (QED), or $\alpha(t)$ decreases with t (or Q^2) if β is negative at small α (QCD). A theory like QCD where the running coupling vanishes asymptotically at large Q^2 is called (ultraviolet) "asymptotically free". An important result that has been proven is that in four spacetime dimensions all and only non-abelian gauge theories are asymptotically free.

Going back to Eq. (4.28) we replace $\beta(\alpha) \sim \pm b\alpha^2$, do the integral and perform a simple algebra. We find

$$\text{QED}: \quad \alpha(t) \sim \frac{\alpha}{1 - b\alpha t} \tag{4.46}$$

and

$$\text{QCD}: \quad \alpha(t) \sim \frac{\alpha}{1 + b\alpha t} \tag{4.47}$$

A slightly different form is often used in QCD. Defining $1/\alpha = b \log \mu^2/\Lambda^2_{QCD}$ we can write:

$$\alpha(t) \sim \frac{1}{\frac{1}{\alpha} + bt} = \frac{1}{b \log \frac{\mu^2}{\Lambda^2_{QCD}} + b \log \frac{Q^2}{\mu^2}} = \frac{1}{b \log \frac{Q^2}{\Lambda^2_{QCD}}} \tag{4.48}$$

We see that $\alpha(t)$ decreases logarithmically with Q^2 and that one can introduce a dimensional parameter Λ_{QCD} that replaces μ. Often in the following we will simply write Λ for Λ_{QCD}. Note that it is clear that Λ depends on the particular definition of α, not only on the defining scale μ but also on the renormalisation scheme (see, for example, the discussion in the next session). Through the parameter b, and in general through the β function, it also depends on the number n_f of coupled flavours. It is very important to note that QED and QCD are theories with "decoupling": up to the scale Q only quarks with masses $m << Q$ contribute to the running of α. This is clearly very important, given that all applications of perturbative QCD so far apply to energies below the top quark mass m_t. For the validity of the decoupling theorem [11] it is necessary that the theory where all the heavy particle internal lines are eliminated is still renormalisable and that the coupling constants do not vary with the mass. These requirements are true for the mass of heavy quarks in QED and QCD, but are not true in the electroweak theory where the elimination of the top would violate $SU(2)$ symmetry (because the t and b left quarks are in a doublet) and the quark couplings to the Higgs multiplet (hence to the longitudinal gauge bosons) are proportional to the mass. In conclusion, in QED and QCD, quarks with $m >> Q$ do not contribute to n_f in the coefficients of the relevant β function. The effects of heavy quarks are power suppressed and can be taken separately into account. For example, in e^+e^- annihilation for $2m_c < Q < 2m_b$ the relevant asymptotics is for $n_f = 4$, while for $2m_b < Q < 2m_t$ $n_f = 5$. Going accross the b threshold the β function coefficients change, so the $\alpha(t)$ slope changes. But $\alpha(t)$ is continuous, so that Λ changes so as to keep constant $\alpha(t)$ at the matching point at $Q \sim o(2m_b)$. The effect on Λ is large: approximately $\Lambda_5 \sim 0.65\Lambda_4$.

Note the presence of a pole in Eqs. (4.46, 4.47) at $\pm b\alpha t = 1$, called the Landau pole, who realised its existence in QED already in the '50's. For $\mu \sim m_e$ (in QED) the pole occurs beyond the Planck mass. In QCD the Landau pole is located for negative t or at $Q < \mu$ in the region of light hadron masses. Clearly the issue of the definition and the behaviour of the physical coupling (which is always finite, when defined in terms of some physical process) in the region around the perturbative Landau pole is a problem that lies outside the domain of perturbative QCD.

The non leading terms in the asymptotic behaviour of the running coupling can in principle be evaluated going back to Eq. (4.36) and computing b' at 2-loops and so on. But in general the perturbative coefficients of $\beta(\alpha)$ depend on the definition of the renormalised coupling α (the renormalisation scheme), so one wonders whether it is worthwhile to do a complicated calculation to get b' if then it must be repeated for a different definition or scheme. In this respect it is interesting to remark that actually both b and b' are independent of the definition of α, while higher order coefficients do depend on that. Here is the simple proof. Two different perturbative definitions of α are related by $\alpha' \sim \alpha(1 + c_1\alpha + \ldots)$. Then we have:

$$\begin{aligned}\beta(\alpha') &= \frac{d\alpha'}{d\log\mu^2} = \frac{d\alpha}{d\log\mu^2}(1 + 2c_1\alpha + \ldots)\\ &= \pm b\alpha^2(1 + b'\alpha + \ldots)(1 + 2c_1\alpha + \ldots)\\ &= \pm b\alpha'^2(1 + b'\alpha' + \ldots)\end{aligned} \tag{4.49}$$

which shows that, up to the first subleading order, $\beta(\alpha')$ has the same form as $\beta(\alpha)$.

In QCD ($N_C = 3$) one has calculated:

$$b' = \frac{153 - 19n_f}{2\pi(33 - 2n_f)} \tag{4.50}$$

By taking b' into account one can write the expression of the running coupling at next to the leading order (NLO):

$$\alpha(Q^2) = \alpha_{LO}(Q^2)[1 - b'\alpha_{LO}(Q^2)\log\log\frac{Q^2}{\Lambda^2} + \ldots] \tag{4.51}$$

where $\alpha_{LO}^{-1} = b\log Q^2/\Lambda^2$ is the LO result (actually at NLO the definition of Λ is modified according to $b\log\mu^2/\Lambda^2 = 1/\alpha + b'\log b\alpha$).

Summarizing, we started from massless classical QCD which is scale invariant. But we have seen that the procedure of quantisation, regularisation and renormalisation necessarily breaks scale invariance. In the quantum QCD theory there is a scale of energy, Λ, which from experiment is of the order of a few hundred MeV, its precise value depending on the definition, as we shall see in detail. Dimensionless quantities depend on the energy scale through the running coupling which is a logarithmic function of Q^2/Λ^2. In QCD the running coupling decreases logarithmically at large Q^2 (asymptotic freedom), while in QED the coupling has the opposite behaviour.

4.4 More on the Running Coupling

In the previous section we have introduced the renormalised coupling α in terms of the 3-gluon vertex at $p^2 = -\mu^2$ (momentum subtraction). The Ward identities of QCD then ensure that the coupling defined from other vertices like the $\bar{q}qg$ vertex are renormalised in the same way and the finite radiative corrections are related. But at present the universally adopted definition of α_s is in terms of dimensional regularisation because of computational simplicity which is essential given the great complexity of present day calculations. So we now briefly review the principles of dimensional regularisation and the definition of Minimal Subtraction (MS) and Modified Minimal Subtraction ($\overline{MS}$). The $\overline{MS}$ definition of α_s is the one most commonly adopted in the literature and a value quoted for it is nomally referring to this definition.

Dimensional Regularisation (DR) is a gauge and Lorentz invariant regularisation that consists in formulating the theory in $D < 4$ spacetime dimensions in order to make loop integrals ultraviolet finite. In DR one rewrites the theory in D dimensions (D is integer at the beginning, but then we will see that the expression of diagrams makes sense at all D except for isolated singularities). The metric tensor is extended into a $D \times D$ matrix $g_{\mu\nu} = diag(1, -1, -1, ..., -1)$ and 4-vectors are given by $k^\mu = (k^0, k^1, \ldots, k^{D-1})$. The Dirac γ^μ are $f(D) \times f(D)$ matrices and it is not important what is the precise form of the function $f(D)$. It is sufficient to extend the usual algebra in a straightforward way like $\{\gamma_\mu, \gamma_\nu\} = 2g_{\mu,\nu}I$, with I the D-dimensional identity matrix, $\gamma^\mu\gamma^\nu\gamma_\mu = -(D-2)\gamma^\nu$ or $Tr(\gamma^\mu\gamma^\nu) = f(D)g_{\mu\nu}$.

The physical dimensions of fields change in D dimensions and, as a consequence, the gauge couplings become dimensional $e_D = \mu^\epsilon e$, where e is dimensionless, $D = 4 - 2\epsilon$ and μ is a scale of mass (this is how a scale of mass is introduced in the DR of massless QCD!). The dimension of fields is determined by requiring that the action $S = \int d^D x \mathcal{L}$ is dimensionless. By inserting for $\mathcal{L}$ terms like $m\bar{\Psi}\Psi$ or $m^2\phi^\dagger\phi$ or $e\bar{\Psi}\gamma^\mu\Psi A_\mu$ the dimensions of the fields and coupling are determined as: $m, \Psi, \phi, A_\mu, e = 1, (D-1)/2, (D-2)/2, (D-2)/2, (4-D)/2$, respectively. The formal expression of loop integrals can be written for any D. For example:

$$\int \frac{d^D k}{(2\pi)^D} \frac{1}{(k^2 - m^2)^2} = \frac{\Gamma(2 - D/2)(-m^2)^{D/2-2}}{(4\pi)^{D/2}} \tag{4.52}$$

For $D = 4 - 2\epsilon$ one can expand using:

$$\Gamma(\epsilon) = \frac{1}{\epsilon} - \gamma_E + o(\epsilon), \qquad \gamma_E = 0.5772\ldots.. \tag{4.53}$$

For some Green function G, normalised to one in lowest order, (like V/e with V the 3-g vertex function at the symmetric point $p^2 = q^2 = r^2$, considered in the previous section) we typically find at 1-loop:

$$G_{bare} = 1 + \alpha_0(\frac{-\mu^2}{p^2})^\epsilon\, [B(\frac{1}{\epsilon} + \log 4\pi - \gamma_E) + A + o(\epsilon)] \tag{4.54}$$

In $\overline{MS}$ one rewrites this at 1-loop accuracy (diagram by diagram: this is a virtue of the method):

$$\begin{aligned} G_{bare} &= Z G_{ren} \\ Z &= 1 + \alpha\, [B(\frac{1}{\epsilon} + \log 4\pi - \gamma_E)] \\ G_{ren} &= 1 + \alpha\, [B \log \frac{-\mu^2}{p^2} + A] \end{aligned} \tag{4.55}$$

Here Z stands for the relevant product of renormalisation factors. In the original MS prescription only $1/\epsilon$ was subtracted (that clearly plays the role of a cutoff) and not also $\log 4\pi$ and γ_E. Later, since these constants always appear from the expansion of Γ functions it was decided to modify MS into $\overline{MS}$. Note that the $\overline{MS}$ definition of α is different than that in the momentum subtraction scheme because the finite terms (those beyond logs) are different. In particular here δG_{ren} does not vanish at $p^2 = -\mu^2$.

The third [12] and fourth [13] coefficients of the QCD β function are also known in the $\overline{MS}$ prescription (recall that only the first two coefficients are scheme independent). The calculation of the last term involved the evaluation of some 50,000 4-loop diagrams. Translated in numbers, for $n_f = 5$ one obtains :

$$\beta(\alpha) = -0.610\alpha^2[1 + 1.261\ldots\frac{\alpha}{\pi} + 1.475\ldots(\frac{\alpha}{\pi})^2 + 9.836\ldots(\frac{\alpha}{\pi})^3\ldots] \tag{4.56}$$

It is interesting to remark that the expansion coefficients are all of order 1 or (10 for the last one), so that the $\overline{MS}$ expansion looks reasonably well behaved.

It is important to keep in mind that the QED and QCD perturbative series, after renormalisation, have all their coefficients finite, but the expansion does not converge. Actually the perturbative series are not even Borel summable. After Borel resummation for a given process one is left with a result which is ambiguous by terms typically down by $\exp -n/(b\alpha)$, with n an integer and b the first β function coefficient. In QED these corrective terms are extremely small and not very important in practice. On the contrary in QCD $\alpha = \alpha_s(Q^2) \sim 1/(b \log Q^2/\Lambda^2)$ and the ambiguous terms are of order $(1/Q^2)^n$, that is are power suppressed. It is interesting that, through this mechanism, the perturbative version of the theory is able to somehow take into account the power suppressed corrections. A sequence of diagrams with factorial growth at large order n is made up by dressing gluon propagators by any number of quark bubbles together with their gauge completions (renormalons).The problem of the precise relation between the ambiguities of the perturbative expansion and the higher twist corrections has been discussed in recent years [14].

4.5 Application to Hard Processes

4.5.1 $R_{e^+e^-}$ and Related Processes

The simplest hard process is $R_{e^+e^-}$ that we have already started to discuss. R is dimensionless and in perturbation theory is given by $R = N_C \sum_i Q_i^2 F(t, \alpha_s)$, where $F = 1 + o(\alpha_S)$. We have already mentioned that for this process the "anomalous dimension" function vanishes: $\gamma(\alpha_s) = 0$ because of electric charge non renormalisation by strong interactions. Let us review how this happens in detail. The diagrams that are relevant for charge renormalisation in QED at 1-loop are shown in Fig. 4.8. The Ward identity that follows from gauge invariance in QED imposes that the vertex (Z_V) and the self-energy (Z_f) renormalisation factors cancel and the only divergence remains in Z_γ, the vacuum polarization of the photon. So the charge is only renormalised by the photon blob, hence it is universal (the same factor for all fermions, independent of their charge) and is not affected by QCD at 1-loop. It is true that at higher orders the photon vacuum polarization diagram is affected by QCD (for example, at 2-loops we can exchange a gluon between the quarks in the photon loop) but the renormalisation induced by the vacuum polarisation diagram remains independent of the nature of the fermion to which the photon line is attached. The gluon contributions to the vertex (Z_V) and to the

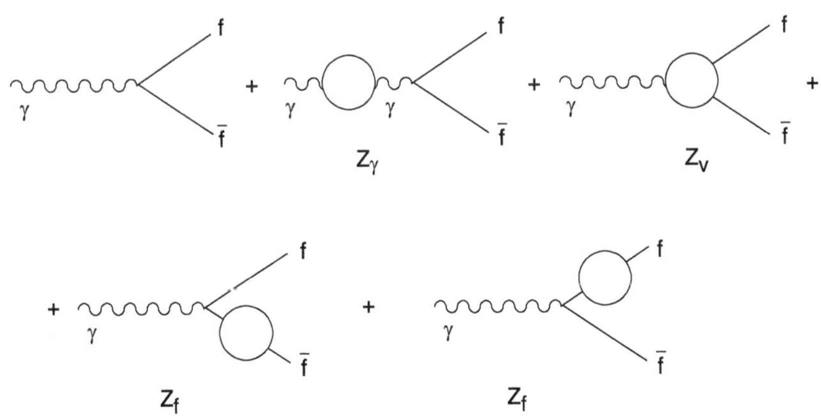

Fig. 4.8 Diagrams for charge renormalisation in QED at 1-loop

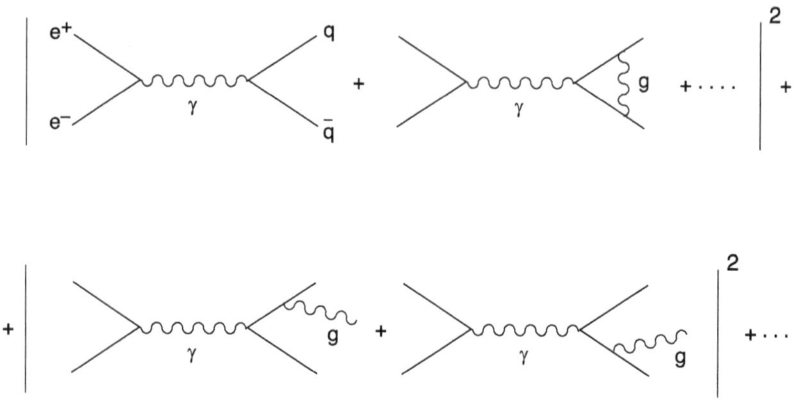

Fig. 4.9 Real and virtual diagrams relevant for the computation of R at 1-loop accuracy

self-energy (Z_f) cancel because they have exactly the same structure as in QED, so that $\gamma(\alpha_s) = 0$.

At 1-loop the diagrams relevant for the computation of R are shown in Fig. 4.9. There are virtual diagrams and real diagrams with one additional gluon in the final state. Infrared divergences cancel between the interference term of the virtual diagrams and the absolute square of the real diagrams, according to the Bloch-Nordsieck theorem. Similarly there are no mass singularities, in agreement with the Kinoshita-Lee-Nauenberg theorem, because the initial state is purely leptonic and all degenerate states that can appear at the given order are included in the final state. Given that $\gamma(\alpha_s) = 0$ the RGE prediction is simply given, as we have already seen, by $F(t, \alpha_s) = F[0, \alpha_s(t)]$. This means that if we do, for example, a 2-loop calculation, we must obtain a result of the form:

$$F(t, \alpha_s) = 1 + c_1\alpha_s(1 - b\alpha_s t) + c_2\alpha_s^2 + o(\alpha_s^3) \tag{4.57}$$

In fact we see that this form, taking into account that from Eq. (4.47) we have:

$$\alpha_s(t) \sim \frac{\alpha_s}{1 + b\alpha_s t} \sim \alpha_s(1 - b\alpha_s t + \ldots.) \tag{4.58}$$

can be rewritten as

$$F(t, \alpha_s) = 1 + c_1\alpha_s(t) + c_2\alpha_s^2(t) + o(\alpha_s^3(t)) = F[0, \alpha_s(t)] \tag{4.59}$$

The content of the RGE prediction is, at this order, that there are no $\alpha_s t$ and $(\alpha_s t)^2$ terms (the leading log sequence must be absent) and the term of order $\alpha_s^2 t$ has the coefficient that allows to reabsorb it in the transformation of α_s into $\alpha_s(t)$.

At present the first three coefficients have been computed in the $\overline{MS}$ scheme [15]. Clearly $c_1 = 1/\pi$ does not depend on the definition of α_s but c_2 and c_3 do. The subleading coefficients also depend on the scale choice: if instead of expanding in $\alpha_s(Q)$ we decide to choose $\alpha_s(Q/2)$ the coefficients c_2 and c_3 change. In the $\overline{MS}$ scheme, for γ-exchange and $n_f = 5$, which are good approximations for $2m_b << Q << m_Z$, one has:

$$F[0, \alpha_s(t)] = 1 + \frac{\alpha_s(t)}{\pi} + 1.409 \ldots (\frac{\alpha_s(t)}{\pi})^2 - 12.8 \ldots (\frac{\alpha_s(t)}{\pi})^3 + \ldots \quad (4.60)$$

Similar perturbative results at 3-loop accuracy also exist for $R_Z = \Gamma(Z \rightarrow hadrons)/\Gamma(Z \rightarrow leptons)$, $R_\tau = \Gamma(\tau \rightarrow \nu_\tau + hadrons)/\Gamma(\tau \rightarrow \nu_\tau + leptons)$, etc. We will discuss these results later when we deal with measurements of α_s.

The perturbative expansion in powers of $\alpha_s(t)$ takes into account all contributions that are suppressed by powers of logarithms of the large scale Q^2 ("leading twist" terms). In addition there are corrections suppressed by powers of the large scale Q^2 ("higher twist" terms). The pattern of power corrections is controlled by the light-cone Operator Product Expansion (OPE) [16] which (schematically) leads to:

$$F = \text{pert.} + r_2 \frac{m^2}{Q^2} + r_4 \frac{< 0|Tr[\mathbf{F}_{\mu\nu}\mathbf{F}^{\mu\nu}]|0 >}{Q^4} + \ldots + r_6 \frac{< 0|O_6|0 >}{Q^6} + \ldots \quad (4.61)$$

Here m^2 generically indicates mass corrections, notably from b quarks, for example (t quark mass corrections only arise from loops, vanish in the limit $m_t \rightarrow \infty$ and are included in the coefficients as those in Eq. (4.60) and the analogous ones for higher twist terms), $\mathbf{F}_{\mu\nu} = \sum_A F^A_{\mu\nu} t^A$, O_6 is typically a 4-fermion operator, etc. For each possible gauge invariant operator the corresponding power of Q^2 is fixed by dimensions.

We now consider the light-cone OPE in some more detail. $R_{e^+e^-} \sim \Pi(Q^2)$ where $\Pi(Q^2)$ is the scalar spectral function related to the hadronic contribution to the imaginary part of the photon vacuum polarization $T_{\mu\nu}$:

$$T_{\mu\nu} = (-g_{\mu\nu}Q^2 + q_\mu q_\nu)\Pi(Q^2) = \int \exp iqx < 0|J^\dagger_\mu(x) J_\nu(0)|0 > dx =$$
$$= \sum_n < 0|J^\dagger_\mu(0)|n >< n|J_\nu(0)|0 > (2\pi)^4 \delta^4(q - p_n) \quad (4.62)$$

For $Q^2 \rightarrow \infty$ the $x^2 \rightarrow 0$ region is dominant. To all orders in perturbation theory the OPE can be proven. Schematically, dropping Lorentz indices, for simplicity, near $x^2 \sim 0$ we have:

$$J^\dagger(x) J(0) = I(x^2) + E(x^2) \sum_{n=0}^{\infty} c_n(x^2) x^{\mu_1} \ldots x^{\mu_n} \cdot O^n_{\mu_1 \ldots \mu_n}(0)$$
$$+ \text{less sing. terms} \quad (4.63)$$

Here $I(x^2)$, $E(x^2)$,..., $c_n(x^2)$ are c-number singular functions, O^n is a string of local operators. $E(x^2)$ is the singularity of free field theory, $I(x^2)$ and $c_n(x^2)$ contain powers of $\log \mu^2 x^2$ in interaction. Some O^n are already present in free field theory, other ones appear when interactions are switched on. Given that $\Pi(Q^2)$ is related to the Fourier transform of the vacuum expectation value of the product of currents, less singular terms in x^2 lead to power suppressed terms in $1/Q^2$. The perturbative terms come from $I(x^2)$ which is the leading twist term. The logarithmic scaling violations induced by the running coupling are the logs in $I(x^2)$.

4.5.2 The Final State in e^+e^- Annihilation

Experiments on e^+e^- annihilation at high energy provide a remarkable possibility of systematically testing the distinct signatures predicted by QCD for the structure of the final state averaged over a large number of events. Typical of asymptotic freedom is the hierarchy of configurations emerging as a consequence of the smallness of $\alpha_s(Q^2)$. When all corrections of order $\alpha_s(Q^2)$ are neglected one recovers the naive parton model prediction for the final state: almost collinear events with two back-to-back jets with limited transverse momentum and an angular distribution as $(1 + \cos^2\theta)$ with respect to the beam axis (typical of spin 1/2 parton quarks: scalar quarks would lead to a $\sin^2\theta$ distribution). At order $\alpha_s(Q^2)$ a tail of events is predicted to appear with large transverse momentum $p_T \sim Q/2$ with respect to the thrust axis (the axis that maximizes the sum of the absolute values of the longitudinal momenta of the final state particles). This small fraction of events with large p_T mostly consists of three-jet events with an almost planar topology. The skeleton of a three-jet event, at leading order in $\alpha_s(Q^2)$, is formed by three hard partons $q\bar{q}g$, the third being a gluon emitted by a quark or antiquark line. The distribution of three-jet events is given by:

$$\frac{1}{\sigma}\frac{d\sigma}{dx_1 dx_2} = \frac{2\alpha_s}{3\pi}\frac{x_1^2 + x_2^2}{(1-x_1)(1-x_2)} \tag{4.64}$$

here $x_{1,2}$ refer to energy fractions of massless quarks: $x_i = 2E_i/\sqrt{s}$ with $x_1 + x_2 + x_3 = 2$. At order $\alpha_s^2(Q^2)$ a hard perturbative non planar component starts to build up and a small fraction of four-jet events $q\bar{q}gg$ or $q\bar{q}q\bar{q}$ appear, and so on.

A quantitatively specified definition of jet counting must be introduced for precise QCD tests and for measuring α_s, which must be infrared safe (i.e. not altered by soft particle emission or collinear splittings of massless particles) in order to be computable at parton level and as much as possible insensitive to the transformation of partons into hadrons. One introduces a resolution parameter y_{cut} and a suitable pair variable; for example [17]:

$$y_{ij} = \frac{min(E_i^2, E_j^2)(1-\cos\theta_{ij})}{s} \tag{4.65}$$

The particles i,j belong to different jets for $y_{ij} > y_{cut}$. Clearly the number of jets becomes a function of y_{cut}: there are more jets for smaller y_{cut}. Measurements of $\alpha_s(Q^2)$ have been performed starting from jet multiplicities, the largest error coming from the necessity of correcting for non-perturbative hadronisation effects.

4.5.3 Deep Inelastic Scattering

Deep Inelastic Scattering (DIS) processes have played and still play a very important role for our understanding of QCD and of nucleon structure. This set of processes actually provides us with a rich laboratory for theory and experiment. There are several structure functions that can be studied, $F_i(x, Q^2)$, each a function of two variables. This is true separately for different beams and targets and different polarizations. Depending on the charges of l and l' (see Eq. (4.13)) we can have neutral currents (γ,Z) or charged currents in the l-l' channel (Fig. 4.6). In the past DIS processes were crucial for establishing QCD as the theory of strong interactions and quarks and gluons as the QCD partons. At present DIS remains very important for quantitative studies and tests of QCD. The theory of scaling violations for totally inclusive DIS structure functions, based on operator expansion or diagrammatic techniques and renormalisation group methods, is crystal clear and the predicted Q^2 dependence can be tested at each value of x. The measurement of quark and gluon densities in the nucleon, as functions of x at some reference value of Q^2, which is an essential starting point for the calculation of all relevant hadronic hard processes, is performed in DIS processes. At the same time one measures $\alpha_s(Q^2)$ and the DIS values of the running coupling can be compared with those obtained from other processes. At all times new theoretical challenges arise from the study of DIS processes. Recent examples are the so-called "spin crisis" in polarized DIS and the behaviour of singlet structure functions at small x as revealed by HERA data. In the following we will review the past successes and the present open problems in the physics of DIS.

The cross-section $\sigma \sim L^{\mu\nu} W_{\mu\nu}$ is given in terms of the product of a leptonic ($L^{\mu\nu}$) and a hadronic ($W_{\mu\nu}$) tensor. While $L^{\mu\nu}$ is simple and easily obtained from the lowest order electroweak (EW) vertex plus QED radiative corrections, the complicated strong interaction dynamics is contained in $W_{\mu\nu}$. The latter is proportional to the Fourier transform of the forward matrix element between the nucleon target states of the product of two EW currents:

$$W_{\mu\nu} = \int dx \, \exp iqx \, < p|J_\mu^\dagger(x) J_\nu(0)|p > \tag{4.66}$$

Structure functions are defined starting from the general form of $W_{\mu\nu}$ given Lorentz invariance and current conservation. For example, for EW currents between unpolarized nucleons we have:

$$W_{\mu\nu} = (-g_{\mu\nu} + \frac{q_\mu q_\nu}{q^2})\, W_1(\nu, Q^2) + (p_\mu - \frac{m\nu}{q^2} q_\mu)(p_\nu - \frac{m\nu}{q^2} q_\nu)\, \frac{W_2(\nu, Q^2)}{m^2} - $$
$$- \frac{i}{2m^2} \epsilon_{\mu\nu\lambda\rho} p^\lambda q^\rho\, W_3(\nu, Q^2)$$

W_3 arises from VA interference and is absent for pure vector currents. In the limit $Q^2 >> m^2$, x fixed, the structure functions obey approximate Bjorken scaling which in reality is broken by logarithmic corrections that can be computed in QCD:

$$\begin{aligned} m W_1(\nu, Q^2) &\to F_1(x) \\ \nu W_{2,3}(\nu, Q^2) &\to F_{2,3}(x) \end{aligned} \tag{4.67}$$

The $\gamma - N$ cross-section is given by ($W_i = W_i(Q^2, \nu)$):

$$\frac{d\sigma^\gamma}{dQ^2 d\nu} = \frac{4\pi\alpha^2 E'}{Q^4 E} \cdot [2\sin^2\frac{\theta}{2} W_1 + \cos^2\frac{\theta}{2} W_2] \tag{4.68}$$

while for the $\nu - N$ or $\bar{\nu} - N$ cross-section one has:

$$\frac{d\sigma^{\nu,\bar{\nu}}}{dQ^2 d\nu} = \frac{G_F^2 E'}{2\pi E} (\frac{m_W^2}{Q^2 + m_W^2})^2 \cdot [2\sin^2\frac{\theta}{2} W_1 + \cos^2\frac{\theta}{2} W_2 \pm \frac{E + E'}{m} \sin^2\frac{\theta}{2} W_3] \tag{4.69}$$

(W_i for photons, ν and $\bar{\nu}$ are all different, as we shall see in a moment).

In the scaling limit the longitudinal and transverse cross sections are given by:

$$\begin{aligned} \sigma_L &\sim \frac{1}{s}[\frac{F_2(x)}{2x} - F_1(x)] \\ \sigma_{RH,LH} &\sim \frac{1}{s}[F_1(x) \pm F_3(x)] \\ \sigma_T &= \sigma_{RH} + \sigma_{LH} \end{aligned} \tag{4.70}$$

where L, RH, LH refer to the helicity 0, 1, -1, respectively, of the exchanged gauge vector boson.

In the '60's the demise of hadrons from the status of fundamental particles to that of bound states of constituent quarks was the breakthrough that made possible the construction of a renormalisable field theory for strong interactions. The presence of an unlimited number of hadrons species, many of them with large spin values, presented an obvious dead-end for a manageable field theory. The evidence for constituent quarks emerged clearly from the systematics of hadron spectroscopy.

The complications of the hadron spectrum could be explained in terms of the quantum numbers of spin 1/2, fractionally charged, u, d and s quarks. The notion of colour was introduced to reconcile the observed spectrum with Fermi statistics. But confinement that forbids the observation of free quarks was a clear obstacle towards the acceptance of quarks as real constituents and not just as fictitious entities describing some mathematical pattern (a doubt expressed even by Gell-Mann at the time). The early measurements at SLAC of DIS dissipated all doubts: the observation of Bjorken scaling and the success of the "naive" (not so much after all) parton model of Feynman imposed quarks as the basic fields for describing the nucleon structure (parton quarks).

In the language of Bjorken and Feynman the virtual γ (or, in general, any gauge boson) sees the quark partons inside the nucleon target as quasi-free, because their (Lorentz dilated) QCD interaction time is much longer than $\tau_\gamma \sim 1/Q$, the duration of the virtual photon interaction. Since the virtual photon 4-momentum is spacelike, we can go to a Lorentz frame where $E_\gamma = 0$ (Breit frame). In this frame $q = (E_\gamma = 0; 0, 0, Q)$ and the nucleon momentum, neglecting the mass $m << Q$, is $p = (Q/2x; 0, 0, -Q/2x)$ (note that this correctly gives $q^2 = -Q^2$ and $x = Q^2/2(p{\cdot}q)$). Consider (Fig. 4.10) the interaction of the photon with a quark carrying a fraction y of the nucleon 4-momentum: $p_q = yp$ (we are neglecting the transverse components of p_q which are of order m). The incoming parton with $p_q = yp$ absorbs the photon and the final parton has 4-momentum p'_q. Since in the Breit frame the photon carries no energy but only a longitudinal momentum Q, the photon can only be absorbed by those partons with $y = x$: then the longitudinal component of $p_q = yp$ is $-yQ/2x = -Q/2$ and can be flipped into $+Q/2$ by the photon. As a result, the photon longitudinal momentum $+Q$ disappears, the parton quark momentum changes of sign from $-Q/2$ into $+Q/2$ and the energy is not changed. So the structure functions are proportional to the density of partons with fraction x of the nucleon momentum, weighted with the squared charge. Also, recall that the helicity of a massless quark is conserved in a vector (or axial vector) interaction. So when the momentum is reversed also the spin must flip. Since the process is collinear there is no orbital contribution and only a photon with helicity ± 1 (transverse photon) can be absorbed. Alternatively, if partons were spin zero only longitudinal photons would instead contribute.

Using these results, which are maintained in QCD at leading order, the quantum numbers of the quarks were confirmed by early experiments. The observation that $R = \sigma_L/\sigma_T \to 0$ implies that the charged partons have spin 1/2. The quark charges

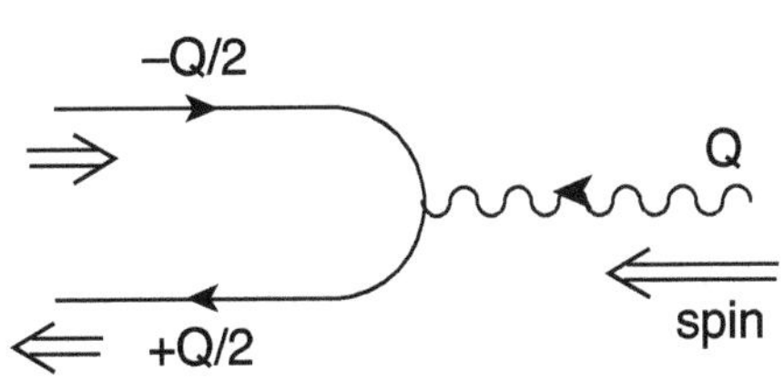

Fig. 4.10 Schematic diagram for the interaction of the virtual photon with a parton quark in the Breit frame

were derived from the data on the electron and neutrino structure functions:

$$F_{ep} = 4/9u(x) + 1/9d(x) + \ldots..; \qquad F_{en} = 4/9d(x) + 1/9u(x) + \ldots.$$
$$F_{\nu p} = F_{\bar{\nu} n} = 2d(x) + \ldots..; \qquad F_{\nu n} = F_{\bar{\nu} p} = 2u(x) + \ldots.. \quad (4.71)$$

where $F \sim 2F_1 \sim F_2/x$ and $u(x)$, $d(x)$ are the parton number densities in the proton (with fraction x of the proton longitudinal momentum), which, in the scaling limit, do not depend on Q^2. The normalisation of the structure functions and the parton densities are such that the charge relations hold:

$$\int_0^1 [u(x) - \bar{u}(x)]dx = 2, \quad \int_0^1 [d(x) - \bar{d}(x)]dx = 1, \quad \int_0^1 [s(x) - \bar{s}(x)]dx = 0 \quad (4.72)$$

Also it was proven by experiment that at values of Q^2 of a few GeV^2, in the scaling region, about half of the nucleon momentum, given by the momentum sum rule:

$$\int_0^1 [\sum_i (q_i(x) + \bar{q}_i(x)) + g(x)]xdx = 1 \quad (4.73)$$

is carried by neutral partons (gluons).

In QCD there are calculable log scaling violations induced by $\alpha_s(t)$. The parton rules just introduced can be summarised in the formula:

$$F(x,t) = \int_x^1 dy \frac{q_0(y)}{y} \sigma_{point}(\frac{x}{y}, \alpha_s(t)) + o(\frac{1}{Q^2}) \quad (4.74)$$

Before QCD corrections $\sigma_{point} = e^2\delta(x/y - 1)$ and $F = e^2 q_0(x)$ (here we denote by e the charge of the quark in units of the positron charge, i.e. $e = 2/3$ for the u quark). QCD modifies σ_{point} at order α_s via the diagrams of Fig. 4.11. Note that the integral is from x to 1, because the energy can only be lost by radiation before interacting with the photon (which eventually wants to find a fraction x, as we have

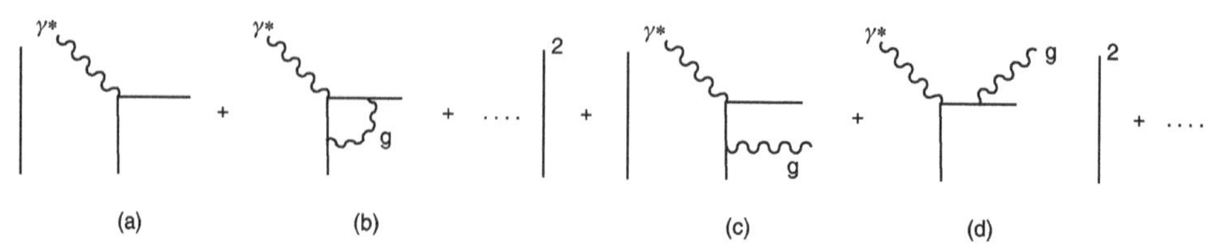

Fig. 4.11 First order QCD corrections to the virtual photon-quark cross-section: (a) leading order with (b) one-loop virtual correction; (c-d) next-to-leading order real emission

explained). From a direct computation of the diagrams one obtains a result of the following form:

$$\sigma_{point}(z, \alpha_s(t)) \simeq e^2[\delta(z-1) + \frac{\alpha_s}{2\pi}(t \cdot P(z) + f(z))] \tag{4.75}$$

For $y > x$ the correction arises from diagrams with real gluon emission. Only the sum of the two real diagrams in Fig. 4.11 is gauge invariant, so that the contribution of one given diagram is gauge dependent. There is a special form of axial gauge, called physical gauge, where, among real diagrams, the diagram of Fig. 4.11c gives the whole t-proportional term. It is obviously not essential to go to this gauge, but this diagram has a direct physical interpretation: a quark in the proton has a fraction $y > x$ of the parent 4-momentum; it then radiates a gluon and looses energy down to a fraction x before interacting with the photon. The log arises from the virtual quark propagator, according to the discussion of collinear mass singularities in Eq. (4.12). In fact in the massless limit one has:

$$\begin{aligned} \text{propagator} &= \frac{1}{r^2} = \frac{1}{(k-h)^2} = \frac{-1}{2E_k E_h} \cdot \frac{1}{1-\cos\theta} \\ &= \frac{-1}{4E_k E_h} \cdot \frac{1}{\sin^2\theta/2} \propto \frac{-1}{p_T^2} \end{aligned} \tag{4.76}$$

where p_T is the transverse momentum of the virtual quark. So the square of the propagator goes like $1/p_T^4$. But there is a p_T^2 factor in the numerator, because in the collinear limit, when $\theta = 0$ and the initial and final quarks and the emitted gluon are all aligned, the quark helicity cannot flip (vector interaction) so that the gluon should carry helicity zero but a real gluon can only have ± 1 helicity. Thus the numerator vanishes as p_T^2 in the forward direction and the cross-section behaves as:

$$\sigma \sim \int^{Q^2} \frac{1}{p_T^2} dp_T^2 \sim \log Q^2 \tag{4.77}$$

Actually the log should be read as $\log Q^2/m^2$ because in the massless limit a genuine mass singularity appears. In fact the mass singularity connected with the initial quark line is not cancelled because we do not have the sum of all degenerate initial states, but only a single quark. But in correspondence to the initial quark we have the (bare) quark density $q_0(y)$ that appear in the convolution integral. This is a non perturbative quantity that is determined by the nucleon wave function. So we can factorize the mass singularity in a redefinition of the quark density: we replace $q_0(y) \to q(y,t) = q_0(y) + \Delta q(y,t)$ with:

$$\Delta q(x,t) = \frac{\alpha_s}{2\pi} t \int_x^1 dy \frac{q_0(y)}{y} \cdot P(\frac{x}{y}) \tag{4.78}$$

Here the factor of t is a bit symbolic: it stands for log Q^2/km^2 and what we exactly put below Q^2 depends on the definition of the renormalised quark density, which also fixes the exact form of the finite term $f(z)$ in Eq. (4.75).

The effective parton density $q(y,t)$ that we have defined is now scale dependent. In terms of this scale dependent density we have the following relations, where we have also replaced the fixed coupling with the running coupling according to the prescription derived from the RGE:

$$F(x,t) = \int_x^1 dy \frac{q(y,t)}{y} e^2 [\delta(\frac{x}{y} - 1) + \frac{\alpha_s(t)}{2\pi} f(\frac{x}{y})] = e^2 q(x,t) + o(\alpha_s(t))$$

$$\frac{d}{dt} q(x,t) = \frac{\alpha_s(t)}{2\pi} \int_x^1 dy \frac{q(y,t)}{y} \cdot P(\frac{x}{y}) + o(\alpha_s(t)^2) \tag{4.79}$$

We see that in lowest order we reproduce the naive parton model formulae for the structure functions in terms of effective parton densities that are scale dependent. The evolution equations for the parton densities are written down in terms of kernels (the "splitting functions") that can be expanded in powers of the running coupling. At leading order, we can interpret the evolution equation by saying that the variation of the quark density at x is given by the convolution of the quark density at y times the probability of emitting a gluon with fraction x/y of the quark momentum.

It is interesting that the integro-differential QCD evolution equation for densities can be transformed into an infinite set of ordinary differential equations for Mellin moments [2]. The moment f_n of a density $f(x)$ is defined as:

$$f_n = \int_0^1 dx x^{n-1} f(x) \tag{4.80}$$

By taking moments of both sides of the second of Eqs. (4.79) one finds, with a simple interchange of the integration order, the simpler equation for the n-th moment:

$$\frac{d}{dt} q_n(t) = \frac{\alpha_s(t)}{2\pi} \cdot P_n \cdot q_n(t) \tag{4.81}$$

To solve this equation we observe that:

$$\log \frac{q_n(t)}{q_n(0)} = \frac{P_n}{2\pi} \int_0^t \alpha_s(t) dt = \frac{P_n}{2\pi} \int_{\alpha_s}^{\alpha_s(t)} \frac{d\alpha'}{-b\alpha'} \tag{4.82}$$

where we used Eq. (4.31) to change the integration variable from dt to $d\alpha(t)$ (denoted as $d\alpha'$) and $\beta(\alpha) \simeq -b\alpha^2 + \ldots$. Finally the solution is:

$$q_n(t) = [\frac{\alpha_s}{\alpha_s(t)}]^{\frac{P_n}{2\pi b}} \cdot q_n(0) \tag{4.83}$$

The connection of these results with the RGE general formalism occurs via the light cone OPE (recall Eq. (4.66) for $W_{\mu\nu}$ and Eq. (4.63) for the OPE of two currents). In the case of DIS the c-number term $I(x^2)$ does not contribute, because we are interested in the connected part $< p|\ldots|p> - <0|\ldots|0>$. The relevant terms are:

$$J^\dagger(x)J(0) = E(x^2)\sum_{n=0}^{\infty} c_n(x^2)x^{\mu_1}\ldots x^{\mu_n}\cdot O^n_{\mu_1\ldots\mu_n}(0) + \text{less sing. terms} \tag{4.84}$$

A formally intricate but conceptually simple argument (Ref. [6], page 28) based on the analiticity properties of the forward virtual Compton amplitude shows that the Mellin moments M_n of structure functions are related to the individual terms in the OPE, precisely to the Fourier transform $c_n(Q^2)$ (we will write it as $c_n(t,\alpha)$) of the coefficient $c_n(x^2)$ times a reduced matrix element h_n from the operators O^n: $< p|O^n_{\mu_1\ldots\mu_n}(0)|p >= h_n p_{\mu_1}\cdots p_{\mu_n}$:

$$c_n < p|O^n|p > \rightarrow M_n = \int_0^1 dx x^{n-1}F(x) \tag{4.85}$$

Since the matrix element of the products of currents satisfy the RGE so do the moments M_n. Hence the general form of the Q^2 dependence is given by the RGE solution (see Eq. (4.33)):

$$M_n(t,\alpha) = c_n[0,\alpha(t)]\exp\int_\alpha^{\alpha(t)}\frac{\gamma_n(\alpha')}{\beta(\alpha')}d\alpha'\cdot h_n(\alpha) \tag{4.86}$$

In lowest order, identifying in the simplest case M_n with q_n, we have:

$$\gamma_n(\alpha) = \frac{P_n}{2\pi}\alpha + \ldots, \qquad \beta(\alpha) = -b\alpha^2 + \ldots \tag{4.87}$$

and

$$q_n(t) = q_n(0)\exp\int_\alpha^{\alpha(t)}\frac{\gamma_n(\alpha')}{\beta(\alpha')}d\alpha' = [\frac{\alpha_s}{\alpha_s(t)}]^{\frac{P_n}{2\pi b}}\cdot q_n(0) \tag{4.88}$$

which exactly coincides with Eq. (4.83).

Up to this point we have implicitly restricted our attention to non-singlet (under the flavour group) structure functions. The Q^2 evolution equations become non diagonal as soon as we take into account the presence of gluons in the target. In fact the quark which is seen by the photon can be generated by a gluon in the target (Fig. 4.12).

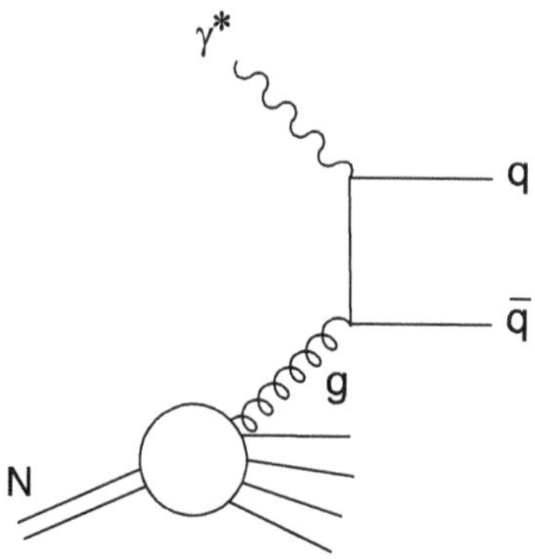

Fig. 4.12 Lowest order diagram for the interaction of the virtual photon with a parton gluon

The quark evolution equation becomes:

$$\frac{d}{dt}q_i(x,t) = \frac{\alpha_s(t)}{2\pi}[q_i \otimes P_{qq}] + \frac{\alpha_s(t)}{2\pi}[g \otimes P_{qg}] \tag{4.89}$$

where we introduced the shorthand notation:

$$[q \otimes P] = [P \otimes q] = \int_x^1 dy \frac{q(y,t)}{y} \cdot P(\frac{x}{y}) \tag{4.90}$$

(it is easy to check that the convolution, like an ordinary product, is commutative). At leading order, the interpretation of Eq. (4.89) is simply that the variation of the quark density is due to the convolution of the quark density at a higher energy times the probability of finding a quark in a quark (with the right energy fraction) plus the gluon density at a higher energy times the probability of finding a quark (of the given flavour i) in a gluon. The evolution equation for the gluon density, needed to close the system, can be obtained by suitably extending the same line of reasoning to a gedanken probe sensitive to colour charges, for example a virtual gluon. The resulting equation is of the form:

$$\frac{d}{dt}g(x,t) = \frac{\alpha_s(t)}{2\pi}[\sum_i (q_i + \bar{q}_i) \otimes P_{gq}] + \frac{\alpha_s(t)}{2\pi}[g \otimes P_{gg}] \tag{4.91}$$

The explicit form of the splitting functions in lowest order [18, 19] can be directly derived from the QCD vertices [19]. They are a property of the theory and do not depend on the particular process the parton density is taking part in. The results are:

$$P_{qq} = \frac{4}{3}[\frac{1+x^2}{(1-x)_+} + \frac{3}{2}\delta(1-x)] + o(\alpha_s)$$

$$P_{gq} = \frac{4}{3}\frac{1+(1-x)^2}{x} + o(\alpha_s)$$

$$P_{qg} = \frac{1}{2}[x^2 + (1-x)^2] + o(\alpha_s)$$

$$P_{gg} = 6[\frac{x}{(1-x)_+} + \frac{1-x}{x} + x(1-x)] + \frac{33-2n_f}{6}\delta(1-x) + o(\alpha_s) \quad (4.92)$$

For a generic non singular weight function $f(x)$, the "+" distribution is defined as:

$$\int_0^1 \frac{f(x)}{(1-x)_+}dx = \int_0^1 \frac{f(x)-f(1)}{1-x}dx \quad (4.93)$$

The $\delta(1-x)$ terms arise from the virtual corrections to the lowest order tree diagrams. Their coefficient can be simply obtained by imposing the validity of charge and momentum sum rules. In fact, from the request that the charge sum rules in Eq. (4.72) are not affected by the Q^2 dependence one derives that

$$\int_0^1 P_{qq}(x)dx = 0 \quad (4.94)$$

which can be used to fix the coefficient of the $\delta(1-x)$ terms of P_{qq}. Similarly, by taking the t-derivative of the momentum sum rule in Eq. (4.73) and imposing its vanishing for generic q_i and g, one obtains:

$$\int_0^1 [P_{qq}(x) + P_{gq}(x)]xdx = 0, \qquad \int_0^1 [2n_f P_{qg}(x) + P_{gg}(x)]xdx = 0. \quad (4.95)$$

At higher orders the evolution equations are easily generalised but the calculation of the splitting functions rapidly becomes very complicated. For many years the splitting functions were only completely known at NLO accuracy [20]: $\alpha_s P \sim \alpha_s P_1 + \alpha_s^2 P_2 + \ldots$. Then in recent years the NNLO results P_3 have been first derived in analytic form for the first few moments and, then the full NNLO analytic calculation, a really monumental work, was completed in 2004 by Moch, Vermaseren and Vogt [21]. Beyond leading order a precise definition of parton densities should be specified. One can take a physical definition (for example, quark densities can be defined as to keep the LO expression for the structure function F_2 valid at all orders, the so called DIS definition [22], and the gluon density could be defined starting from F_L, the longitudinal structure function, or a more abstract specification (for example, in terms of the $\overline{MS}$ prescription). Once the definition of parton densities is fixed, the coefficients that relate the different structure functions to the parton densities at each fixed order can be computed. Similarly the higher order splitting functions also depend, to some extent, from the definition of parton densities, and a consistent set of coefficients and splitting functions must be used at each order.

The scaling violations are clearly observed by experiment and their pattern is very well reproduced by QCD fits at NLO. Examples are seen in Fig. 4.13a–d [23].

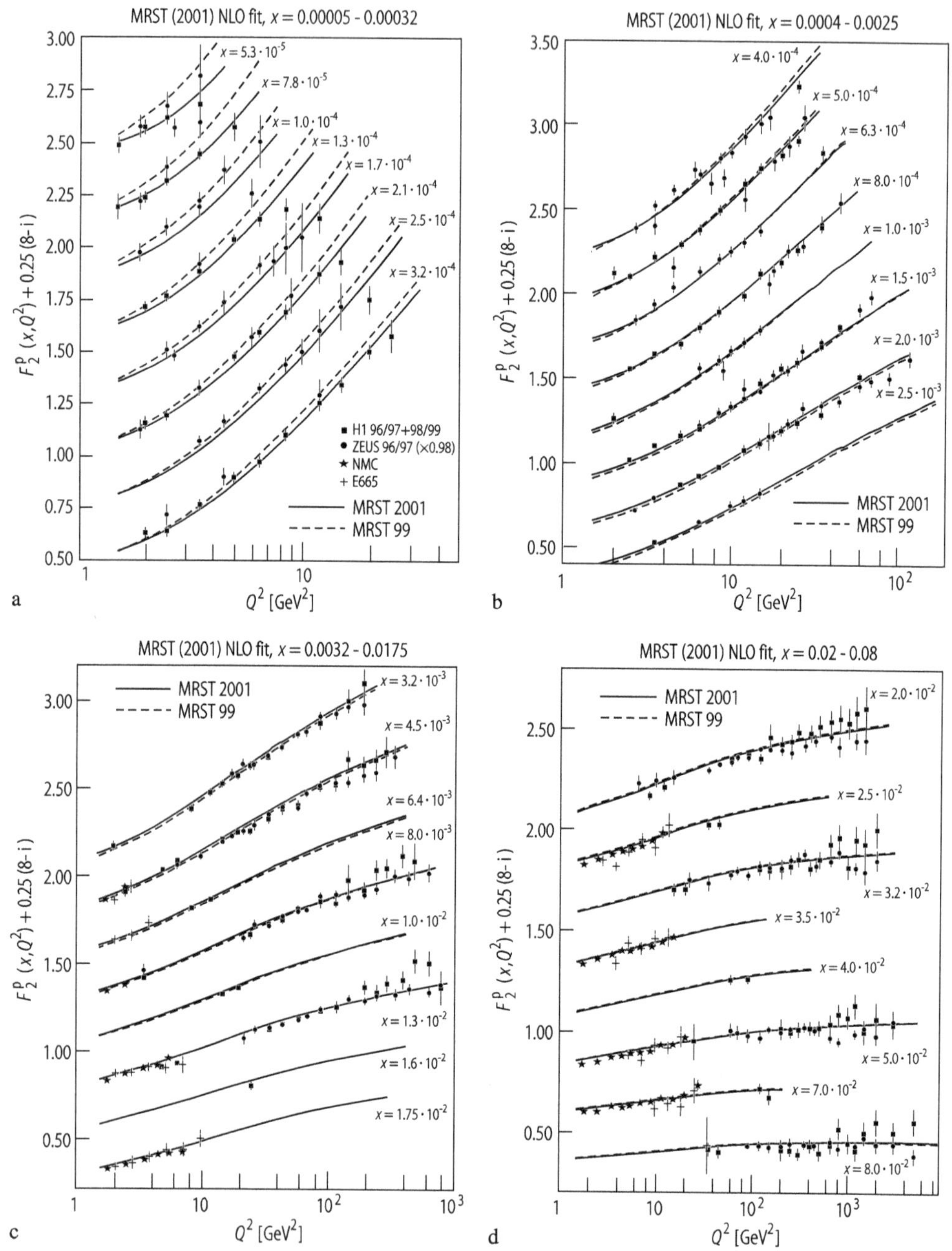

Fig. 4.13 A recent NLO fit of scaling violations from Ref. [23], for different x ranges, as functions of Q^2

These fits provide an impressive confirmation of a quantitative QCD prediction, a measurement of $q_i(x, Q_0^2)$ and $g(x, Q_0^2)$ at some reference value Q_0^2 of Q^2 and a precise measurement of $\alpha_s(m_Z^2)$.

4.5.3.1 Resummation for Deep Inelastic Structure Functions

At small or at large values of x (with Q^2 large) those terms of higher order in α_s in either the coefficients or the splitting functions which are multiplied by powers of $\log 1/x$ or $\log(1-x)$ eventually become important and should be taken into account. Fortunately the sequences of leading and subleading logs can be evaluated at all orders by special techniques and resummed to all orders.

For large x resummation [24] I refer to the recent papers [25, 26] (the latter also involving higher twist corrections, which are important at large x) where a list of references to previous work can be found.

Here we will briefly summarise the small-x case for the singlet structure function which is the dominant channel at HERA, dominated by the sharp rise of the gluon and sea parton densities at small x. The small x data collected by HERA can be fitted reasonably well even at the smallest measured values of x by the NLO QCD evolution equations, so that there is no dramatic evidence in the data for departures. This is surprising also in view of the fact that the NNLO effects in the evolution have recently become available and are quite large. Resummation effects have been shown to resolve this apparent paradox. For the singlet splitting function the coefficients of all LO and NLO corrections of order $[\alpha_s(Q^2)\log 1/x]^n$ and $\alpha_s(Q^2)[\alpha_s(Q^2)\log 1/x]^n$, respectively, are explicitly known from the BFKL analysis of virtual gluon-virtual gluon scattering [27, 28]. But the simple addition of these higher order terms to the perturbative result (with subtraction of all double counting) does not lead to a converging expansion (the NLO logs completely overrule the LO logs in the relevant domain of x and Q^2). A sensible expansion is only obtained by a proper treatment of momentum conservation constraints, also using the underlying symmetry of the BFKL kernel under exchange of the two external gluons, and especially, of the running coupling effects (see the recent papers [29, 30] and refs. therein). In Fig. 4.14 we present the results for the dominant singlet splitting function $xP(x, \alpha_s(Q^2))$ for $\alpha_s(Q^2) \sim 0.2$. We see that while the NNLO perturbative splitting function sharply deviates from the NLO approximation at small x, the resummed result only shows a moderate dip with respect to the NLO perturbative splitting function in the region of HERA data, and the full effect of the true small x asymptotics is only felt at much smaller values of x. The related effects are not very important for processes at the LHC but could become relevant for next generation hadron colliders.

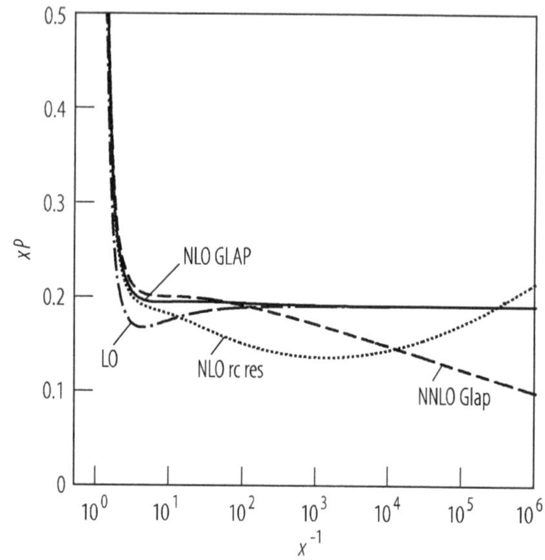

Fig. 4.14 The dominant singlet splitting function $xP(x, \alpha_s(Q^2))$ for $\alpha_s(Q^2) \sim 0.2$. The resummed result from Ref. [29] is compared with the LO, NLO and NNLO perturbative results

4.5.3.2 Polarized Deep Inelastic Scattering

In polarized DIS one main question is how the proton helicity is distributed among quarks, gluons and orbital angular momentum: $1/2\Delta\Sigma + \Delta g + L_z = 1/2$ (for a recent review, see, for example, [31]). For a parton density p (either a quark or a gluon) Δp indicates the first moment of the net helicity difference $p_+ - p_-$ in a polarized proton with helicity $+1/2$ or:

$$\Delta p(Q^2) = \int_0^1 dx[p_+(x, Q^2) - p_-(x, Q^2)] \tag{4.96}$$

Experiments have shown that the quark moment $\Delta\Sigma$ is small (the "spin crisis"): values from a recent fit [32] are $\Delta\Sigma_{exp} \sim 0.21 \pm 0.14$ and $\Delta g_{exp} \sim 0.50 \pm 1.27$ at $Q^2 = 1\,\mathrm{GeV}^2$ (see also [33]). This is surprising because $\Delta\Sigma$ is conserved in perturbation theory at LO (i.e. it does not evolve in Q^2). For conserved quantities we would expect that they are the same for constituent and for parton quarks. But actually the conservation of $\Delta\Sigma$ is broken by the axial anomaly. In perturbation theory the conserved density is actually $\Delta\Sigma' = \Delta\Sigma + n_f/2\pi\alpha_s\,\Delta g$ [34]. Note that also $\alpha_s\Delta g$ is conserved in LO, that is $\Delta g \sim logQ^2$. This behaviour is not controversial but it will take long before the log growth of Δg will be confirmed by experiment! But to establish this behaviour would show that the extraction of Δg from the data is correct and that the QCD evolution works as expected. If Δg is large enough it could account for the difference between partons ($\Delta\Sigma$) and constituents ($\Delta\Sigma'$). From the spin sum rule it is clear that the log increase should cancel between Δg and L_z. This cancelation is automatic as a consequence of helicity conservation in the basic QCD vertices. From the spin sum rule one obtains that either $\Delta g + L_z$ is large or there are contributions to $\Delta\Sigma$ at very small x outside of the measured region.

Δg can be measured indirectly by scaling violations and directly from asymmetries, e.g. in $c\bar{c}$ production. Existing measurements by Hermes, Compass, and at RHIC are still crude but show no hint of a large Δg at accessible value of x and Q^2. Present data are consistent with Δg large enough to sizeably contribute to the spin sum rule but there is no indication that $\alpha_s \Delta g$ can explain the difference between constituents and parton quarks. The perspectives of better measurements are good at Compass and RHIC in the near future.

4.5.4 Factorisation and the QCD Improved Parton Model

The parton densities defined and measured in DIS are instrumental to compute hard processes initiated by hadronic collisions via the Factorisation Theorem (FT). Suppose you have a hadronic process of the form $h_1 + h_2 \to X + all$ where h_i are hadrons and X is some triggering particle or pair of particles which specify the large scale Q^2 relevant for the process, in general somewhat, but not much, smaller than s, the total c.o.m. squared mass. For example, in pp or $p\bar{p}$ collisions, X can be a W or a Z or a virtual photon with large Q^2, or a jet at large transverse momentum p_T, or a pair of heavy quark-antiquark of mass M. By "all" we mean a totally inclusive collection of gluons and light quark pairs. The FT states that for the total cross-section or some other sufficiently inclusive distribution we can write, apart from power suppressed corrections, the expression:

$$\sigma(s,\tau) = \sum_{AB} \int dx_1 dx_2 p_{1A}(x_1, Q^2) p_{2B}(x_2, Q^2) \sigma_{AB}(x_1 x_2 s, \tau) \tag{4.97}$$

Here $\tau = Q^2/s$ is a scaling variable, p_{iC} are the densities for a parton of type C inside the hadron h_i, σ_{AB} is the partonic cross-section for parton-A + parton-B$\to$ $X + all'$. This result is based on the fact that the mass singularities that are associated with the initial legs are of universal nature, so that one can reproduce the same modified parton densities, by absorbing these singularities into the bare parton densities, as in deep inelastic scattering. Once the parton densities and α_s are known from other measurements, the prediction of the rate for a given hard process is obtained with not much ambiguity (e.g from scale dependence or hadronisation effects). The NLO calculation of the reduced partonic cross-section is needed in order to correctly specify the scale and in general the definition of the parton densities and of the running coupling in the leading term. The residual scale and scheme dependence is often the most important source of theoretical error. In the following we consider a few examples.

A comparison of data and predictions on the production of jets at large $\sqrt{s}$ and p_T in pp or $p\bar{p}$ collisions is shown in Fig. 4.15 [9, 35].

This is a particularly significant test because the rates at different c.o.m. energies and, for each energy, at different values of p_T span over many orders of magnitude.

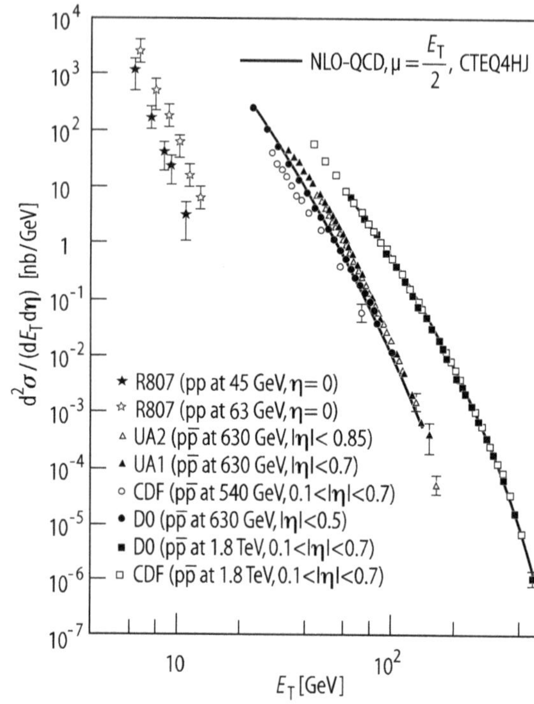

Fig. 4.15 Jet production cross-section at pp or $p\bar{p}$ colliders, as function of p_T [9]

This steep behaviour is determined by the sharp falling of the parton densities at large x. Also the corresponding values of $\sqrt{s}$ and p_T are large enough to be well inside the perturbative region. The overall agreement of the data from ISR, UA1,2 and CDF and D0 is spectacular. Similar results also hold for the production of photons at large p_T. The collider data [36], shown in Fig. 4.16 [9], are in fair agreement with the theoretical predictions. For the same process less clear a situation is found with fixed target data. Here, first of all, the experimental results show some internal discrepancies. Also, the p_T accessible values being smaller, the theoretical uncertainties are larger. But it is true that the agreement is poor, so that the necessity of an "intrinsic" transverse momentum of partons inside the hadron of over 1 GeV has been claimed, which theoretically is not justified (rather, given the sharp falling down at large p_T, it could be interpreted as a correction for p_T calibration errors).

For heavy quark production at colliders [42] the agreement is very good for the top crosssection at the Tevatron (Fig. 4.17) [43, 44]. The bottom production at the Tevatron has for some time represented a problem [45]. The total rate and the p_T distribution of b quarks observed at CDF appeared in excess of the prediction, up to the largest measured values of p_T. But this is a complicated case, with different scales being present at the same time: $\sqrt{s}$, p_T, m_b. Finally the problem has been solved (Fig. 4.18) by better taking into account a number of small effects from resummation of large logarithms, the difference between b hadrons and b partons, the inclusion of better fragmentation functions etc. [46].

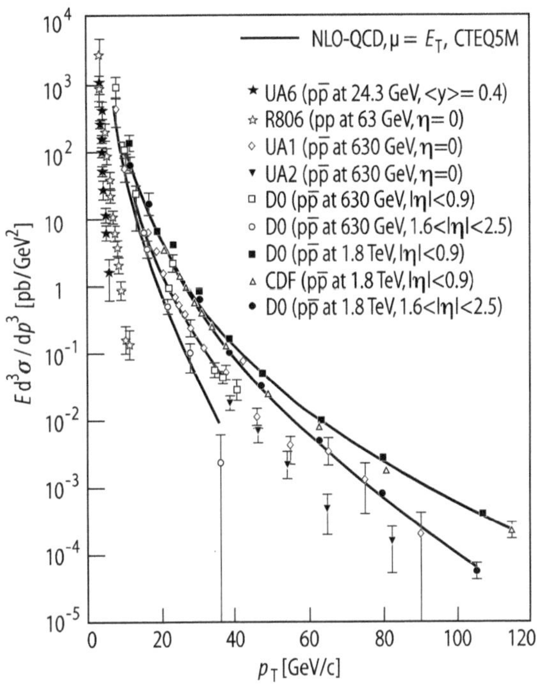

Fig. 4.16 Single photon production in $p\bar{p}$ colliders as function of p_T [9]

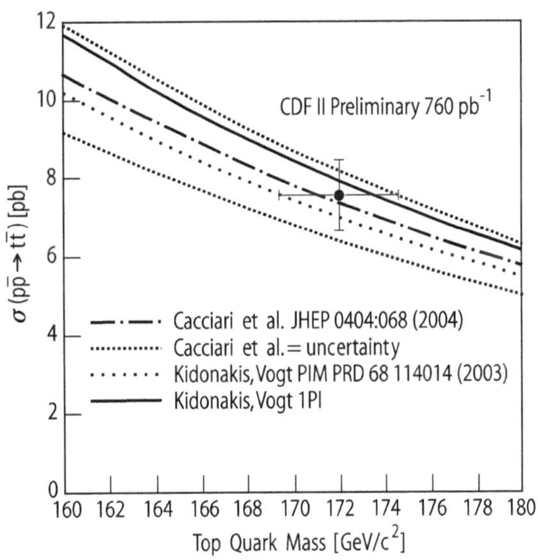

Fig. 4.17 The t production cross-section at the Tevatron $p\bar{p}$ collider [44]

Drell-Yan processes, including lepton pair production via virtual γ, W or Z exchange, offer a particularly good opportunity to test QCD. The process is quadratic in parton densities, and the final state is totally inclusive, while the large scale is specified and measured by the invariant mass squared Q^2 of the lepton pair which itself is not strongly interacting (so there no dangerous hadronisation effects). The QCD improved parton model leads directly to a prediction for the total rate as a function of Q^2. The value of the LO cross-section is inversely proportional to the number of colours N_C because a quark of given colour can only annihilate with an

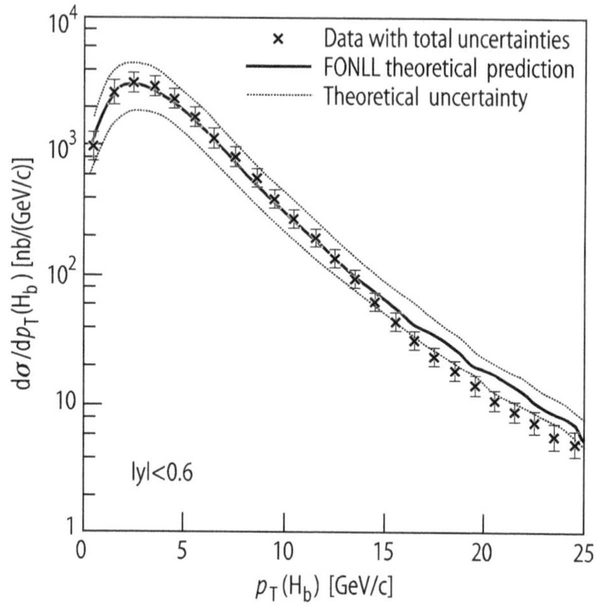

Fig. 4.18 The b production p_T distribution at the Tevatron $p\bar{p}$ collider [47]. The data from CDF also include systematics and correlations. The theoretical curve with the uncertainty range is from Ref. [46]

antiquark of the same colour to produce a colourless lepton pair. The order $\alpha_s(Q^2)$ corrections to the total rate were computed long ago and found to be particularly large [22, 38], when the quark densities are defined from the structure function F_2 measured in DIS at $q^2 = -Q^2$. The ratio $\sigma_{corr}/\sigma_{LO}$ of the corrected and the Born cross-sections, was called K-factor, because it is almost a constant in rapidity. In recent years also the NLO full calculation of the K-factor was completed, a very remarkable calculation [37]. The QCD predictions have been best tested for W and Z production at CERN $Sp\bar{p}S$ and Tevatron energies. $Q \sim m_{W,Z}$ is large enough to make the prediction reliable (with a not too large K-factor) and the ratio $\sqrt{\tau} = Q/\sqrt{s}$ is not too small. Recall that in lowest order $x_1 x_2 s = Q^2$ so that the parton densities are probed at x values around $\sqrt{\tau}$. We have $\sqrt{\tau} = 0.13 - 0.15$ (for W and Z production, respectively) at $\sqrt{s} = 630\,\text{GeV}$ (CERN $Sp\bar{p}S$ Collider) and $\sqrt{\tau} = 0.04 - 0.05$ at the Tevatron. In this respect the prediction is more delicate at the LHC, where $\sqrt{\tau} \sim 5.7 - 6.5 \cdot 10^{-3}$. One comparison of the experimental total rates at the Tevatron [48] with the QCD predictions is shown in Fig. 4.19, together with the expected rates at the LHC (based on the structure functions obtained in [23]).

The calculation of the W/Z p_T distribution has been a classic problem in QCD. For large p_T, for example $p_T \sim o(m_W)$, the p_T distribution can be reliably computed in perturbation theory, which was done up to NLO in the late '70's and early '80's. A problem arises in the intermediate range $\Lambda_{QCD} << p_T << m_W$, where the bulk of the data is concentrated, because terms of order $\alpha_s(p_T^2)\log m_W^2/p_T^2$ become of order one and should included to all orders [39]. At order α_s we have:

$$\frac{1}{\sigma_0}\frac{d\sigma_0}{dp_T^2} = (1+A)\delta(p_T^2) + \frac{B}{p_T^2}\log\frac{m_W^2}{p_T^2}_+ + \frac{C}{(p_T^2)_+} + D(p_T^2) \tag{4.98}$$

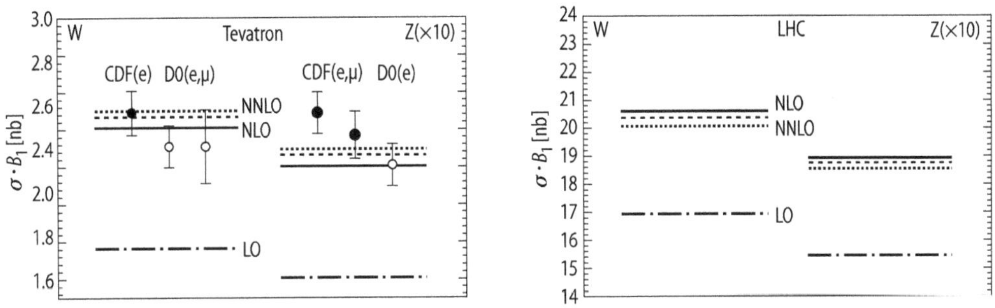

Fig. 4.19 Data vs. theory for W and Z production at the Tevatron ($\sqrt{s}$ = 1.8 TeV) together with the corresponding predictions for the LHC ($\sqrt{s}$= 1.4 TeV) [48]

where A, B, C, D are coefficients of order α_s. The "+" distribution is defined in complete analogy with Eq. (4.93):

$$\int_0^{p^2_{TMAX}} g(z)f(z)_+ dz \;=\; \int_0^{p^2_{TMAX}} [g(z)-g(0)]f(z)dz \tag{4.99}$$

The content of this, at first sight mysterious, definition is that the singular "+" terms do not contribute to the total cross-section. In fact for the cross-section the weight function $g(z) = 1$ and we obtain:

$$\sigma \;=\; \sigma_0[(1+A) \;+\; \int_0^{p^2_{TMAX}} D(z)dz] \tag{4.100}$$

The singular terms, of infrared origin, are present at the non completely inclusive level but disappear in the total cross-section. Arguments have been given that these singularities are expected to exponentiate. Explicit calculations in low order support the exponentiation which leads to the following expression:

$$\frac{1}{\sigma_0}\frac{d\sigma_0}{dp_T^2} \;=\; \int \frac{d^2b}{4\pi}\exp(-ib\cdot p_T)(1+A)\exp S(b) \tag{4.101}$$

with:

$$S(b) \;=\; \int_0^{p_{TMAX}} \frac{d^2k_T}{2\pi}[\exp ik_T\cdot b - 1][\frac{B}{k_T^2}\log\frac{m_W^2}{k_T^2} \;+\; \frac{C}{k_T^2}] \tag{4.102}$$

At large p_T the LO perturbative expansion is recovered. At intermediate p_T the infrared p_T singularities are resummed (the Sudakov log terms, which are typical of vector gluons, are related to the fact that for a charged particle in acceleration it is impossible not to radiate, so that the amplitude for no soft gluon emission is exponentially suppressed). However this formula has problems at small p_T, for example, because of the presence of α_s under the integral for $S(b)$:

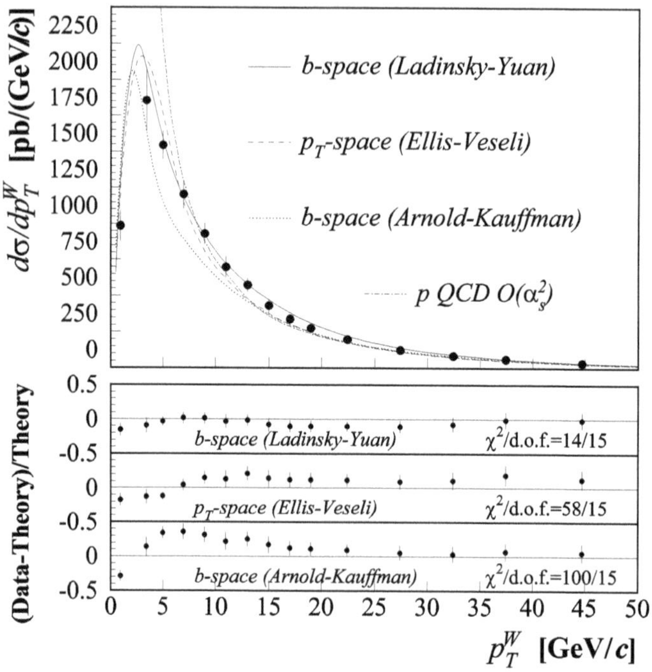

Fig. 4.20 QCD predictions for the W p_T distribution compared with recent D0 data at the Tevatron ($\sqrt{s} = 1.8\,\text{TeV}$)[49] [40]

presumably the relevant scale is of order k_T^2. So it must be completed by some non perturbative ansatz or an extrapolation into the soft region. All the formalism has been extended to NLO accuracy, where one starts from the perturbative expansion at order α_s^2, and generalises the resummation to also include NLO terms of order $\alpha_s(p_T^2)^2 \log m_W^2/p_T^2$ (see, for example, [40]). The comparison with the data is very impressive. In Fig. 4.20 we see the p_T distribution as predicted in QCD (with a number of variants that mainly differ in the approach to the soft region) compared with some recent data at the Tevatron [49].

A great effort is being devoted to the preparation to the LHC. Calculations for specific processes are being completed. A very important example is Higgs production via $g + g \to H$. The amplitude is dominated by the top quark loop, as discussed in Chap. 3 [51]. The NLO corrections turn out to be particularly large [52], as seen in Fig. 4.21. Higher order corrections can be computed either in the effective lagrangian approach, where the heavy top is integrated away and the loop is shrunk down to a point [53] [the coefficient of the effective vertex is known to α_s^4 accuracy [54]], or in the full theory. At the NLO the two approaches agree very well for the rate as a function of m_H [55]. The NNLO corrections have been computed in the effective vertex approximation [56] (see Fig. 4.21). Beyond fixed order resummation of large logs were carried out [57]. Also the NLO EW contributions have been computed [58]. Rapidity (at NNLO) [59] and p_T distributions (at NLO) [60] have also been evaluated. At smaller p_T the large logarithms $[log(p_T/m_H)]^n$ have been

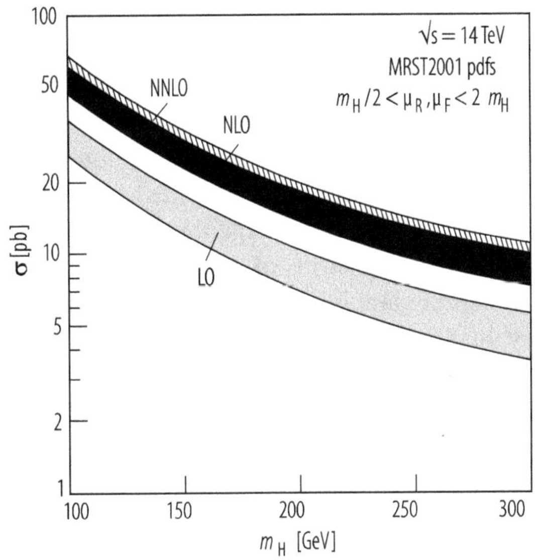

Fig. 4.21 The Higgs gluon fusion cross section in LO, NLO and NLLO

resummed in analogy with what was done long ago for W and Z production [61]. For additional recent works on Higgs physics at colliders see, for example, [62].

The activity on event simulation also received a big boost from the LHC preparation (see, for example, the review [50]). General algorithms for performing NLO calculations numerically (requiring techniques for the cancellation of singularities between real and virtual diagrams) have been developed (see, for example, [65]). The matching of matrix element calculation of rates together with the modeling of parton showers has been realised in packages, as for example in the MC@NLO [63] or POWHEG [64] based on HERWIG. The matrix element calculation, improved by resummation of large logs, provides the hard skeleton (with large p_T branchings) while the parton shower is constructed by a sequence of factorized collinear emissions fixed by the QCD splitting functions. In addition, at low scales a model of hadronisation completes the simulation. The importance of all the components, matrix element, parton shower and hadronisation can be appreciated in simulations of hard events compared with the Tevatron data.

At different places in the previous pages we have seen examples of resummation of large logs. This is a very important chapter of modern QCD. The resummation of soft gluon logs enter in different problems and the related theory is subtle. I refer the reader here to some recent papers where additional references can be found [66]. A particularly interesting related development has to do with the so called non global logs (see, for example, [67]). If in the measurement of an observable some experimental cuts are introduced, which is a very frequent case, then a number of large logs can arise from the corresponding breaking of inclusiveness. The discussion of this problem has led to rethinking the theory of final state observables. It is also important to mention the development of software for the automated implementation of resummation (see, for example, [68]).

Before closing this section I would like to mention some very interesting developments at the interface between string theory and QCD, twistor calculus. A precursor work was the Parke-Taylor result in 1986 [69] on the amplitudes for n incoming gluons with given helicities [70]. Inspired by dual models, they derived a compact formula for the maximum non vanishing helicity violating amplitude (with n−2 plus and 2 minus helicities) in terms of spinor products. Using the relation between strings and gauge theories in twistor space Witten developed in '03 [71] a formalism in terms of effective vertices and propagators that allows to compute all helicity amplitudes. The method, alternative to other modern techniques for the evaluation of Feynman diagrams [73], leads to very compact results. Since then rapid progress followed (for reviews, see [72]): for tree level processes powerful recurrence relations were established [74], the method was extended to include massless external fermions [75] and also external EW vector bosons [76] and Higgs particles [77]. The level already attained is already important for multijet events at the LHC. And the study of loop diagrams has been started. In summary, this road looks new and promising.

4.6 Measurements of α_s

Very precise and reliable measurements of $\alpha_s(m_Z^2)$ are obtained from e^+e^- colliders (in particular LEP) and from deep inelastic scattering.

4.6.1 α_s from e^+e^- Colliders

The main methods at e^+e^- colliders are: (a) Inclusive hadronic Z decay, R_l, σ_h, σ_l, Γ_Z. (b) Inclusive hadronic τ decay. (c) Event shapes and jet rates.

As we have seen, for a quantity like R_l we can write a general expression of the form:

$$R_l = \frac{\Gamma(Z, \tau \rightarrow hadrons)}{\Gamma(Z, \tau \rightarrow leptons)} \sim R^{EW}(1 + \delta_{QCD} + \delta_{NP}) + \ldots \qquad (4.103)$$

where R^{EW} is the electroweak-corrected Born approximation, δ_{QCD}, δ_{NP} are the perturbative (logarithmic) and non perturbative (power suppressed) QCD corrections. For a measurement of α_s at the Z (in the following we always refer to the $\overline{MS}$ definition of α_s) one can use all info from R_l, $\Gamma_Z = 3\Gamma_l + \Gamma_h$ and (f=h or l) $\sigma_f = 12\pi\Gamma_l\Gamma_f/(m_Z^2\Gamma_Z^2)$. In the past the measurement from R_l was preferred (by itself it leads to $\alpha_s(m_Z) = 0.1226 + 0.0058 - 0.0036$) but at LEP there is no reason for that. In all these quantities α_s enters through Γ_h, but the measurements of, say, Γ_Z, R_l and σ_l are really independent (they are affected by entirely different systematics: Γ_Z is extracted from the line shape, R_l and σ_l are measured at the peak but R_l does not depend on the absolute luminosity while σ_l does).The most sensitive single quantity is σ_l. The combined value from the measurements at the Z (assuming the validity of the SM and a light Higgs mass) is [78]:

$$\alpha_s(m_Z) = 0.119 \pm 0.003 \tag{4.104}$$

For a relatively light Higgs (even if not as light as from the fit to EW observables) the final error is mainly experimental with a theoretical component from our ignorance of m_H, of higher orders in the QCD expansion [79] and also from uncertainties on the Bhabha luminometer (which affect $\sigma_{h,l}$) [80]. By adding all other electroweak precision electroweak tests (in particular m_W) one similarly finds [41]:

$$\alpha_s(m_Z) = 0.1185 \pm 0.003 \tag{4.105}$$

We now consider the measurement of $\alpha_s(m_Z)$ from τ decay. R_τ has a number of advantages that, at least in part, tend to compensate for the smallness of $m_\tau = 1.777$ GeV. First, R_τ is maximally inclusive, more than $R_{e^+e^-}(s)$, because one also integrates over all values of the invariant hadronic squared mass:

$$R_\tau = \frac{1}{\pi}\int_0^{m_\tau^2} \frac{ds}{m_\tau^2}(1 - \frac{s}{m_\tau^2})^2 Im\Pi_\tau(s) \tag{4.106}$$

The perturbative contribution is known at NNLO. Analyticity can be used to transform the integral into one on the circle at $|s| = m_\tau^2$:

$$R_\tau = \frac{1}{2\pi i}\oint_{|s|=m_\tau^2} \frac{ds}{m_\tau^2}(1 - \frac{s}{m_\tau^2})^2 \Pi_\tau(s) \tag{4.107}$$

Also, the factor $(1 - \frac{s}{m_\tau^2})^2$ is important to kill the sensitivity the region $Re[s] = m_\tau^2$ where the physical cut and the associated thresholds are located. Still the quoted result [81] looks a bit too precise:

$$\alpha_s(m_\tau) = 0.345 \pm 0.010 \tag{4.108}$$

or

$$\alpha_s(m_Z) = 0.1215 \pm 0.0012 \tag{4.109}$$

This precision is obtained by taking for granted that corrections suppressed by $1/m_\tau^2$ are negligible. This is because, in the massless theory, the light cone expansion is given by:

$$\delta_{NP} = \frac{ZERO}{m_\tau^2} + c_4 \cdot \frac{< O_4 >}{m_\tau^4} + c_6 \cdot \frac{< O_6 >}{m_\tau^6} + \cdots \tag{4.110}$$

In fact there are no dim-2 Lorentz and gauge invariant operators. For example, $g_\mu g^\mu$ is not gauge invariant. In the massive theory, the ZERO is replaced by light quark mass-squared m^2. This is still negligible if m is taken as a lagrangian mass of a few MeV. If on the other hand the mass were taken to be the constituent mass of order Λ_{QCD}, this term would not be at all negligible and would substantially affect the result (note that $\alpha_s(m_\tau)/\pi \sim 0.1 \sim (0.6\,\text{GeV}/m_\tau)^2$ and that Λ_{QCD} for three flavours is large). For example, the PDG value and estimate of the error is [9]:

$$\alpha_s(m_Z) = 0.120 \pm 0.003. \tag{4.111}$$

Most people believe the optimistic version. I am not convinced that the gap is not filled up by ambiguities of $0(\Lambda^2_{QCD}/m^2_\tau)$ from δ_{pert} [82]. In any case, one can discuss the error, but it is true and remarkable, that the central value from τ decay, obtained at very small Q^2, is in good agreement with all other precise determinations of α_s at more typical LEP values of Q^2.

Important determinations of $\alpha_s(m_Z)$ are obtained from different infrared safe observables related to event rates and jet shapes in e^+e^- annihilation. The main problem of these measurements is the large impact of non perturbative hadronization effects on the result and therefore on the theoretical error. The perturbative part is known at NLO. One advantage is that the same measurements can be repeated at different $\sqrt{s}$ values (e.g. with the same detectors at LEP1 or LEP2) allowing for a direct observation of the energy dependence. A typical result, from jets and event shapes at LEP, quoted in Ref. [83], is given by:

$$\alpha_s(m_Z) = 0.121 \pm 0.005. \tag{4.112}$$

Recently the rate of 4-jet events (proportional to α_s^2) at LEP as function of y_{cut} has been used [84], for which a NLO theoretical calculation exists [85]. The quoted result is $\alpha_s(m_Z) = 0.1176 \pm 0.0022$ (the actual error could be somewhat larger because the ambiguity from hadronisation modeling is always debatable).

4.6.2 α_s *from Deep Inelastic Scattering*

QCD predicts the Q^2 dependence of $F(x, Q^2)$ at each fixed x, not the x shape. But the Q^2 dependence is related to the x shape by the QCD evolution equations. For each x-bin the data allow to extract the slope of an approximately straight line in $dlogF(x, Q^2)/dlogQ^2$: the log slope. The Q^2 span and the precision of the data are not much sensitive to the curvature, for most x values. A single value of Λ_{QCD} must be fitted to reproduce the collection of the log slopes. For the determination of α_s the scaling violations of non-singlet structure functions would be ideal, because of the minimal impact of the choice of input parton densities. We can write the non-singlet evolution equations in the form:

$$\frac{d}{dt} log F(x,t) = \frac{\alpha_s(t)}{2\pi} \int_x^1 \frac{dy}{y} \frac{F(y,t)}{F(x,t)} P_{qq}(\frac{x}{y}, \alpha_s(t)) \tag{4.113}$$

where P_{qq} is the splitting function. At present NLO and NNLO corrections are known. It is clear from this form that, for example, the normalisation error on the input density drops away, and the dependence on the input is reduced to a minimum (indeed, only a single density appears here, while in general there are quark and gluon densities). Unfortunately the data on non-singlet structure functions are not very accurate. If we take the difference of data on protons and neutrons, $F_p - F_n$, experimental errors add up in the difference and finally are large. The $F_{3\nu N}$ data are directly non-singlet but are not very precise. A determination of α_s from the CCFR data on $F_{3\nu N}$ has led to [86]:

$$\alpha_s(m_Z) = 0.119 \pm 0.006 \tag{4.114}$$

A recent analysis of the same data leads to $\alpha_s(m_Z) = 0.119 \pm 0.002$ [87], but the theoretical error associated with the method and with the choice adopted for the scale ambiguities is not considered. A fit to non singlet structure functions in electro- or muon-production extracted from proton and deuterium data at the NNLO level was performed in Ref. [88] with the result:

$$\alpha_s(m_Z) = 0.114 \pm 0.002 \tag{4.115}$$

When one measures α_s from scaling violations on F_2 from e or μ beams, the data are abundant, the errors small but there is an increased dependence on input parton densities and especially a strong correlation between the result on α_s and the input on the gluon density. There are complete and accurate derivations of α_s from scaling violations in F_2. In a well known analysis by Santiago and Yndurain [89], the data on protons from SLAC, BCDMS, E665 and HERA are used with NLO kernels plus the NNLO first few moments. The analysis is based on an original method that uses projections on a specially selected orthogonal basis, the Bernstein polynomials. The quoted result is given by:

$$\alpha_s(m_Z) = 0.1163 \pm 0.0014 \tag{4.116}$$

(these authors also quote $\alpha_s(m_Z) = 0.115 \pm 0.006$ for F_3 data in νN scattering). A different analysis by Alekhin [90] of existing data off proton and deuterium targets with NNLO kernels and a more conventional method leads to

$$\alpha_s(m_Z) = 0.114 \pm 0.002 \tag{4.117}$$

In both analyses the dominant error is theoretical and, in my opinion, should be somewhat larger than quoted. An interesting perspective on theoretical errors can be obtained by comparing analyses with different methods. We add the following examples. From truncated moments (but with a limited set of proton data and NLO kernels) [91]: $\alpha_s(m_Z) = 0.122 \pm 0.006$, from Nachtmann moments (which take into account some higher twist terms) and proton data [92]: $\alpha_s(m_Z) = 0.1188 \pm 0.0017$. A combination of measurements at HERA by H1 and Zeus, also including final state jet observables, leads to $\alpha_s(m_Z) = 0.1186 \pm 0.0051$ [93], most of the error being theoretical. Finally, to quote a number that appears to me as a good summary of the situation of $\alpha_s(m_Z)$ from DIS one can take the result from a NNLO analysis of available data by the MRST group [94] as quoted by Particle Data Group, W.-M. Yao et al. [9]:

$$\alpha_s(m_Z) = 0.1167 \pm 0.0036 \tag{4.118}$$

If we compare these results on α_s from DIS with the findings at the Z, given by Eq. (4.105), we see that the agreement is good, with the value of α_s from the most precise DIS measurements a bit on the low side with respect to e^+e^-.

4.6.3 Summary on α_s

There are a number of other determinations of α_s which are important because they arise from qualitatively different observables and methods. For example [9, 83], some are obtained from the Bjorken sum rule and the scaling violations in polarized DIS, from Υ decays, from the 4-jet rate in e^+e^-. A special mention deserves the "measurement" of α_s from lattice QCD [95]. A number of hadronic observables, in particular $\Upsilon' - \Upsilon$ splitting, pion and kaon decay constants, the B_s mass and the Ω baryon mass are used to fix the lattice spacing and to accurately tune the QCD simulation. The value of α_s is then obtained by computing non perturbatively a number of quantities related to Wilson loops that can also be given in perturbation theory. The result is then evolved with state of the art beta functions to obtain $\alpha_s(m_Z) = 0.1170 \pm 0.0012$. This result is interesting for its really special nature but it is not clear that the systematics due to the lattice technology is as small as claimed.

Summarising: there is very good agreement among many different measurements of α_s. In Fig. 4.22 [83], a compilation of the data is reported with each measurement plotted at the scale of the experiment, which shows the consistency of the measurements and the running of α_s. This is a very convincing, quantitative test of QCD. If I take the values of $\alpha_s(m_Z)$ from precision electroweak data, Eq. (4.105), from τ decay with the central value as in Eq. (4.109) but the larger error as in Eq. (4.111), from jets in e^+e^-, Eq. (4.112), and from DIS, Eq. (4.118), the average is :

$$\alpha_s(m_Z) = 0.119 \pm 0.002 \tag{4.119}$$

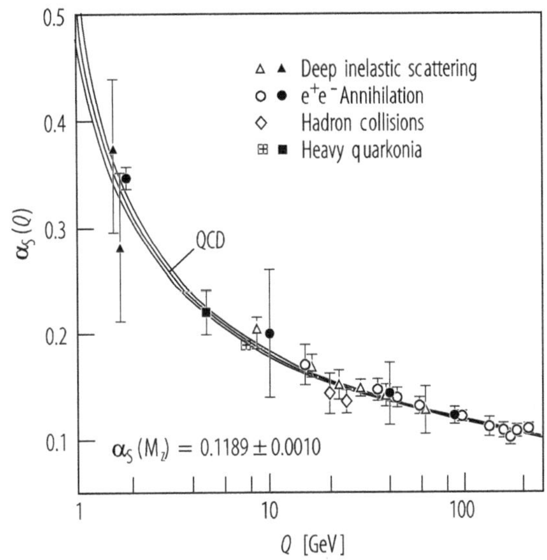

Fig. 4.22 The running of α_s as determined from present data [83]

For comparison, the average value quoted by PDG 2006 is $\alpha_s(m_Z) = 0.1176 \pm 0.0020$ while Ref. [83] gives $\alpha_s(m_Z) = 0.1189 \pm 0.0010$.

The value of Λ (for $n_f = 5$) which corresponds to Eq. (4.119) is:

$$\Lambda_5 = 221 \pm 25 \text{ MeV} \tag{4.120}$$

Λ is the scale of mass that finally appears in massless QCD. It is the scale where $\alpha_s(\Lambda)$ is of order one. Hadron masses are determined by Λ. Actually the ρ mass or the nucleon mass receive little contribution from the quark masses (the case of pseudoscalar mesons is special, as they are the pseudo Goldstone bosons of broken chiral invariance). Hadron masses would be almost the same in massless QCD.

4.7 Conclusion

We have seen that perturbative QCD based on asymptotic freedom offers a rich variety of tests and we have described some examples in detail. QCD tests are not as precise as for the electroweak sector. But the number and diversity of such tests has established a very firm experimental foundation for QCD as a theory of strong interactions. The field of QCD appears as one of great maturity but also of robust vitality with many rich branches and plenty of new blossoms. The physics content of QCD is very large and our knowledge, especially in the non perturbative domain, is still very limited but progress both from experiment (Tevatron, RHIC, LHC......) and from theory is continuing at a healthy rate. And all the QCD predictions that we were able to formulate and to test appear to be in very good agreement with experiment.

Acknowledgments I am very grateful to Richard Ball, Keith Ellis, Stefano Forte, Ashutosh Kotwal, Vittorio Lubicz, Lorenzo Magnea, Michelangelo Mangano, Silvano Simula, Hartmut Wittig for their help and advise.

References for 4

Disclaimer: This list of references is by far incomplete. It is just meant to provide a guide to the vast literature on QCD. So it includes only a few old papers and, for most of the issues, some relatively recent works where to look for a more complete list of references.

References

1. M. Gell-Mann, Acta Phys. Austriaca Suppl. IX (1972) 733; H. Fritzsch and M. Gell-Mann, Proc. XVI Int. Conf. on High Energy Physics, Chicago-Batavia, 1972; H. Fritzsch, M. Gell-Mann, H. Leutwyler, Phys. Lett. B47 (1973) 365.
2. D.J. Gross, F. Wilczek, Phys. Rev. Lett. 30 (1973) 1343; Phys. Rev. D8 (1973) 3633; H.D. Politzer, Phys. Rev. Lett. 30 (1973) 1346.
3. S. Weinberg, Phys. Rev. Lett. 31 (1973) 494.
4. G. Altarelli, Phys. Rep. 81(1982)1.
5. G. Altarelli, Ann. Rev. Nucl. Part. Sci. 39(1989)357.
6. G. Altarelli, *The Development of Perturbative QCD*, World Scientific, (1994); W. Greiner, S. Schramm and E. Stein, *Quantum Chromodynamics*, Springer, 1994; *Handbuch of QCD*. ed. by M. Shifman, Vol.1–4, World Sci. (2001); R. K. Ellis, W. J. Stirling, and B. R. Webber, *QCD and Collider Physics*, Cambridge Monographs, (2003); G. Dissertori, I. Knowles and M. Schmelling, *Quantum Chromodynamics: High Energy Experiments and Theory*, Oxford Univ. Press, 2003.
7. M. Gell-Mann, Phys. Lett. 8 (1964) 214 ; G. Zweig, CERN TH 401 and 412 (1964). O.W. Greenberg,Phys. Rev. Lett. 13 (1964) 598.
8. F. Bloch and H. Nordsieck, Phys. Rev. 52(1937)54.
9. Particle Data Group, W.-M. Yao et al., Journal of Physics G 33, 1 (2006).
10. T. Kinoshita, J. Math. Phys. 3(1962)650; T.D. Lee and M. Nauenberg, Phys. Rev. 133(1964)1549.
11. T. Applequist and J. Carazzone, Phys. Rev. D11 (1975)2856.
12. O.V. Tarasov, A.A. Vladimirov and A. Yu. Zharkov, Phys. Lett. 93B(1980)429.
13. T. van Ritbergen, J.A.M. Vermaseren and S.A. Larin, Phys.Lett.B400(1997)379; see also M. Czakon, Nucl. Phys. B710 (2005) 485.
14. G. Altarelli, Proceedings of the E. Majorana Summer School, Erice, 1995, Plenum Press, ed. by A. Zichichi; M. Beneke and V. M. Braun, in *Handbuch of QCD*. ed. by M. Shifman, Vol.3, pag. 1719, World Sci. (2001).
15. S.G. Gorishny, A.L. Kataev and S.A. Larin, Phys. Lett. 259B (1991) 144; L. R. Surguladze and M. A. Samuel, Phys. Rev. Lett. 66 (1991) 560.
16. R. Brandt and G. Preparata,Nucl. Phys. 27B (1971) 541.
17. Yu. Dokshitzer, J. Phys.G17 (1991)1572; N. Brown and W. J. Stirling, Z. Phys. C53 (1992) 629.
18. V.N. Gribov and L.N. Lipatov, Sov. J. Nucl. Phys. 15 (1972) 438; see also Yu. L. Dokshitzer, Sov. Phys. JETP 46 (1977) 641.
19. G.Altarelli and G. Parisi, Nucl. Phys. B126 (1977) 298.
20. E. G. Floratos, D.A. Ross and C. T. Sachrajda, Nucl. Phys. B129 (1977) 66, E B139 (1978) 545; B152 (1979) 493; A. Gonzales-Arroyo, C. Lopez and F. J. Yndurain, Nucl. Phys. B153 (1979) 161; G. Curci, W. Furmanski and R. Petronzio, Nucl. Phys. B175 (1980) 27; W. Furmanski

and R. Petronzio, Phys. Lett. 97B (1980) 438; E.G. Floratos, R. Lacaze and C. Kounnas, Phys. Lett. 99B (1981) 89, 285; R. T. Herrod and S. Wada, Z. Phys. C9 (1981) 351.
21. S. Moch, J. A. M. Vermaseren and A. Vogt, Nucl. Phys. B **688**, 101 (2004), [arXiv:hep-ph/0403192]; A. Vogt, S. Moch and J. A. M. Vermaseren, Nucl. Phys. B **691**, 129 (2004), [arXiv:hep-ph/0404111].
22. 5) G. Altarelli, R.Keith Ellis, G. Martinelli, Nucl.Phys.B143:521,1978, E.ibid.B146:544,1978; Nucl.Phys.B157:461,1979.
23. A. D. Martin, R. G. Roberts, W. J. Stirling and R. S. Thorne, hep-ph/0110215, hep-ph/0201127.
24. G. Sterman, Nucl. Phys. B281 (1987) 310; S. Catani and L. Trentadue, Nucl. Phys. B327 (1989) 323; N. Kidonakis and G. Sterman, Nucl. Phys. B505 (1997) 321.
25. S. Forte and G. Ridolfi, Nucl.Phys. B650:229,2003; S. Moch, J. A. M. Vermaseren and A. Vogt, Nucl. Phys. B **726** (2005) 317 [arXiv:hep-ph/0506288]; T. Becher, M. Neubert and B. D. Pecjak, JHEP **0701** (2007) 076 [arXiv:hep-ph/0607228].
26. E. Gardi, G. P. Korchemsky, D. A. Ross and S. Tafat, Nucl. Phys. B **636** (2002) 385 [arXiv:hep-ph/0203161]; E. Gardi and R. G. Roberts, Nucl. Phys. B **653** (2003) 227 [arXiv:hep-ph/0210429].
27. L.N. Lipatov, Sov. Jour. Nucl. Phys. 23 (1976) 338; V.S. Fadin, E.A. Kuraev and L.N. Lipatov, Phys. Lett. 60B (1975) 50; Sov. Phys. JETP 44 (1976) 443; 45(1977)199; Y.Y. Balitski and L.N.Lipatov, Sov. Jour. Nucl. Phys. 28 (1978) 822.
28. V.S. Fadin and L.N. Lipatov, Phys. Lett. B429 (1998) 127; B429 (1998) 127; V.S. Fadin et al, Phys. Lett. B359 (1995)181; B387 (1996)593; Nucl. Phys. B406 (1993) 259; Phys. Rev. D50 (1994) 5893; Phys. Lett. B389 (1996) 737; Nucl. Phys. B477 (1996) 767; Phys. Lett. B415 (1997) 97; B422 (1998) 287; G. Camici and M. Ciafaloni, Phys. Lett. B412 (1997) 396; Phys. Lett. B430 (1998) 349. V. del Duca, Phys. Rev. D54 (1996) 989; D54 (1996) 4474; V. del Duca and C.R. Schmidt, Phys. Rev. D57 (1998) 4069; Z. Bern, V. del Duca and C.R. Schmidt, Phys. Lett. B445 (1998) 168.
29. G. Altarelli, R. Ball and S. Forte, Nucl. Phys. B742 (2006) 1.
30. M. Ciafaloni, D. Colferai, G. P. Salam and A. M. Stasto, Phys. Lett. B587 (2004) 87; see also G. P. Salam, hep-ph/0501097; C. D. White and R. S. Thorne, Phys. Rev. D **75** (2007) 034005 [hep-ph/0611204].
31. S. Forte and Y. Goto, *Proceedings of DIS 2006*, Tsukuba, Japan, World Scientific, (2007), p.908 [hep-ph/0609127].
32. M. Hirai, S. Kumano and N. Saito, Phys.Rev. D69 (2004) 05402.
33. G. Mallot, hep-ex/0612055.
34. G.Altarelli and G. G. Ross, Phys. Lett. B212 (1988) 391; A. V. Efremov and O. V. Terayev, *Proceedings of the Czech Hadron Symposium*, 1988, p. 302; R. D. Carlitz, J. C. Collins and A. H. Mueller, Phys. Lett. B214 (1988) 229; G. Altarelli and B. Lampe, Z. Phys. C47 (1990) 315.
35. CDF Run II Collaboration, Phys.Rev.D74:071103,2006; e-Print: hep-ex/0512020 the D0 Collaboration, Phys.Rev.Lett.86:2523,2001; Phys.Lett.B639:151,2006.
36. The CDF Collaboration, Phys.Rev.Lett.73:2662–2666,1994, Erratum-ibid.74:1891–1893,1995; Phys.Rev.D65:112003,2002; Phys.Rev.D70:074008,2004;
37. R. Ha the D0 Collaboration, Phys.Rev.Lett.87:251805,2001; Phys.Rev.Lett.84:2786,2000.
38. J. Kubar-Andre' and F. Paige, Phys. Rev. D19(1979) 221. mberg, W. L. van Neerven and T. Matsuura, Nucl. Phys. B359 (1991) 343; W. L. van Neerven and E. B. Zijlstra, Nucl. Phys. B382(1992) 11.
39. V.V. Sudakov, Sov. Phys. JETP, 3 (1956) 75; Yu. Dokshitzer, D. Dyakonov and S. Troyan, Phys. Lett. 76B (1978) 290, Phys. Rep. 58 (1980 269; G. Parisi and R. Petronzio, Nucl. Phys. B154 ((1979) 427; G. Curci, M. Greco and Y. Srivastava, Phys. Rev. Lett. 43 (1979), Nucl. Phys. B159 (1979) 451; J. Collins and D. Soper, Nucl. Phys. B139 (1981) 381, B194 (1982) 445, B197 (1982)446; J. Kodaira and L. Trentadue, Phys. Lett. 112B (1982) 66, 123B (1983)335; J. Collins, D. Soper and G. Sterman, Nucl. Phys. B250 (1985) 199; C. Davies, B. Webber and J. Stirling, Nucl. Phys. B256 (1985) 413.
40. P. Arnold and R. Kauffman, Nucl.Phys.B349:381,1991; G.A. Ladinsky and C.P. Yuan, Phys.Rev.D50:4239,1994; R.K. Ellis, S. Veseli, Nucl.Phys.B511:649,1998.

41. For up-to-date results see, The LEP Electroweak Group, http://lepewwg.web.cern.ch/LEPEWWG/.
42. P. Nason, S. Dawson and R. K. Ellis, Nucl. Phys. B **303** (1988) 607; W. Beenakker, H. Kuijf, W. L. van Neerven and J. Smith, Phys. Rev. D **40** (1989) 54; P. Nason, S. Dawson and R. K. Ellis, Nucl. Phys. B **327** (1989) 49 [Erratum-ibid. B **335** (1989) 260]; M. L. Mangano, P. Nason and G. Ridolfi, Nucl. Phys. B **373** (1992) 295; R. Bonciani, S. Catani, M. L. Mangano and P. Nason, Nucl.Phys. B529:424,1998; M. Cacciari et al, JHEP 0404:068, 2004; N. Kidonakis and A. Vogt, Phys. Rev. D68 114014(2003); A. Banfi, G. P. Salam and G. Zanderighi, JHEP **0707** (2007) 026 [arXiv:0704.2999 [hep-ph]].
43. The CDF II Collaboration, Phys.Rev.D74:072005,2006; Phys.Rev.D74:072006,2006; Phys.Rev.D71:072005,2005; the CDF Collaboration, Phys.Rev.Lett.97:082004,2006; Phys.Rev.Lett.96:202002,2006; Phys.Rev.D72:052003,2005; Phys.Rev.D71:052003,2005; Phys.Rev.Lett.93:142001,2004. The D0 Collaboration, Phys.Rev.D74:112004,2006; Phys.Lett.B626:55,2005; Phys.Lett.B626:35,2005; Phys.Lett.B626:45,2005.
44. The CDF Collaboration, CDF Note 8148, v. 1.0, 2006.
45. The CDF Collaboration, Phys.Rev.D75:012010,2007; Phys.Rev.D71:032001,2005; Phys.Rev.D65:052005,2002; the D0 Collaboration, Phys.Rev.Lett.94:232001,2005; Phys.Rev.Lett.85:5068,2000; Phys.Lett.B487:264-272,2000.
46. M. Cacciari et al, JHEP 0407:033, 2004.
47. D. Bauer (for the CDF and D0 Collaborations), Nucl. Phys. B (Proc. Suppl.) 156 (2006) 226.
48. The CDF II Collaboration, Phys.Rev.Lett. 94:091803, 2005; the D0 Collaboration, Phys.Rev. D60:052003,1999.
49. The D0 collaboration, Phys. Rev. Lett. 84 (2000) 2792; Phys.Rev. D61 (2000) 032004; see also the CDF Collaboration, Phys.Rev.Lett. 84 (2000) 845.
50. J.M. Campbell, J.W. Huston and W.J. Stirling, Rept.Prog.Phys. 70 (2007) 89
51. H.Georgi, S.Glashow, M.Machacek and D.Nanopoulos,Phys.Rev.Lett. **40**,692(1978)
52. S. Dawson, Nucl. Phys. B **359**, 283 (1991); A. Djouadi, M. Spira and P. M. Zerwas, Phys. Lett. B **264**, 440 (1991); M. Spira, A. Djouadi, D. Graudenz and P. Zerwas, Nucl. Phys. B **453**, 17 (1995).
53. J. Ellis, M. Gaillard and D. Nanopoulos, Nucl. Phys. B **106**, 292 (1976)
54. K. Chetyrkin, B. Kniehl and M. Steinhauser, Phys. Rev. Lett. **79**, 353 (1997)
55. M. Kramer, E. Laenen and M. Spira, Nucl. Phys. B **511**, 523 (1998)
56. S.Catani, D.de Florian and M.Grazzini, JHEP**0105**,025(2001); JHEP**0201**,015(2002); R. V. Harlander, Phys. Lett. B **492**, 74 (2000); V. Ravindran, J. Smith and W. L. van Neerven, Nucl. Phys. B **704**, 332 (2005); R. V. Harlander and W. B. Kilgore, Phys. Rev. D **64**, 013015 (2001); R. V. Harlander and W. B. Kilgore, Phys. Rev. Lett. **88**, 201801 (2002); C. Anastasiou and K. Melnikov, Nucl. Phys. B **646**, 220 (2002); V. Ravindran, J. Smith and W. L. van Neerven, Nucl. Phys. B **665**, 325 (2003).
57. S. Catani, D. de Florian, M. Grazzini and P. Nason, JHEP **0307**, 028 (2003); S. Moch and A. Vogt, Phys. Lett. B **631**, 48 (2005); V. Ravindran, Nucl. Phys. B **752**, 173 (2006) (and references therein).
58. U. Aglietti, R. Bonciani, G. Degrassi and A. Vicini, Phys. Lett. B **595**, 432 (2004); G. Degrassi and F. Maltoni, Phys. Lett. B **600**, 255 (2004).
59. C. Anastasiou, K. Melnikov and F. Petriello, Nucl. Phys. B **724**, 197 (2005); V. Ravindran, J. Smith and W. L. van Neerven, Nucl. Phys. B **767** (2007) 100 [hep-ph/0608308].
60. D. de Florian, M. Grazzini and Z. Kunszt, Phys. Rev. Lett. **82**, 5209 (1999); V. Ravindran, J. Smith and W. L. Van Neerven, Nucl. Phys. B **634**, 247 (2002); C. J. Glosser and C. R. Schmidt, JHEP **0212**, 016 (2002).
61. G. Bozzi, S. Catani, D. de Florian and M. Grazzini, Phys. Lett. B **564**, 65 (2003); Nucl. Phys. B **737**, 73 (2006); A. Kulesza, G. Sterman and W. Vogelsang, Phys. Rev. D **69**, 014012 (2004).
62. R.K. Ellis, W.T. Giele and G. Zanderighi, Phys.Rev. D72 (2005) 054018; E. D74 (2006) 079902; J. M. Campbell, R. K. Ellis and G. Zanderighi, JHEP 0610 (2006) 028; C. Anastasiou et al, JHEP 0701 (2007) 082; G. Davatz et al, JHEP 0607 (2006) 037; S. Catani and M. Grazzini, Phys. Rev. Lett. 98 (2007) 222002; C. Anastasiou, G. Dissertori and F. Stoeckli,

JHEP **0709** (2007) 018 [arXiv:0707.2373 [hep-ph]]; P. M. Nadolsky, C. Balazs, E. L. Berger and C.-P. Yuan, Phys. Rev. D **76** (2007) 013008 [hep-ph/0702003 [HEP-PH]]; C. Balazs, E. L. Berger, P. M. Nadolsky and C.-P. Yuan, Phys. Rev. D **76** (2007) 013009 [arXiv:0704.0001 [hep-ph]]; M. Ciccolini, A. Denner and S. Dittmaier, Phys. Rev. Lett. **99** (2007) 161803 [arXiv:0707.0381 [hep-ph]].
63. S. Frixione and B. Webber, hep-ph/0612272 and references therein.
64. S. Frixione, P. Nason and G. Ridolfi, arXiv:0707.3081; S. Frixione, P. Nason and G. Ridolfi, JHEP **0709** (2007) 126 [arXiv:0707.3088 [hep-ph]]
65. K. Fabricius, I. Schmitt, G. Kramer and G. Schierholz, Z. Phys. C **11** (1981) 315; G. Kramer and B. Lampe, Fortsch. Phys. **37** (1989) 161; R. K. Ellis, D. A. Ross and A. E. Terrano, Nucl. Phys. B **178** (1981) 421; S. Frixione, Z. Kunszt and A. Signer, Nucl.Phys.B467:399,1996; D. Kosower, Phys.Rev.D71:045016,2005; S. Catani, S. Dittmaier, M. H. Seymour and Z. Trócsányi, Nucl.Phys.B627,189 (2002).
66. A. Banfi and M. Dasgupta, JHEP **0401** (2004) 027; P Bolzoni, S. Forte and G. Ridolfi, Nucl.Phys.B731:85,2005; Y. Delenda, R. Appleby, M. Dasgupta and A. Banfi, JHEP **0612** (2006) 044; P. Bolzoni, Phys.Lett.B643:325,2006; S. Mert Aybat, L. J. Dixon and G. Sterman, Phys. Rev. D **74** (2006) 074004; Yu. L. Dokshitzer and G. Marchesini, JHEP **0601** (2006) 007; E. Laenen and L. Magnea, Phys. Lett. B **632** (2006) 270; C. Lee and G. Sterman, Phys. Rev. D **75**, 014022 (2007); D. de Florian and W. Vogelsang, Phys. Rev. D **76** (2007) 074031 [arXiv:0704.1677 [hep-ph]]; R. Abbate, S. Forte and G. Ridolfi, Phys. Lett. B **657** (2007) 55 [arXiv:0707.2452 [hep-ph]].
67. M. Dasgupta and G. P. Salam, Phys. Lett. B **512** (2001) 323; M. Dasgupta and G. P. Salam, J. Phys. G **30** (2004) R143; A. Banfi, G. Corcella, M. Dasgupta, Y. Delenda, G. P. Salam and G. Zanderighi, hep-ph/0508096; J. R. Forshaw, A. Kyrieleis and M. H. Seymour, JHEP **0608** (2006) 059.
68. A. Banfi, G. P. Salam and G. Zanderighi, JHEP **0503** (2005) 073.
69. S.J. Parke and T.R. Taylor, Phys.Rev.Lett.56:2459,1986.
70. F.A. Berends and W. Giele, Nucl. Phys. B294 (1987) 700; M. L. Mangano, S.J. Parke and Z. Xu, Nucl. Phys. B298 (1988) 653.
71. E. Witten, Commun.Math.Phys.252:189,2004, [hep-th/0312171].
72. F. Cachazo and P. Svrcek, PoS RTN2005:004,2005, [hep-th/0504194]; L. J. Dixon, *Proceedings of the EPS International Europhysics Conference on High Energy Physics*, Lisbon, Portugal, 2005, PoS HEP2005:405,2006, [hep-ph/0512111] and references therein.
73. M. Dinsdale, M. Ternick and S. Weinzierl, JHEP 0603:056,2006, [hep-ph/0602204].
74. R. Roiban, M. Spradlin and A. Volovich, Phys. Rev. Lett. 94 102002 (2005); R. Britto, F. Cachazo and B. Feng, Nucl. Phys. B715 (2005) 499; R. Britto, F. Cachazo, B. Feng and E. Witten, Phys. Rev. Lett. 94 181602 (2005).
75. G. Georgiou and V. V. Khoze, JHEP 0405 (2004) 015; J. B. Wu and C.J. Zhu, JHEP 0409 (2004) 063; G. Georgiou, E. W. N. Glover and V. V. Khoze, JHEP 0407 (2004) 048.
76. Z. Bern, D. Forde, D. A. Kosower and P. Mastrolia, Phys.Rev.D72:025006,2005.
77. L. J. Dixon, E. W. N. Glover and V. V. Khoze, JHEP 0412 (2004) 015; S. D. Badger, E. W. N. Glover and V. V. Khoze, JHEP 0503 (2005) 023; S.D. Badger, E.W.N. Glover and K. Risager, Acta Phys.Polon.B38:2273–2278,2007.
78. S. Kluth, Rept. Prog. Phys. 69 (2006) 1771.
79. H. Stenzel, JHEP 0507:0132,2005.
80. W. de Boer and C. Sander, Phys.Lett.B585:276,2004.
81. M. Davier, A. Hocker and Z. Zhang, Rev.Mod.Phys.78:1043,2006, [hep-ph/0507078].
82. G Altarelli, P. Nason and G. Ridolfi, Z.Phys.C68:257,1995.
83. S. Bethke, Prog.Part.Nucl.Phys.58:351,2007, [hep-ex/0606035].
84. ALEPH Collaboration, A. Heister et al, Eur. J. Phys. C 27 (2003) 1; DELPHI Collaboration, J. Abdallah et al, Eur. J. Phys. C 38 (2005) 413; OPAL Collaboration, G. Abbiendi et al, hep-ex/0601048.
85. Z. Nagy and Z. Trocsanyi, Phys. Rev. D59 (1999) 14020, E. D62 (2000) 099902.

86. A.L. Kataev, G. Parente and A.V. Sidorov, Nucl.Phys. B573 (2000) 405; Nucl.Phys.Proc.Suppl.116:105,2003,[hep-ph/0211151].
87. P. M. Brooks and C.J. Maxwell, Nucl.Phys.B780:76, 2007, [hep-ph/0610137].
88. J. Blumlein, H. Bottcher and A. Guffanti, Nucl.Phys.B774:182,2007, [hep-ph/0607200].
89. J. Santiago and F.J. Yndurain, Nucl.Phys. B563 (1999) 45; Nucl. Phys. Proc. Suppl. 86 (2000)69; Nucl.Phys. B611 (2001) 45.
90. S.I. Alekhin, Phys.Rev. D59:114016,1999.
91. S. Forte, J. I. Latorre, L. Magnea and A. Piccione, Nucl.Phys.B643:477,2002, [hep-ph/0205286].
92. S. Simula, M. Osipenko, Nucl.Phys.B675:289,2003, [hep-ph/0306260].
93. C. Glasman, hep-ex/0506035.
94. A.D. Martin, R.G. Roberts, W.J. Stirling and R.S. Thorne, Eur.Phys.J.C35:325, 2004, [hep-ph/0308087].
95. Q. Mason et al, Phys. Rev. Lett. 95 (2005)052002, hep-lat/0503005.

5

The Lattice Approach to Quantum Chromodynamics

Hartmut Wittig

5.1 Introduction and Outline

Since Wilson's seminal papers of the mid-1970s, the lattice approach to Quantum Chromodynamics has become increasingly important for the study of the strong interaction at low energies, and has now turned into a mature and established technique. In spite of the fact that the lattice formulation of Quantum Field Theory has been applied to virtually all fundamental interactions, it is appropriate to discuss this topic in a chapter devoted to QCD, since by far the largest part of activity is focused on the strong interaction. Lattice QCD is, in fact, the only known method which allows ab initio investigations of hadronic properties, starting from the QCD Lagrangian formulated in terms of quarks and gluons.

5.1.1 Historical Perspective

In order to illustrate the wide range of applications of the lattice formulation, we give a brief historical account below.

First applications of the lattice approach in the late 1970s employed analytic techniques, predominantly the strong coupling expansion, in order to investigate colour confinement and also the spectrum of glueballs. While these attempts gave valuable insights, it soon became clear that in the case of non-Abelian gauge theories such expansions were not sufficient to produce quantitative results.

H. Wittig (✉)
Johannes-Gutenberg-Universität Mainz, Mainz, Germany
e-mail: hartmut.wittig@uni-mainz.de

First numerical investigations via Monte Carlo simulations, focusing in particular on the confinement mechanism in pure Yang–Mills theory, were carried out around 1980. The following years saw already several valiant attempts to study QCD numerically, yet it was realized that the available computer power was grossly inadequate to incorporate the effects of dynamical quarks. It was then that the so-called "quenched approximation" of QCD was proposed as a first step to solving full QCD numerically. This approximation rests on the *ad hoc* assumption that the dominant non-perturbative effects are mediated by the gluon field. Hadronic observables can then be computed on a pure gauge background with far less numerical effort compared to the real situation where quarks have a feedback on the gluon field. The main focus of activity during the 1980s was on bosonic theories: numerical simulations were used to compute the glueball spectrum in pure Yang–Mills theory. Another important result during this period concerned ϕ^4-theory and the implications of its supposed "triviality" for the Higgs-Yukawa sector of the Standard Model. Using a combination of analytic and numerical techniques, the triviality of ϕ^4 theory could be rigorously established.

Except for a brief spell of activity around the turn of the decade to simulate QCD with dynamical fermions, most projects in the 1990s were devoted to explore quenched QCD. Having recognized that the available computers and the efficiency of known algorithms were by far not sufficient to perform "realistic" simulations of QCD with controlled errors, lattice physicists resorted to exploring the quenched approximation and its limitations for a number of phenomenologically interesting quantities. Although the systematic error that arises by neglecting dynamical quarks could not be quantified reliably, many important quantities, such as quark and hadron masses, the strong coupling constant and weak hadronic matrix elements, were computed for the first time. One of the icons of that period was surely a plot of the masses of the lightest hadrons in the continuum limit of quenched QCD, produced by the CP-PACS Collaboration: their results indicated that the quenched approximation works surprisingly well (at least for these quantities), since the computed spectrum agreed with experimental determinations at the level of 10%. Simultaneously, a number of sophisticated techniques have been developed during the 1990s, thereby helping to control systematic effects, mainly pertaining to the influence of lattice artefacts, as well as the renormalization of local operators in the lattice regularized theory and their relation to continuum schemes such as $\overline{\text{MS}}$. Perhaps the most significant development at the end of the 1990s was the clarification of the issue of chiral symmetry and lattice regularization. Following this work it is now understood under which conditions the lattice formulation is compatible with chiral symmetry. The importance of this development extends far beyond QCD and implies new prospects for the non-perturbative study of chiral gauge theories.

Since 2000 the focus has decidedly shifted from the quenched approximation to serious attempts to simulate QCD with dynamical quarks, thereby tackling the biggest remaining systematic uncertainty. Progress in this area has not just been determined by the vast increase in computer power since the very first Monte Carlo simulations, but rather by the development of new algorithmic ideas, combined with the use of alternative discretizations that are numerically more efficient. At the time of writing this contribution (2007), the whole field is actually in a state of transition: although the quenched approximation is being abandoned, the latest results from simulations with dynamical quarks have not yet reached the same level of accuracy in regard to controlling systematic errors due to lattice artefacts and effects from renormalization, as compared to earlier quenched calculations. It can thus be expected that many of the results discussed later in this chapter will soon be superseded by more accurate numbers. In turn, the quenched approximation will be completely obsolete in a few years time, except perhaps to test new ideas or for exploratory studies of more complex quantities.

5.1.2 Outline

We begin with an introduction of the basic concepts of the lattice formulation of QCD. This shall include the field theoretical foundations, discretizations of the QCD Lagrangian, as well as simulation algorithms and other technical aspects related to the actual calculation of physical observables from suitable correlation functions. The following sections deal with various applications. Lattice calculations of the hadron spectrum are described in Sect. 5.3. Section 5.4 is devoted to lattice investigations of the confinement phenomenon. Determinations of the fundamental parameters of QCD, namely the strong coupling constant and quark masses are a major focus of this article, and are presented in Sect. 5.5. Another important property of QCD, namely the spontaneously broken chiral symmetry, is discussed in some detail in Sect. 5.6, which also includes a brief introduction into analytical non-perturbative approaches to the strong interaction, based on effective field theories. Lattice calculations of weak hadronic matrix elements, which serve to pin down the elements of the Cabibbo–Kobayashi–Maskawa matrix, are covered in Sect. 5.7. We end this contribution with a few concluding remarks.

In addition to the topics listed above, lattice simulations of QCD have also made important contributions to the determination of the phase structure of QCD, including results for the critical temperature of the deconfinement phase transition. Nevertheless, in this chapter we restrict the discussion to QCD at zero temperature and refer the reader to other parts of this volume.

5.2 The Lattice Approach to QCD

The essential features of the lattice formulation can be summarized by the following statement:

> Lattice QCD is the non-perturbative approach to the gauge theory of the strong interaction through regularized, Euclidean functional integrals. The regularization is based on a discretization of the QCD action which preserves gauge invariance at all stages.

This definition includes all basic ingredients: starting from the functional integral itself avoids any particular reference to perturbation theory. This is what we mean when we call lattice QCD an ab initio method. The Euclidean formulation, which is obtained by rotating to imaginary time, reveals the close relation between Quantum Field Theory and Statistical Mechanics. In particular, the Euclidean functional integral is equivalent to the partition function of the corresponding statistical system. This equivalence is particularly transparent if the field theory is formulated on a discrete space-time lattice. Via this relation, the whole toolkit of condensed matter physics, including high-temperature expansions, and, perhaps most importantly, Monte Carlo simulations, are at the disposal of the field theorist.

Many of the basic concepts introduced in this section are discussed in several common textbooks on the subject [1–4], which can be consulted for further details.

5.2.1 Euclidean Quantization

The generic steps in the Euclidean quantization procedure of a lattice field theory are the following:

1. Define the classical, Euclidean field theory in the continuum;
2. Discretize the corresponding Lagrangian;
3. Quantize the theory by defining the functional integral;
4. Determine the particle spectrum from Euclidean correlation functions.

We shall now illustrate this procedure for a simple example, namely the theory for a neutral scalar field.

Step 1 Consider a real, classical field $\phi(x)$, with $x = (x^0, x^1, x^2, x^3)$, whose time variable x^0 is obtained by analytically continuing t to $-\mathrm{i}x^0$. The Euclidean action $S_{\mathrm{E}}[\phi]$ is defined as

$$S_{\mathrm{E}}[\phi] = \int \mathrm{d}^4x \left\{ \frac{1}{2}\partial_\mu\phi(x)\partial_\mu\phi(x) + V(\phi) \right\}, \qquad \partial_\mu \equiv \frac{\partial}{\partial x_\mu}, \tag{5.1}$$

where

$$V(\phi) = \frac{1}{2}m^2\phi(x)^2 + \frac{\lambda}{4!}\phi(x)^4. \tag{5.2}$$

Step 2 In order to discretize the theory, a hyper-cubic lattice, Λ_{E}, is introduced as the set of discrete space-time points, i.e.

$$\begin{aligned} \Lambda_{\mathrm{E}} &= \left\{ x \in \mathbb{R}^4 \,\middle|\, x^0/a = 1, \ldots, N_{\mathrm{t}};\; x^j/a = 1, \ldots, N_{\mathrm{s}}\,,\; j = 1, 2, 3 \right\}, \\ T &= N_{\mathrm{t}}a,\; L = N_{\mathrm{s}}a. \end{aligned} \tag{5.3}$$

Thus, any space-time point is an integer multiple of the lattice spacing a. The total number of lattice sites is $N_t \times N_s^3$, while the physical space-time volume is $T \times L^3$. The discretized action is then given by

$$S_E[\phi] = a^4 \sum_{x \in \Lambda_E} \left\{ \frac{1}{2} d_\mu \phi(x) d_\mu \phi(x) + \frac{1}{2} m^2 \phi(x)^2 + \frac{\lambda}{4!} \phi(x)^4 \right\}, \tag{5.4}$$

where the lattice derivatives can be defined as

$$d_\mu \phi(x) := \frac{1}{a} \left(\phi(x + a\hat{\mu}) - \phi(x) \right) \qquad \text{“forward” derivative,} \tag{5.5}$$

$$d_\mu^* \phi(x) := \frac{1}{a} \left(\phi(x) - \phi(x - a\hat{\mu}) \right) \qquad \text{“backward” derivative.} \tag{5.6}$$

Here and below $\hat{\mu}$ denotes a unit vector in direction of μ. Via a Fourier transform, the Euclidean lattice Λ_E is related to the dual lattice, Λ_E^*, defined by

$$\Lambda_E^* = \left\{ p \in \mathbb{R}^4 \,\middle|\, p_0 = \frac{2\pi}{T} n^0,\ p_j = \frac{2\pi}{L} n^j \right\}$$

$$n^0 = -\frac{N_t}{2}, -\frac{N_t}{2} + 1, \ldots, \frac{N_t}{2} - 1, \quad n^j = -\frac{N_s}{2}, -\frac{N_s}{2} + 1, \ldots, \frac{N_s}{2} - 1. \tag{5.7}$$

This not only implies that the momenta p_0 and p_j are quantized in units of $2\pi/T$ and $2\pi/L$, respectively, but also that a momentum cutoff has been introduced, since

$$-\frac{\pi}{a} \leq p_\mu \leq \frac{\pi}{a}. \tag{5.8}$$

As we shall see below, this way of introducing a momentum cutoff can be extended to gauge theories in such a way that gauge invariance is respected. An important point to realize is that the lattice action is not unique: it is only required that the discretized expression for S_E reproduces the continuum result as the lattice spacing a is taken to zero.

Step 3 The theory is quantized via the Euclidean functional integral

$$Z_E := \int D[\phi]\, e^{-S_E[\phi]}, \qquad D[\phi] = \prod_{x \in \Lambda_E} d\phi(x). \tag{5.9}$$

Here one sees explicitly that the discretization procedure has given a mathematical meaning to the integration measure, which reduces to that of an ordinary, multiple-dimensional integration.

One can now define Euclidean correlation functions of local fields through

$$\langle\phi(x_1)\cdots\phi(x_n)\rangle = \frac{1}{Z_\mathrm{E}}\int D[\phi]\phi(x_1)\cdots\phi(x_n)\mathrm{e}^{-S_\mathrm{E}[\phi]}. \tag{5.10}$$

In the continuum limit, these correlation functions approach the Schwinger functions, which encode the physical information about the spectrum within the Euclidean formulation. Osterwalder and Schrader [5] have laid down the general criteria which must be satisfied such that the information in Minkowskian space-time can be reconstructed from the Schwinger functions.

Step 4 The particle spectrum is extracted from the exponential fall-off of the Euclidean two-point correlation function. To this end, one must define the Euclidean time evolution operator. The *transfer matrix* T describes time propagation by a finite Euclidean time interval a. The functional integral can be expressed in terms of the transfer matrix as

$$Z_\mathrm{E} = \mathrm{Tr}\,\mathsf{T}^{N_\mathrm{t}}, \tag{5.11}$$

where the trace is taken over the basis $|\alpha\rangle$ of the Hilbert space of physical states. In order to obtain expressions which are more reminiscent of those in Minkowski space-time, one can define a Hamiltonian H_E by

$$\mathsf{T} =: \mathrm{e}^{-a\mathsf{H}_\mathrm{E}}. \tag{5.12}$$

If $|\alpha\rangle$ denotes an eigenstate of the transfer matrix with eigenvalue λ_α, i.e.

$$\mathsf{T}|\alpha\rangle = \lambda_\alpha|\alpha\rangle = \mathrm{e}^{-aE_\alpha}|\alpha\rangle, \tag{5.13}$$

then one can work out the spectral decomposition of the two-point correlation function, viz.

$$\langle\phi(x)\phi(y)\rangle = \frac{1}{Z_\mathrm{E}}\int D[\phi]\phi(x)\phi(y)\mathrm{e}^{-S_\mathrm{E}[\phi]} \tag{5.14}$$

$$= \sum_\alpha \mathrm{e}^{-(E_\alpha-E_0)(x_0-y_0)}\big\langle\alpha\big|\hat{\phi}(0,\vec{y})\big|0\big\rangle\big\langle 0\big|\hat{\phi}(0,\vec{x})\big|\alpha\big\rangle. \tag{5.15}$$

Here, the quantity $(E_\alpha - E_0)$ is the so-called mass gap, i.e. the energy of the state $|\alpha\rangle$ above the vacuum. For large Euclidean time separations $(x_0 - y_0)$ the lowest state dominates the two-point function, i.e. all higher states die out exponentially. The spectral decomposition of the two-point function forms the basis for numerical simulations of lattice field theories, as the mass (or energy) of a given state is given by the dominant exponential fall-off at large Euclidean times (see Sect. 5.2.3).

5.2.2 Lattice Actions for QCD

Our goal now is to find a lattice transcription of the Euclidean QCD action in the continuum, i.e.

$$S_{\rm QCD} = \int d^4x \left\{ -\frac{1}{2g_0^2} {\rm Tr}\,(F_{\mu\nu}F_{\mu\nu}) + \sum_{f=u,d,s\ldots} \bar{\psi}_f \left(\gamma_\mu D_\mu + m_f\right) \psi_f \right\}, \qquad (5.16)$$

where g_0 denotes the gauge coupling, and our conventions are chosen such that the covariant derivative is defined through

$$D_\mu = \partial_\mu + A_\mu, \qquad (5.17)$$

while the field tensor reads

$$F_{\mu\nu} = \partial_\mu A_\nu - \partial_\nu A_\mu + [A_\mu, A_\nu], \qquad A_\mu^\dagger = -A_\mu. \qquad (5.18)$$

Before attempting to write down a discretized version, we must first elucidate the notion of a lattice gauge field in a non-Abelian theory. In fact, in this case it turns out that the gauge potential A_μ must be abandoned when the theory is discretized. The reason is that the familiar non-Abelian transformation law, i.e.

$$A_\mu(x) \to g(x)A_\mu(x)g(x)^{-1} + g(x)\partial_\mu(x)g(x)^{-1}, \quad g(x) \in {\rm SU}(3), \qquad (5.19)$$

no longer holds exactly when ∂_μ is replaced by its discrete counterpart d_μ of Eq. (5.5). Strict gauge invariance at the level of the regularized theory cannot be maintained in this fashion.

The definition of a lattice gauge field relies on the concept of the parallel transporter. If a quark moves in the presence of a background gauge field from y to x, it picks up a non-Abelian phase factor, given by

$$U(x,y) = {\rm P.O.}\exp\left\{ -\int_y^x dz_\mu\, A_\mu(z) \right\}, \qquad (5.20)$$

where "P.O." denotes path ordering, as a consequence of the non-Abelian nature of the gauge field. By contrast to the gauge potential A_μ, which is an element of the Lie algebra of SU(3), the parallel transporter $U(x,y)$ is an element of the gauge group itself. On the lattice, the parallel transporter between neighbouring lattice sites x and $x + a\hat{\mu}$ is called *link variable:*

$$U(x, x+a\hat{\mu}) \equiv U_\mu(x), \qquad U(x+a\hat{\mu}, x) = U(x, x+a\hat{\mu})^{-1} = U_\mu(x)^{-1}. \qquad (5.21)$$

A consistent and manifestly gauge invariant discretization of QCD is obtained by identifying the gauge degrees of freedom with the link variables $U_\mu(x)$, which transform under the gauge group as

$$U_\mu(x) \to g(x)\, U_\mu(x)\, g(x+a\hat{\mu})^{-1}, \quad g(x),\, g(x+a\hat{\mu}) \in \mathrm{SU}(3). \tag{5.22}$$

The connection with the gauge potential $A_\mu(x)$ is somewhat subtle: if $U_\mu(x)$ denotes a *given* link variable in the discretized theory, it can be used to *define* a vector field $A_\mu(x)$ as an element of the Lie algebra of SU(3) via

$$\mathrm{e}^{aA_\mu(x)} \equiv U_\mu(x). \tag{5.23}$$

In turn, if A_μ^{c} is a *given* gauge potential in the continuum theory, one can always find a link variable which approximates A_μ^{c} up to cutoff effects.

Now we turn to the problem of defining a discretized version of the Yang–Mills action. To this end we define the *plaquette* $P_{\mu\nu}(x)$ as the product of link variables around an elementary square of the lattice:

$$P_{\mu\nu}(x) \equiv U_\mu(x) U_\nu(x+a\hat{\mu}) U_\mu(x+a\hat{\nu})^{-1} U_\nu(x)^{-1}. \tag{5.24}$$

A graphical representation is shown in Fig. 5.1. Using the transformation property in Eq. (5.22), it is easy to convince oneself that this object is manifestly gauge invariant. Moreover, it serves to define the simplest discretization of the Yang–Mills action, the Wilson plaquette action [6]

$$S_{\mathrm{G}}[U] = \beta \sum_{x\in\Lambda_{\mathrm{E}}} \sum_{\mu<\nu} \left(1 - \frac{1}{3}\mathrm{Re\,Tr}\; P_{\mu\nu}(x)\right). \tag{5.25}$$

It has become a standard textbook exercise to verify that for small lattice spacings

$$S_{\mathrm{G}}[U] \longrightarrow -\frac{1}{2g_0^2}\int \mathrm{d}^4x\, \mathrm{Tr}\,(F_{\mu\nu}F_{\mu\nu}) + \mathrm{O}(a), \tag{5.26}$$

provided that one relates the parameter β to the bare gauge coupling via $\beta = 6/g_0^2$ in Eq. (5.25). We have remarked already that the discretization of a field theory is not

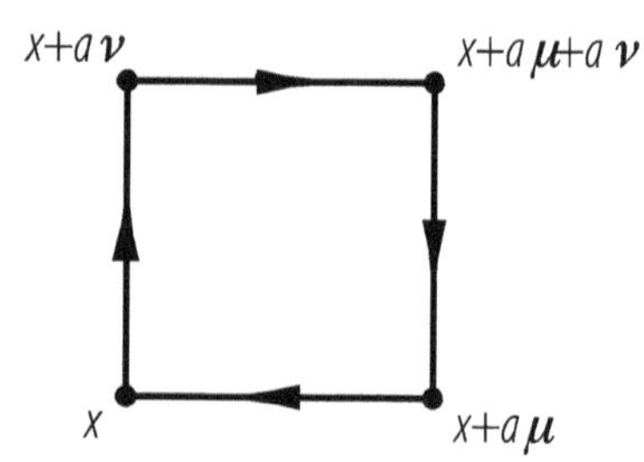

Fig. 5.1 Graphical representation of the plaquette $P_{\mu\nu}(x)$ in the (μ, ν)-plane. The arrow between sites $x + a\hat{\mu}$ and x denotes the link variable $U_\mu(x)$

unique, and hence one is free to add further gauge invariant terms to the plaquette action which formally vanish as $a \to 0$, but which produce a discretization with an accelerated rate of convergence to the continuum limit. The most widely chosen alternatives are the Symanzik [7] and Iwasaki [8] actions.

Quark and antiquark fields, $\psi(x)$ and $\bar{\psi}(x)$, are associated with the lattice sites and transform under the gauge group as

$$\psi(x) \to g(x)\psi(x), \qquad \bar{\psi}(x) \to \bar{\psi}(x)g(x)^{-1}. \tag{5.27}$$

Using the transformation property of the link variables, it is straightforward to write down a discretized version of the covariant derivative, i.e.

$$\begin{aligned} \nabla_\mu \psi(x) &:= \frac{1}{a}\left(U_\mu(x)\psi(x+a\hat{\mu}) - \psi(x)\right) \\ \nabla^*_\mu \psi(x) &:= \frac{1}{a}\left(\psi(x) - U_\mu(x-a\hat{\mu})^{-1}\psi(x-a\hat{\mu})\right), \end{aligned} \tag{5.28}$$

where ∇_μ and ∇^*_μ denote the "forward" and "backward" derivatives, respectively. Finally, we note that in Euclidean space-time, the Dirac matrices can be defined to satisfy $\{\gamma_\mu, \gamma_\nu\} = 2\delta_{\mu\nu}$.

Before we attempt to construct the fermionic part of the action of lattice QCD, it is useful to identify the basic properties that the discretized, massless Dirac operator, D, should satisfy:

(a) D is local;
(b) $\widetilde{D}(p) = \mathrm{i}\gamma_\mu p_\mu + \mathrm{O}(ap^2)$;
(c) $\widetilde{D}(p)$ is invertible for $p \neq 0$;
(d) $\gamma_5 D + D\gamma_5 = 0$.

Locality, i.e. the absence of long-ranged interactions, is a basic property of any quantum field theory describing elementary particles. Property (b) implies that the correct continuum behaviour of the quark-gluon interaction is reproduced. Furthermore, condition (c) ensures that the correct fermion spectrum is obtained: fermion masses are associated with poles of $\{\widetilde{D}(p)\}^{-1}$, which, in the continuum theory, only occur at vanishing four-momentum. Finally, property (d) ensures that the massless theory respects chiral symmetry.

Using the definition of the covariant derivative and the conventions for the Dirac matrices in Euclidean space-time, we can now write down the simplest discretized version of the massless lattice Dirac operator:

$$D_{\text{disc}} = \tfrac{1}{2}\gamma_\mu(\nabla_\mu + \nabla^*_\mu). \tag{5.29}$$

It turns out, however, that this "naïve" discretization violates condition (c) and therefore produces spurious fermionic degrees of freedom. This is the so-called fermion doubling problem, which is most easily explained by considering D_{disc} in momentum space for the free theory. The Fourier transform yields

$$\widetilde{D}_{\rm disc}(p) = \mathrm{i}\gamma_\mu \frac{1}{a}\sin(ap_\mu) = \mathrm{i}\gamma_\mu p_\mu + \mathrm{O}(a^2). \tag{5.30}$$

The discretization procedure has thus replaced p_μ by a sine function. While the Taylor expansion guarantees that condition (b) is satisfied, the occurrence of $\sin(ap_\mu)$ implies that $\widetilde{D}_{\rm disc}(p)$ vanishes not only at $p_\mu = 0$, but also at π/a for $\mu = 0, \ldots, 3$ in the permitted range of momenta, thereby violating condition (c). The massless propagator $\{\widetilde{D}_{\rm disc}(p)\}^{-1}$ therefore has $2^4 = 16$ poles, and thus there is a 16-fold degeneracy of the fermion spectrum.

As we shall see below, the fermion doubling problem is closely linked with the issue of chiral symmetry on the lattice. For now we simply list the various methods that have been devised to address fermion doubling. Historically the first was due to Wilson ("Wilson fermions") [6]. Here, the degeneracy is lifted completely, but the price to pay is the explicit breaking of chiral symmetry at the level of the regularized theory. Another method, due to Kogut and Susskind ("staggered fermions") [9], is based on the idea of spreading individual spinor components over the corners of an elementary hypercube of the lattice. Although the degeneracy is only lifted partially (from 16 to 4), this formulation has the advantage of leaving a subgroup of chiral symmetry unbroken. More recent developments include the use of so-called "domain wall" [10, 11] or "overlap" [12] fermions. These formulations leave chiral symmetry unbroken in principle, and also succeed in lifting the degeneracy completely. Finally, there are the so-called "perfect" actions [13], which are based on a renormalization group approach and which are in principle completely free of lattice artefacts. An exact realization of the perfect action which can be used in simulations is, however, difficult to obtain. In practice, one typically uses a so-called truncated fixed point action. Domain wall and overlap fermions, as well as perfect actions are particular realizations of a class of discretizations dubbed "Ginsparg-Wilson fermions". They have the remarkable feature that chiral symmetry is preserved, while the fermion doubling problem is completely avoided. We shall come back to this issue in more detail below.

For now we turn specifically to Wilson's treatment of the fermion doubling problem. It exploits the fact that the discretization is not unique. Thus, one can add a term to $D_{\rm disc}$, which formally vanishes as $a \to 0$, but which pushes the masses of the unwanted doubler states to the cutoff scale at any non-zero value of the lattice spacing. Explicitly, the massless Wilson-Dirac operator $D_{\rm w}$ reads

$$D_{\rm w} = \tfrac{1}{2}\gamma_\mu(\nabla_\mu + \nabla_\mu^*) + ar\nabla_\mu^*\nabla_\mu, \tag{5.31}$$

where r is the so-called Wilson parameter, which is usually set to one. The Fourier transform of $D_{\rm w}$ for a trivial gauge field reads

$$\widetilde{D}_{\rm w}(p) = \mathrm{i}\gamma_\mu \frac{1}{a}\sin(ap_\mu) + \frac{2r}{a}\sin^2\left(\frac{ap_\mu}{2}\right), \tag{5.32}$$

which explicitly demonstrates (for the free theory, at least) that the poles at $p_\mu = \pi/a$ receive additional contributions proportional to r/a, which is of order of the cutoff for $r = \mathrm{O}(1)$. Although this procedure leads to a complete lifting of the degeneracy,[1] it has a number of unwanted features: first, it should be noted that the Wilson fermion action differs from the classical action in the continuum by terms of order a, as a result of adding the counterterm proportional to r. By contrast, the leading discretization effects of the Wilson plaquette action for Yang–Mills theory are only $\mathrm{O}(a^2)$. The Wilson fermion formulation will thus have a reduced rate of convergence towards the continuum limit. Secondly, the addition of the Wilson term results in an explicit breaking of chiral symmetry, since the massless theory is no longer invariant under global axial rotations, such as

$$\psi(x) \to \mathrm{e}^{\mathrm{i}\alpha\gamma_5}\psi(x), \qquad \bar{\psi}(x) \to \bar{\psi}(x)\mathrm{e}^{\mathrm{i}\alpha\gamma_5}, \tag{5.33}$$

which implies that property (d) is violated. While the rate of convergence to the continuum limit can be accelerated by employing what is known as "$\mathrm{O}(a)$ improvement" (see below), the explicit breaking of chiral symmetry cannot be cured within the Wilson theory. Thus, quantities like the quark condensate, which arises from the spontaneous breaking of chiral symmetry, cannot be studied in a conceptually "clean" manner using Wilson fermions. A detailed discussion how this can be achieved with the help of a more sophisticated fermionic discretization ("Ginsparg-Wilson fermions") is presented in Sect. 5.6. However, for most applications of lattice QCD, explicit chiral symmetry breaking is merely an inconvenience, but no serious obstacle.

We have already remarked when discussing the discretized Yang–Mills part of the QCD action that the non-uniqueness of the discretization opens the possibility to construct lattice actions with an accelerated rate of convergence towards the continuum limit. A systematic way how to do this is the so-called Symanzik improvement programme [14], in which lattice artefacts can be removed order by order in the lattice spacing. In a nutshell, the improvement programme amounts to extending the renormalization procedure of a field theory to the level of irrelevant operators, i.e. operators that formally vanish as $a \to 0$. In this sense one adds suitable counterterms, which for any non-zero value of a produce a cancellation of the cutoff effects at a given order, provided that their coefficients are tuned appropriately. For QCD with Wilson fermions, Sheikholeslami and Wohlert [15] have shown that the Symanzik improvement programme to lowest order is realized by adding one $\mathrm{O}(a)$ counterterm to the Wilson-Dirac operator D_w. The resulting expression in the massless case reads

$$D_\mathrm{sw} = D_\mathrm{w} + \frac{ia}{4} c_\mathrm{sw}\sigma_{\mu\nu}\widehat{F}_{\mu\nu}, \tag{5.34}$$

[1] That the degeneracy is indeed completely lifted in the presence of a non-trivial gauge field can be verified in numerical simulations.

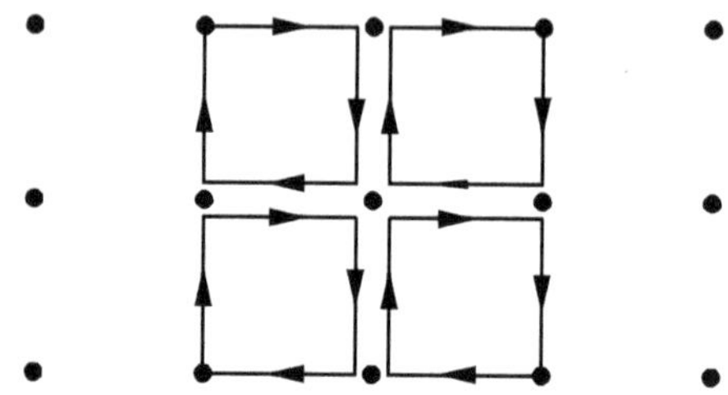

Fig. 5.2 Four plaquettes that must be summed over to yield the quantity $Q_{\mu\nu}(x)$ in the lattice definition of the field strength tensor. The site x is at the center of the "clover" leaf

where $\sigma_{\mu\nu} = \frac{i}{2}[\gamma_\mu, \gamma_\nu]$, and $\widehat{F}_{\mu\nu}$ is a lattice transcription of the gluon field strength tensor $F_{\mu\nu}$. A suitable representation of $\widehat{F}_{\mu\nu}$ in terms of plaquette variables is given by

$$\widehat{F}_{\mu\nu}(x) = \frac{1}{8a^2}\left(Q_{\mu\nu}(x) - Q_{\nu\mu}(x)\right), \tag{5.35}$$

where $Q_{\mu\nu}(x)$ is the sum of the four plaquettes emanating from the site x, as depicted in Fig. 5.2. The object $Q_{\mu\nu}(x)$ is aptly called "clover" leaf. In order to remove all lattice artefacts of order a in hadron masses, the improvement coefficient c_{sw} must be fixed by imposing a suitable improvement condition. Without going into details here, we note that it is possible to find such a condition, which can also be evaluated at the non-perturbative level [16, 17]. The resulting, non-perturbatively $\text{O}(a)$ improved Wilson action can then be used to compute, say, hadron masses whose values differ from the continuum result by terms of only $\text{O}(a^2)$.

The Wilson-Dirac operator for a quark with bare mass m_0 is simply $(D_{\text{w}} + m_0)$. However, the form of the Wilson fermion action, $S_{\text{F}}^{\text{W}}[U, \bar{\psi}, \psi]$, which is found in the literature is usually expressed in terms of the "hopping parameter" κ rather than m_0. By rescaling the fermion fields according to

$$\psi(x) \to \sqrt{2\kappa}\,\psi(x), \qquad \bar{\psi}(x) \to \bar{\psi}(x)\sqrt{2\kappa}, \tag{5.36}$$

one obtains

$$\begin{aligned} S_{\text{F}}^{\text{W}}[U, \bar{\psi}, \psi] &\equiv a^4 \sum_{x\in\Lambda_{\text{E}}} \bar{\psi}(x)(D_{\text{w}} + m_0)\psi(x) \\ &= a^4 \sum_{x\in\Lambda_{\text{E}}} \Bigg\{ -\kappa \sum_{\mu=0}^{3} \frac{1}{a}\Big[\bar{\psi}(x)(r - \gamma_\mu)U_\mu(x)\psi(x + a\hat{\mu}) \\ &\quad + \bar{\psi}(x + a\hat{\mu})(r + \gamma_\mu)U_\mu(x)^{-1}\psi(x)\Big] \\ &\quad + \bar{\psi}(x)\psi(x)\Bigg\}. \end{aligned} \tag{5.37}$$

The hopping parameter κ is related to the bare mass m_0 via

$$\kappa = \frac{1}{2am_0 + 8r}, \tag{5.38}$$

while the dimensionless parameter r is usually set to one. Taken together with the plaquette action of Eq. (5.25), the Wilson action for QCD is thus conveniently parameterized in terms of the bare parameters (β, κ), with $\beta = 6/g_0^2$ and κ as above, instead of the bare gauge coupling and quark mass (g_0, m_0).

Another consequence of adding the Wilson term to the naïve lattice action is the resulting additive renormalization of the quark mass. In other words, the point where the quark mass vanishes is a priori unknown. The value that must be subtracted is called the critical quark mass, which corresponds to the critical value of the hopping parameter, κ_c. The bare subtracted quark mass is then given by

$$m = \frac{1}{2a}\left(\frac{1}{\kappa} - \frac{1}{\kappa_c}\right). \tag{5.39}$$

From Eq. (5.38) one easily infers that the critical value of κ in the free theory occurs at

$$\kappa_c = \frac{1}{8}, \qquad r = 1, \tag{5.40}$$

while for non-zero g_0 the value of κ_c must be determined, for instance, by adjusting κ to the point where the pion mass vanishes.

We now turn to discussing one alternative to using Wilson's solution to the fermion doubling problem, namely the so-called "staggered" (or Kogut-Susskind) fermions. One might think that the doubling problem arises since there are too many fermion degrees of freedom in the discretized theory, if one associates a four-component Dirac spinor with each individual lattice site. Pictorially, the main idea of Kogut and Susskind was to "thin out" the degrees of freedom by distributing single spinor components over different lattice sites. In their particular formulation, the 16 corners of a four-dimensional hypercube serve to accommodate the individual components of four Dirac spinors. Therefore, if these hypercubes are regarded as the main building blocks for the fermionic discretization, rather than the lattice sites themselves, this procedure will result in a partial lifting of the degeneracy from 16 fermion species down to four. It is clear, though, that a simple distribution of spinor components is not sufficient to define the action, since the Dirac matrices mix different spinor components. Thus, the staggered fermion action is only obtained after performing a diagonalization in spinor space, which then decouples the individual components.

Rather than describing the details of this procedure, which can be found in most textbooks, we simply state the result. Starting from the usual four-component spinor and performing a spin-diagonalization, the lattice action for staggered fermions with bare mass m_0 coupled to the gauge field is derived as

$$S_{\mathrm{F}}^{\mathrm{stagg}}[U,\bar{\chi},\chi] = a^4 \sum_{x\in\Lambda_{\mathrm{E}}} \sum_{\alpha=1}^{4} \Big\{ m_0 \bar{\chi}_\alpha(x)\chi_\alpha(x)$$

$$+ \frac{1}{2a}\sum_{\mu=0}^{3} \eta_\mu(x)\Big[\bar{\chi}_\alpha(x)U_\mu(x)\chi_\alpha(x+a\hat{\mu}) - \bar{\chi}_\alpha(x+a\hat{\mu})U_\mu(x)^{-1}\chi_\alpha(x)\Big]\Big\}, \quad (5.41)$$

where χ_α denotes a one-component Grassmann variables. The spin-diagonalization has thus replaced the Dirac matrices γ_μ by real, position-dependent phase factors $\eta_\mu(x)$, which are given by

$$\eta_0(x) = 1, \qquad \eta_j(x) = (-1)^{n_0+\ldots+n_{j-1}}, \quad n_j = x_j/a. \quad (5.42)$$

At the level of the classical action, the spinor components are completely decoupled, and the action is decomposed into four identical pieces. In order to occupy all 16 corners of a four-dimensional hypercube with one-component Grassmann variables, one needs four Dirac spinors, each of which contributes a term like Eq. (5.41) to the overall action. This produces the fourfold degeneracy of staggered fermions, with the remnant doubler states being referred to as "tastes", in order to distinguish them from physical flavours. The formulation using the one-component fields within a hypercube can be re-expressed in terms of the spin-taste basis [18], from which one can infer directly that the taste symmetry is broken. However, one axial generator of the taste symmetry remains unbroken. The fermion mass in the staggered approach is therefore protected against any additive renormalization through the associated global axial U(1) symmetry, unlike the case of the Wilson action. While the various tastes decouple in the continuum limit, non-vanishing interactions between the tastes at $\mathrm{O}(a^2)$ in the lattice spacing are induced, leading to large lattice artefacts. The Symanzik improvement programme can be employed to reduce these taste-changing interactions [19], and the resulting "improved staggered fermions" (the so-called "Asqtad"-action being one particular example [20]) have been widely used in a series of simulations.

For a long time lattice physicists have struggled to find a fermionic discretization which would both solve the doubling problem and be compatible with chiral symmetry. In fact, physicists grew increasingly doubtful that this could be achieved, following the proof of a "No-Go theorem" by Nielsen and Ninomiya [21], which stated that the conditions (a)–(d) mentioned above could not be satisfied simultaneously. Since one does not want to give up locality and property (b), this would imply that either (c) or (d) must be violated. Indeed, the Wilson and staggered discretizations seem to confirm this expectation: while the Wilson fermion action removes all doublers, it breaks chiral symmetry, leading to an additive renormalization of the quark mass, as well as several other consequences. By contrast, the staggered formulation preserves a U(1) subgroup of chiral symmetry at the price of only partially removing the spurious degrees of freedom.

A way to circumvent the Nielsen–Ninomiya theorem was already pointed out by Ginsparg and Wilson in 1982 [22], when they suggested to relax condition (d) in favour of

$$\gamma_5 D + D\gamma_5 = aD\gamma_5 D. \tag{5.43}$$

However, it was not before 1997 that this condition—now commonly referred to as the Ginsparg-Wilson relation—was confronted with a non-trivial solution. It was shown [23] that the so-called "perfect action" constructed from a renormalization group approach satisfied equation (5.43). It was also realized that any lattice Dirac operator, which is a solution to the Ginsparg-Wilson relation, also satisfies the Atiyah–Singer index theorem, i.e.

$$\{\gamma_5, D\} = aD\gamma_5 D \quad \Leftrightarrow \quad \mathrm{index}(D) = a^5 \sum_{x\in\Lambda_\mathrm{E}} \tfrac{1}{2}\mathrm{Tr}\,(\gamma_5 D) = n_- - n_+, \tag{5.44}$$

such that the operator D exhibits $|n_- - n_+|$ exact chiral zero modes. Finally, it was shown [24] that the Ginsparg-Wilson relation implies an exact symmetry of the associated action, with infinitesimal variations proportional to

$$\delta\psi = \gamma_5(1 - aD)\psi, \qquad \delta\bar{\psi} = \bar{\psi}\gamma_5. \tag{5.45}$$

Moreover, this symmetry reproduces the correct chiral anomaly in the flavour singlet case, and therefore all the hallmarks of the correct chiral behaviour are present in the lattice theory: chiral zero modes, an exact index theorem and the chiral anomaly derived from the Ward identities associated with the exact symmetry.

Another line in the development of lattice fermion actions that preserve chiral symmetry goes back to Kaplan's domain wall fermion approach [10], which was subsequently applied to QCD by Furman and Shamir [11]. Without going into detail, we state that the basic idea is to introduce an extra, fifth dimension and to couple the fermions to a mass defect (the so-called "domain wall height") in that extra dimension. To make this more explicit, let x, y denote the coordinates in the four-dimensional bulk, and s, t the coordinates in the 5th dimension, which has finite length N_5. The gauge fields are trivial in the 5th direction, and the Dirac operator then has the general structure

$$D_\mathrm{dwf}(x, s; y, t) = D^{\|}(x, y)\delta_{st} + \delta(x - y)D^{\perp}_{st} \tag{5.46}$$

where $D^{\|}(x, y)$ is the usual Wilson-Dirac operator with a *negative* mass term, $-M$, which represents the domain wall height. The operator $D^{\perp}_{st}$ couples fermions in the 5th dimension and contains the physical bare quark mass m_0. It can then be shown that for $m_0 = 0$ and in the limit $N_5 \to \infty$ there are no fermion doublers and, more importantly, chiral modes of opposite chirality are trapped in the four-dimensional domain walls at $s = 1,\ N_5$.

However, in a real lattice simulation of domain wall fermions, one has to work with a finite value of N_5, so that the decoupling of chiral modes is not exact. One expects, though, an exponential suppression of the remnant chiral symmetry breaking effects, and this has been confirmed in several simulations. Furthermore, the rate of suppression may be accelerated by optimizing the choice of lattice action for the gauge fields. Hence, the domain wall formulation of QCD offers a method to realize almost exact chiral symmetry at non-zero lattice spacing at the expense of simulating a five-dimensional theory.

Another operator which correctly reproduces the chiral properties of QCD at non-zero lattice spacing was constructed by Neuberger [12]. Its definition is

$$D_{\rm N} = \frac{1}{\overline{a}}\left(1 - \frac{A}{\sqrt{A^\dagger A}}\right), \quad A = 1 + s - aD_{\rm w}, \quad \overline{a} = \frac{a}{1+s}, \tag{5.47}$$

where $D_{\rm w}$ is the massless Wilson-Dirac operator, and $|s| < 1$ is a tunable parameter. By defining $Q = -\gamma_5 A$, one can rewrite Eq. (5.47) as

$$D_{\rm N} = \frac{1}{\overline{a}}\left(1 + \gamma_5 \mathrm{sign}(Q)\right). \tag{5.48}$$

The Neuberger-Dirac operator $D_{\rm N}$ removes all doublers from the spectrum, and can easily be shown to satisfy the Ginsparg-Wilson relation [12]. The occurrence of an inverse square root in $D_{\rm N}$ raises two issues. First, it is a priori not clear whether or not $D_{\rm N}$ is local. Second, the application of $D_{\rm N}$ in a computer program is potentially very costly, since the sign-function of the matrix Q must be implemented using, for instance, a polynomial approximation.

In order to qualify as a viable discretization of the quark action, "strict" locality, meaning that only fields in a local neighbourhood of a given lattice site are coupled, is not actually required. If $D(x, y)$ denotes a generic lattice Dirac operator which couples fields at sites x and y, then a sufficient condition for locality of D is the exponential suppression of non-local interactions, i.e.

$$\|D(x, y)\| \leq \mathrm{e}^{-\gamma|x-y|/a}, \tag{5.49}$$

where $|x - y|$ is the distance between sites and $\|\cdot\|$ denotes a suitably defined matrix norm. In Ref. [25] it was shown that the Neuberger-Dirac operator $D_{\rm N}$ is local in the sense of Eq. (5.49), provided that the lattice spacing in physical units[2] is not larger than about 0.13 fm. As far as the issue of numerical efficiency is concerned, we note that the most widely used approximations of sign(Q) with good convergence properties include Chebysheff or Zolotarev polynomials, as well as rational fractions.

[2]So far we have not discussed how to assign physical units to the lattice spacing a. This is described in Sect. 5.2.4.

The last fermionic discretization we wish to mention here was originally constructed to address another problem of Wilson's discretization, namely the fact that they are not protected against the occurrence of zero modes for *any* non-zero value of the bare quark mass. These unphysical zero modes manifest themselves as "exceptional" configurations, which occur with a certain frequency in numerical simulations with Wilson quarks and which can lead to strong statistical fluctuations. The problem can be cured by introducing a so-called "chirally twisted" mass term, after which the fermionic part of the QCD action in the continuum assumes the form

$$S_{\mathrm{F}}^{\mathrm{tm;\,cont}} = \int \mathrm{d}^4x\, \bar{\psi}(x)(\gamma_\mu D_\mu + m + \mathrm{i}\mu_{\mathrm{q}}\gamma_5\tau^3)\psi(x). \tag{5.50}$$

Here, μ_{q} is the twisted mass parameter, and τ^3 is a Pauli matrix. The standard action in the continuum can be recovered via a global chiral field rotation:

$$\psi'(x) = \mathrm{e}^{\mathrm{i}\alpha\gamma_5\tau^3/2}\psi(x), \qquad \bar{\psi}'(x) = \bar{\psi}(x)\mathrm{e}^{\mathrm{i}\alpha\gamma_5\tau^3/2}. \tag{5.51}$$

Fixing the twist angle α by requiring that $\tan\alpha = \mu_{\mathrm{q}}/m$ one finds

$$S_{\mathrm{F}}' = \int \mathrm{d}^4x\, \bar{\psi}'(x)(\gamma_\mu D_\mu + M)\psi'(x), \qquad M = \sqrt{m^2 + \mu_{\mathrm{q}}^2}, \tag{5.52}$$

which demonstrates the complete equivalence of the twisted formulation with "ordinary" QCD. The lattice action of twisted mass QCD for $N_{\mathrm{f}} = 2$ flavours is defined as [26]

$$S_{\mathrm{F}}^{\mathrm{tm}}[U, \bar{\psi}, \psi] = a^4 \sum_{x\in\Lambda_{\mathrm{E}}} \bar{\psi}(x)(D_{\mathrm{w}} + m_0 + \mathrm{i}\mu_{\mathrm{q}}\gamma_5\tau^3)\psi(x). \tag{5.53}$$

Although this formulation breaks physical parity and flavour symmetries, is has a number of advantages over standard Wilson fermions. In particular, the presence of the twisted mass parameter μ_{q} protects the discretized theory against unphysical zero modes. Another attractive feature of twisted mass lattice QCD is the fact that the leading lattice artefacts are of order a^2 without the need to add the Sheikholeslami-Wohlert term [27], even though the Wilson-Dirac operator is used in Eq. (5.53). Although the problem of explicit chiral symmetry breaking remains, the twisted formulation is particularly useful to circumvent some of the problems that are encountered in connection with the renormalization of local operators on the lattice. Recent review of twisted mass lattice QCD can be found in [28, 29].

We wish to end this part with a few general remarks. Although we have discussed discretizations of the QCD action in some detail, including the most recent developments, many more variants of the basic types of action—including several different combinations of fermionic and pure gauge parts—can be found in the literature. This reflects the fact that the discretization is not unique. The actual choice of lattice action in a particular simulation will influence the convergence rate to the continuum limit, the algorithmic efficiency, the renormalization properties of local operators, or—in the case of domain wall fermions—the extent to which chiral symmetry is realized. Depending on the properties of a particular discretization, the choice of lattice action can be optimized for the physics one wishes to study.

5.2.3 Functional Integral and Observables

The lattice formulation provides a regularization of non-Abelian gauge theories whilst preserving the gauge invariance at all stages of the calculation. This comes at a price, since all continuous space-time symmetries are broken explicitly and must be recovered in the continuum limit. Nevertheless, the lattice regularized theory inherits all consequences of gauge invariance, including renormalizability. Moreover, the lattice regularizes the theory without any reference to perturbation theory. By contrast, in continuum schemes like the $\overline{\text{MS}}$ scheme of dimensional regularization the cutoff is only defined after fixing the order of the perturbative expansion. As we shall see below, observables in lattice QCD are directly given in terms of functional integrals, which can be evaluated stochastically using Monte Carlo integration. In this way, any use of perturbation theory is completely avoided.

For concreteness, let us assume that we have made a particular choice for the Yang–Mills part $S_{\text{G}}[U]$ and the fermionic part $S_{\text{F}}[U, \bar{\psi}, \psi]$, for instance, the Wilson plaquette action and Wilson fermions. Let Ω denote an observable, which is represented by a polynomial in the quark and antiquark fields and the link variables. The expectation value, $\langle\Omega\rangle$, is defined through the Euclidean functional integral[3]

$$\langle\Omega\rangle = \frac{1}{Z}\int D[U]D[\bar{\psi}, \psi]\,\Omega\,\mathrm{e}^{-S_{\text{G}}[U]-S_{\text{F}}[U,\bar{\psi},\psi]}, \tag{5.54}$$

where Z is fixed by the condition $\langle\mathbb{1}\rangle = 1$. The functional integral involves an integration over the gauge group and over all fermionic degrees of freedom, the latter being represented by anti-commuting (Grassmann) variables. Since the fermionic action, $S_{\text{F}}[U, \bar{\psi}, \psi]$ is bilinear in the quark and antiquark fields, the integration over the Grassmann variables is Gaussian and can be performed analytically. This yields

$$\langle\Omega\rangle = \frac{1}{Z}\int \prod_{x\in\Lambda_{\text{E}}}\prod_{\mu=0}^{3} \mathrm{d}U_\mu(x)\,\widetilde{\Omega}\,\{\det D_{\text{lat}}\}^{N_{\text{f}}}\,\mathrm{e}^{-S_{\text{G}}[U]}. \tag{5.55}$$

[3]Here and in the following we drop the subscript "E" on the partition function Z.

Equation (5.55) requires some further explanation:

- $\widetilde{\Omega}$ denotes the representation of Ω in the (effective) theory, where the quark fields have been integrated out and only the link variables remain in the functional integral measure;
- D_{lat} denotes a generic, massive lattice Dirac operator. For instance, for Wilson quarks one has $D_{\text{lat}} = D_{\text{w}} + m_0$. For simplicity we have displayed the expression for QCD with N_{f} flavours of equal mass m_0, which accounts for the power N_{f}. In the case of non-degenerate quarks $\{\det D_{\text{lat}}\}^{N_{\text{f}}}$ must be replaced by a product of determinants, in which each factor represents the contribution from a single flavour:
- The lattice formulation has given a well-defined meaning to the measure $D[U]$. The integration over the gauge degrees of freedom reduces to a finite-dimensional integration over the gauge group, based on the invariant group (Haar) measure.

The numerical evaluation of $\langle\Omega\rangle$ via Monte Carlo integration proceeds as follows. One starts by generating a set of gauge configurations using a computer program. One configuration in the set represents the collection of all link variables on a given lattice, i.e.

$$\left\{U_\mu(x)\,|x \in \Lambda_{\text{E}},\ \mu = 0,\ldots,3\right\}, \tag{5.56}$$

for which we shall use the shorthand $\{U_\mu(x)\}$ below. A collection of an infinite number of configurations is called an *ensemble*. The statistical weight, W, of an individual configuration is given by

$$W = \{\det D_{\text{lat}}\}^{N_{\text{f}}}\ \mathrm{e}^{-S_{\text{G}}[U]}. \tag{5.57}$$

In other words, the composition of the ensemble is determined by a probability distribution, which is given by the negative exponentiated classical action in the integrand of the Euclidean functional integral. Owing to the weight factor, the integrand of the functional integral will be strongly peaked around those configurations for which W is large. This particular feature makes the expectation value amenable to a Monte Carlo treatment. The key idea is to replace the ensemble by a *finite* sample of N_{cfg} gauge configurations, which is dominated by those configurations for which W is large. Provided that one can construct a suitable algorithm, the sample will then consist predominantly of those configurations which give a large contribution to the Euclidean functional integral and thus $\langle\Omega\rangle$. Such a procedure is called *importance sampling*.

Technically, the sample is produced by generating a sequence of configurations via a *Markov process:*

$$\left\{U_\mu(x)\right\}_1 \longrightarrow \left\{U_\mu(x)\right\}_2 \longrightarrow \ldots \longrightarrow \left\{U_\mu(x)\right\}_{N_{\text{cfg}}}. \tag{5.58}$$

One assigns a probability for the transition from $\{U_\mu(x)\}_i$ to $\{U_\mu(x)\}_{i+1}$, which is usually a function of the statistical weights of the two configurations, W_i and W_{i+1}, respectively. For each individual configuration in the sequence one then evaluates the observable, which yields the estimates Ω_i, $i = 1, \ldots, N_{\text{cfg}}$. The expectation value $\langle\Omega\rangle$ is related to the mean value $\overline{\Omega}$ via

$$\langle\Omega\rangle = \lim_{N_{\text{cfg}}\to\infty} \overline{\Omega}, \qquad \overline{\Omega} = \frac{1}{N_{\text{cfg}}} \sum_{i=1}^{N_{\text{cfg}}} \Omega_i. \tag{5.59}$$

In other words, in the limit of infinite statistics the mean value converges to the ensemble average which is identical to the expectation value. An important consequence of approximating the ensemble average by the sample average is a non-zero value of the *variance*. Hence, in order to specify the results from a Monte Carlo integration completely, one must also quote the statistical error which is given by the square root of the variance.

In the standard algorithms that implement Markov processes (such as the Metropolis algorithm [30]), the transition probabilities for going from one configuration to another are determined by comparing the statistical weights for *local* variations in the field variables. This guarantees computational efficiency, since the variation of individual link variables does not involve global information from the entire lattice. In Eq. (5.55) the dynamical effects of the quark fields are incorporated via the determinant of the lattice Dirac operator. The determinant, however, is a non-local object, which is expensive to compute. When the first efforts were made to compute observables in QCD in the 1980s, the available computer power did not allow for the inclusion of the quark determinant. Instead, lattice physicists resorted to what is known as the "quenched approximation", which is based on the assumption that the bulk of non-perturbative contributions is carried by the gauge field, so that the determinant is set to a constant:

$$\text{Quenched approximation:} \qquad \det D_{\text{lat}} = 1 \quad \Leftrightarrow \quad N_{\text{f}} = 0. \tag{5.60}$$

The resulting gain in computer time amounts to several orders of magnitude. In the quenched approximation the effects of virtual quark loops are entirely suppressed. As a consequence, results for observables are afflicted with an unknown systematic error. As we shall see later, there are several quantities (for instance, the masses of the lightest hadrons) for which the quenching error amounts to just 10–15%. Although this justifies the use of the quenched approximation to some extent, it is clear that dynamical quark effects must be taken into account, in order to arrive at reliable, non-perturbative predictions with a total accuracy at the percent level.

Modern algorithms for dynamical quarks, such as the Hybrid Monte Carlo algorithm [31], do not evaluate the quark determinant directly. Rather, one exploits the property that the determinant can be rewritten as a functional integral over bosonic fields, which is then evaluated stochastically. Thereby one avoids computing a global object, but the computational effort involved in the stochastic estimation of the quark determinant is still large compared with the quenched approximation. More details can be found in Sect. 5.2.6 below.

Correlation functions, i.e. the expectation values of polynomials in the quark and gluon fields, are the most important quantities, since they determine implicitly the particle spectrum of the theory. As was discussed already in Sect. 5.2.1, the link between correlation functions and the particle spectrum is provided by the transfer matrix T. For lattice QCD with Wilson fermions, the existence of a positive transfer matrix was rigorously established [32].

As a concrete example we shall discuss the two-point correlation function of a charged kaon. A polynomial of quark fields with the quantum numbers of the kaon is given by

$$\phi_{\mathrm{K}}(x) = (\bar{u}\gamma_5 s)\,(x), \tag{5.61}$$

where the parentheses indicate summation over spinor and colour components of the fields. Mostly one is interested in correlation functions in which all spatial points have been summed over and which therefore only depend on the Euclidean time separation. We define

$$C_{\mathrm{K}}(x_0;\,\vec{p}) = \sum_{\vec{x}} \mathrm{e}^{\mathrm{i}\vec{p}\cdot\vec{x}} \left\langle \phi_{\mathrm{K}}(x)\phi_{\mathrm{K}}^{\dagger}(0)\right\rangle. \tag{5.62}$$

The inclusion of the phase factor in conjunction with the summation over $\vec{x}$ amounts to a projection onto spatial momentum $\vec{p}$. On a finite lattice with periodic boundary conditions $C_{\mathrm{K}}(x_0;\,\vec{p})$ must be symmetric under $x_0 \leftrightarrow T - x_0$. Therefore, the spectral decomposition of $C_{\mathrm{K}}(x_0;\,\vec{p})$ reads

$$C_{\mathrm{K}}(x_0;\,\vec{p}) = \sum_{\alpha} \frac{\left|\langle 0|\phi_{\mathrm{K}}(0)|\alpha\rangle\right|^2}{2\epsilon_\alpha(\vec{p})} \left\{\mathrm{e}^{-\epsilon_\alpha(\vec{p})x_0} + \mathrm{e}^{-\epsilon_\alpha(\vec{p})(T-x_0)}\right\}, \tag{5.63}$$

where the sum runs over all states in the kaon channel with fixed momentum $\vec{p}$, and $\epsilon_\alpha(\vec{p})$ is the mass gap (see Sect. 5.2.1).[4] For large Euclidean times x_0 the ground state dominates. If we further set $\vec{p} = 0$, then the asymptotic form of the two-point function reads

$$\lim_{x_0\to\infty} C_{\mathrm{K}}(x_0;\,\vec{p}) = \frac{\left|\langle 0|\phi_{\mathrm{K}}(0)|K\rangle\right|^2}{m_{\mathrm{K}}} \mathrm{e}^{-m_{\mathrm{K}}T/2} \cosh\left(m_{\mathrm{K}}(T/2 - x_0)\right), \tag{5.64}$$

where $m_{\mathrm{K}} = \epsilon_0(\vec{p})|_{\vec{p}=0}$ is the mass of the kaon, and the sum of the two exponentials has been re-expressed using the cosh function. Owing to the ordering $\epsilon_0(\vec{p}) < \epsilon_1(\vec{p}) < \ldots$, the higher excited states are exponentially suppressed. The functional

[4] In the commonly normalization of hadron states one includes a factor $2\epsilon_\alpha(\vec{p})$ in the denominator.

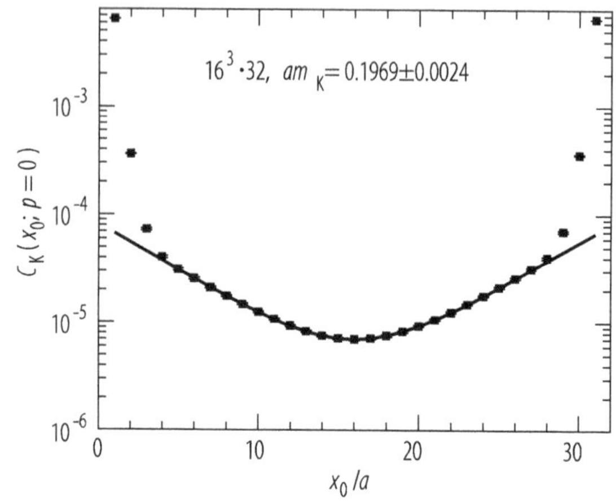

Fig. 5.3 Two-point correlation function for a pseudoscalar meson. The curve denotes a fit to Eq. (5.64) in the interval $6 \le x_0/a \le 26$

form of Eq. (5.64) is nicely illustrated by the plot in Fig. 5.3, where simulation data for $C_K(x_0; \vec{p} = 0)$ are compared to its asymptotic form. The data show indeed the expected cosh-behaviour. Furthermore, one observes how the contributions from higher excited states, which are clearly visible at small values of x_0/a, quickly die out as the time separation increases. From the two-point function we can extract two important quantities: the fall-off of $C_K(x_0; \vec{p} = 0)$ is characteristic of the kaon mass, i.e. the energy of the ground state. Moreover, the pre-factor of the cosh-function yields the transition amplitude between a kaon state and the vacuum, and thus contains information on the kaon's decay properties.

5.2.4 Continuum Limit, Scale Setting and Renormalization

In Sect. 5.2.2 we have discussed how to discretize the QCD action. The main principle for their construction was the condition that the corresponding expressions reproduce the continuum action in the formal limit $a \to 0$, regardless of the values of the bare parameters (such as β and the hopping parameter κ in the case of QCD with Wilson fermions). If one goes beyond the classical theory this is not possible anymore: it is a general property of quantum field theory that the parameters of the regularized theory (masses and couplings) must be adjusted as the regulator is removed. In the context of lattice QCD this implies that the continuum limit, $a \to 0$, is reached by a suitable tuning of the bare parameters.

To make this statement more precise, we shall invoke the close connection between Euclidean lattice field theory and a system in statistical mechanics. Models in statistical physics (think of the Ising model as an example) usually have a phase structure. Depending on the choice of parameters, the different phases may exhibit entirely different physical properties. The analogy with lattice field theory then implies that a particular discretization of QCD also possesses a phase structure in the space of bare parameters (β and κ, for example).[5] We shall now explain that the continuum limit of QCD is associated with a critical point in the phase diagram, which corresponds to a second-order phase transition. In the previous section we have considered hadronic two-point correlation functions, and how the mass in a given channel can be extracted from the asymptotic behaviour at large Euclidean times. Actually, this procedure yields the dimensionless combination (aM), i.e. the hadron mass in lattice units. In order to take the continuum limit, one must take $a \to 0$, while the physical mass M must remain constant. This implies

$$\frac{1}{(aM)} \equiv \xi \to 0. \tag{5.65}$$

In other words, the correlation length ξ diverges in the continuum limit. In the language of statistical physics, a divergent correlation length signals a second-order phase transition. The existence of the continuum limit in lattice QCD is therefore equivalent to the existence of a second-order transition in the space of bare parameters.

For simplicity we shall now consider Yang–Mills theory on the lattice, which we choose to describe by Wilson's plaquette action and the bare coupling parameter $\beta \equiv 6/g_0^2$. The existence of a second-order phase transition corresponds to a critical value of the bare gauge coupling, $g_{0,c}$. Furthermore, it implies that the bare coupling g_0 and the lattice spacing a (or, equivalently, the correlation length ξ) cannot be varied independently when the continuum limit is approached.[6] In this way we may regard the bare coupling as a function of the lattice spacing, $g_0(a)$, such that

$$\lim_{a \to 0} g_0(a) = g_{0,c}. \tag{5.66}$$

Let P be an observable, computed for a particular value of g_0, i.e. $P = P(g_0, a)$. Since P is a physical quantity it must stay constant as the continuum limit is taken, i.e.

$$a \frac{\mathrm{d}}{\mathrm{d}a} P(g_0, a) = 0. \tag{5.67}$$

This leads to the Callan–Symanzik equation

$$\left\{ a\frac{\partial}{\partial a} + a\frac{\partial g_0}{\partial a}\frac{\partial}{\partial g_0} \right\} P(g_0, a) = 0. \tag{5.68}$$

[5]This phase diagram must not be confused with the physical phase diagram of QCD in the plane defined by the temperature and the chemical potential, which is explored at heavy-ion colliders.

[6]Otherwise, an arbitrarily chosen value of g_0 would always correspond the a critical point.

We can define the renormalization group β-function $\beta_{\rm lat}$ as

$$\beta_{\rm lat}(g_0) := -a\frac{\partial g_0}{\partial a}, \tag{5.69}$$

which describes the change in g_0 when a is varied. Note that $\beta_{\rm lat}$ depends on the choice of discretization. In perturbation theory, however, one recovers the familiar universal coefficients at one- and two-loop order. For gauge group SU(N) one has

$$\beta_{\rm lat}(g_0) = -b_0 g_0^3 - b_1 g_0^5 + {\rm O}(g_0^7), \tag{5.70}$$

where

$$b_0 = \frac{1}{(4\pi)^2}\left\{\frac{11}{3}N - \frac{2}{3}N_{\rm f}\right\}, \quad b_1 = \frac{1}{(4\pi)^4}\left\{\frac{34}{3}N^2 - N_{\rm f}\left(\frac{13}{3}N - \frac{1}{N}\right)\right\}, \tag{5.71}$$

and $N_{\rm f} = 0$ in pure Yang–Mills theory. Starting from the perturbative expansion of $\beta_{\rm lat}$ one can integrate the Callan–Symanzik equation, which gives

$$a\Lambda_{\rm lat} = (b_0 g_0)^{-b_1/(2b_0^2)} {\rm e}^{-1/(2b_0 g_0)} \left\{1 + {\rm O}(g_0^2)\right\}, \tag{5.72}$$

where the integration constant $\Lambda_{\rm lat}$ represents a characteristic scale of the theory. The above expression establishes the connection between the lattice spacing and the bare coupling in perturbation theory. One reads off that

$$a \to 0 \quad \Leftrightarrow \quad g_0 \to 0, \tag{5.73}$$

and hence the critical point occurs at $g_{0,c} = 0$. These findings are a consequence of asymptotic freedom. Taking Eq. (5.72) at face value one would conclude that the relation between $P(a, g_0)$ and $P(a', g_0')$, computed for two different values of the bare coupling g_0 and g_0' near the critical point, was simply given by the ratio of Eq. (5.72) evaluated for g_0 and g_0'. However, actual simulations do not confirm this expectation. The reason for the failure to observe "asymptotic scaling", i.e. a variation of $P(a, g_0)$ with g_0 which is consistent with Eq. (5.72), is that the accessible values of g_0 in simulations are by far not near enough the critical point, in order for perturbation theory to be a good approximation.

Let P and P' be two different observables that both satisfy Eq. (5.68). Then, regardless of whether or not asymptotic scaling holds, one would expect the ratio $aP(a, g_0)/aP'(a, g_0)$ to be equal to the physical ratio P/P' for all values of g_0. However, even this weaker scaling criterion is usually not observed, the reason being that the right-hand side of Eq. (5.68) is not strictly zero. Rather one has

$$\left\{a\frac{\partial}{\partial a} - \beta_{\rm lat}(g_0)\frac{\partial}{\partial g_0}\right\} P(g_0, a) = {\rm O}(a^p), \tag{5.74}$$

Table 5.1 Most widely used discretizations of the Dirac operator and some of their properties

Action	Doublers	Leading artefacts	Chiral symmetry
Wilson	None	$O(a)$	Broken
Clover	None	$O(a^2)$	Broken
Staggered	4	$O(a^2)$	$U(1) \otimes U(1)$ subgroup unbroken
Neuberger	None	$O(a^2)$	Preserved
Domain Wall	None	$O(a^2)$	Remnant breaking exponentially suppressed
Twisted Mass Wilson	None	$O(a^2)$	Broken

where p is a positive integer. These so-called scaling violations on the right-hand side depend both on the lattice action and the observable in question. As a consequence, the ratio considered above behaves like

$$\frac{aP(a, g_0)}{aP'(a, g_0)} = O(a^p). \tag{5.75}$$

In other words, as g_0 is tuned towards zero, dimensionless ratios of observables converge to the continuum limit with a rate proportional to a^p, where the power p is characteristic of the particular discretization employed in the lattice calculation. In Table 5.1 we have already listed the leading scaling violations (lattice artefacts) for several widely used fermionic lattice actions. Discretizations of the Yang–Mills part, such as the plaquette action, have leading lattice artefacts of $O(a^2)$. The Symanzik improvement programme allows to construct lattice actions with an accelerated rate of convergence to the continuum limit.

In actual lattice calculations, the continuum limit must be taken by performing simulations at several different values of the lattice spacing and extrapolating the results to $a = 0$. The functional form of the extrapolation is chosen such that it is consistent with the leading discretization errors for a given lattice action. Such a procedure is only viable if the relation between the lattice spacing in physical units and the dimensionless coupling parameter g_0 (which is an input parameter in the simulation) is known with good accuracy. Since the perturbative formula Eq. (5.72) is not of any practical use, the relation between the scale and the coupling must be mapped out non-perturbatively. To this end one picks a value for g_0 and computes in a Monte Carlo simulation a dimensionful quantity Q, whose value is known from experiment. Common choices for Q in the pure gauge theory are the string tension or the hadronic radius r_0 [33, 34], while in full QCD one may choose the mass of the nucleon. The Monte Carlo procedure yields Q in lattice units, (aQ), and the calibration of the lattice spacing is achieved via

$$a^{-1}\,[\text{MeV}] = \frac{Q|_{\text{exp}}\,[\text{MeV}]}{(aQ)|_{g_0}}. \tag{5.76}$$

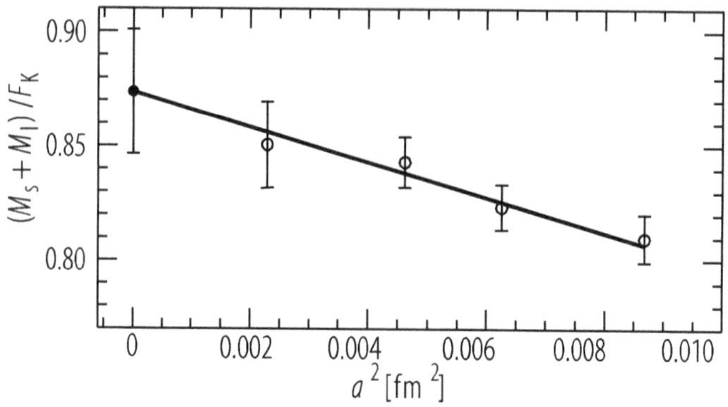

Fig. 5.4 Continuum extrapolation of the dimensionless ratio of quark masses and the kaon decay constant [35]

Knowledge of (aQ) over a range of bare couplings is a prerequisite for performing the continuum extrapolation. In Fig. 5.4 we show a particular example, namely the continuum extrapolation of the combination $M_s + \frac{1}{2}(M_u + M_d)$ of quark masses, normalized by the kaon decay constant, computed using O(a) improved Wilson fermions in the quenched approximation [35]. The expected linear convergence in a^2 is clearly exhibited by the lattice data.

So far we have restricted the discussion to the pure gauge theory which contains only one bare parameter, the gauge coupling g_0 (sometimes expressed in terms of $\beta = 6/g_0^2$). When quarks are incorporated, the set of parameters must be enlarged by the values of the bare masses, one for each flavour. Lattice QCD is thus parameterized by the set of bare parameters

$$\{g_0; m_u, m_d, m_s, m_c, m_b, m_t\}.$$

In order to be predictive, the theory must be renormalized, by expressing the bare parameters in terms of renormalized ones.

A convenient and practical method for lattice QCD is based on so-called *hadronic renormalization schemes*. Here the bare coupling and quark masses are eliminated in favour of renormalized quantities such as hadron masses or decay constants. An example how this works in the pure gauge theory was already given in the preceding discussion on scale setting, where the bare coupling was eliminated by assigning a value in physical units to the lattice spacing. In the process one has to choose a quantity that sets the scale and which cannot be predicted anymore.

Replacing the values of the bare quark masses $m_u, m_d, \ldots$ in favour of hadronic quantities works as follows. Like the bare coupling, the bare quark mass is an input parameter for the simulation and thus freely adjustable. Therefore, simulations yield hadron masses (in lattice units) as a function of the input quark masses. For instance, $am_{\text{PS}}(m_1, m_2)$ denotes the mass in lattice units of a generic pseudoscalar meson composed of a quark and antiquark with bare masses m_1 and m_2, respectively. Let us assume that the lattice spacing a has been calibrated using some input quantity

Q. If we further assume exact isospin symmetry we can then determine the value of the bare isospin-symmetrized light quark mass $\hat{m} = \frac{1}{2}(m_u + m_d)$, by requiring that

$$\frac{m_{\rm PS}(m_1, m_2)}{Q} = \left.\frac{m_\pi}{Q}\right|_{\rm exp}, \qquad m_1 = m_2, \tag{5.77}$$

i.e. the value of $\hat{m}$ is fixed by adjusting the input mass m_1 until $m_{\rm PS}(m_1, m_2)/Q$ coincides with the experimental result. We can extend this procedure to include more massive flavours. The bare strange quark mass is found by tuning m_2 such that

$$\frac{m_{\rm PS}(\hat{m}, m_2)}{Q} = \left.\frac{m_{\rm K}}{Q}\right|_{\rm exp}. \tag{5.78}$$

Alternatively one can fix m_s via the condition $m_{\rm V}(\hat{m}, m_2)/Q = m_{\rm K}^*/Q|_{\rm exp}$, where $m_{\rm V}$ denotes the mass in the vector channel. An example of a particular hadronic renormalization scheme is shown below:

Parameter	Quantity
g_0	f_π
$\frac{1}{2}(m_u + m_d)$	m_π
m_s	$m_{\rm K}$
m_c	$m_{\rm D_s}$
m_b	$m_{\rm B_s}$

All quantities in a lattice calculation are genuine predictions, except for those that are listed in the right-hand column of the table, which are used to eliminate the bare parameters.

Given the multitude of hadronic states, it is obvious that there is considerable freedom in choosing hadronic renormalization schemes. Usually, masses or mass splittings of hadrons that are stable in QCD are suitable to define a scheme. Resonances, such as the ρ, should be avoided, since they do not have a sharply defined energy, owing to their large width.

5.2.5 *Limitations and Systematic Effects*

The lattice formulation is the basis for an exact non-perturbative treatment of QCD. The accuracy of lattice results is chiefly limited by the algorithmic performance and the available computer power. In particular, the set of bare parameters that can be simulated efficiently for a given number of lattice sites is restricted. This has the important consequence that the quark masses at the very extremes of the physical mass scale (i.e. the up/down quarks and the b-quark) cannot be simulated directly with currently available methods and machines. These technical limitations are usually translated into a systematic error, which is quoted alongside the statistical one. The most important systematic effects are due to

- lattice artefacts (cutoff effects),
- finite volume effects, and
- extrapolations in the quark mass.

In order to have sufficient control over these effects, the simulation parameters must be chosen such that the following inequalities are satisfied:

$$\frac{1}{am_{\text{had}}} \ll \frac{L}{a}, \qquad m_{\text{had}} \ll a^{-1}, \tag{5.79}$$

where m_{had} is the mass of a generic hadron in physical units computed in the simulation. The inequality on the left of (5.79) states that the hadron's correlation length must be much smaller than the linear extent of the spatial box (in lattice units), as otherwise its value will be strongly distorted by finite volume effects. The inequality on the right states that the hadron mass must be significantly smaller than the inverse lattice spacing. If this is not the case, lattice artefacts will be uncontrollably large, meaning that the presence of higher-order cutoff effects cannot be excluded, so that a reliable extrapolation to the continuum limit as a linear function of the leading power of lattice artefacts cannot be performed. With current algorithms and machines, lattice sizes of up to $L/a = 48$ and lattice spacings down to 0.05 fm are affordable, even if dynamical quarks are included. Since $a = 0.05$ fm corresponds to $a^{-1} \approx 4$ GeV, it is obvious that the b-quark mass is too large to be simulated directly. Several techniques have been devised to address this problem, and a brief account can be found in Sect. 5.7.2.

In the light quark sector, the primary limitation that forbids making direct contact with the physical values of the up and down quarks is mostly due to algorithmic performance, rather than finite size effects. A detailed discussion of the algorithmic difficulties associated with the simulation of light dynamical quarks is presented separately in the following section. Moreover, it is difficult even in the quenched approximation to reach quark masses significantly smaller than half the physical strange quark mass, in particular with Wilson fermions. This is related to the occurrence of arbitrarily small eigenvalues in the spectrum of the Wilson-Dirac operator, even for small but non-vanishing values of the bare mass. As a result, observables computed on individual, so-called "exceptional" configurations may differ from the Monte Carlo average by orders of magnitude, and thus a reliable determination of the result and its error is virtually impossible. As already mentioned in Sect. 5.2.2, the problem of exceptional configuration can be cured by employing alternative discretizations such as twisted mass QCD or the overlap operator. A related problem arising from the particular spectral properties of the Wilson-Dirac operator is the bad performance of standard algorithms for dynamical quarks, discussed in the next section.

Due to these reasons, many simulations (quenched and unquenched) were restricted to quark masses not much smaller than $m_s/2$. This value translates into a minimum mass of about 490 MeV in the pseudoscalar meson channel, so that in these simulations the pion is as heavy as the physical kaon. In this region of parameter space it is known empirically that a spatial lattice length of 2–3 fm is sufficient to satisfy the first inequality in (5.79) and to rule out significant finite volume effects. Moreover, an important analytic result derived by Lüscher [36], implies that the asymptotic convergence to the result in infinite volume is exponential.

In order to make contact with the chiral regime, lattice results must be extrapolated to the physical values of the up and down quark masses. The functional form for the dependence of observables on the quark mass is usually provided by Chiral Perturbation Theory (ChPT). For instance, at lowest order the relation between the mass of a pseudoscalar meson composed of quarks with masses m_1 and m_2 is

$$m_{\mathrm{PS}}^2 = B_0(m_1 + m_2) + \dots, \tag{5.80}$$

where the ellipses represent higher orders in the chiral expansion. Similar expressions are derived for vector meson and baryon masses, e.g.

$$m_{\mathrm{V}} = m_{\mathrm{V}}^0 + C\,M^2 + \dots, \qquad m_{\mathrm{N}} = m_{\mathrm{N}}^0 + k\,M^2 + \dots, \tag{5.81}$$

and also for other quantities such as pseudoscalar decay constants. In the above formulae, $M^2 \equiv B_0(m_1 + m_2)$, and m_{V}^0 and m_{N}^0 denote the (non-vanishing) masses in the chiral limit. A more formal introduction to the basic concepts of ChPT is presented in Sect. 5.6.1.

It remains largely unknown whether or not the expressions of ChPT considered at a given order in the expansion can be applied in the quark mass range that is accessible in current simulations. Therefore, chiral extrapolations can lead to substantial systematic uncertainties. For instance, lattice predictions for the ratio of decay constants of the B and B_s mesons, $f_{\mathrm{B_s}}/f_{\mathrm{B}}$, may differ by 10%, depending on whether the LO or NLO expressions are used as an *ansatz* for the extrapolation from quark masses around $m_s/2$. Currently it is estimated that pseudoscalar meson masses of 300 MeV and below must be reached in simulations, in order that ChPT at one- or even two-loop order provides an accurate prediction for the quark mass dependence of hadron masses and matrix elements.

In the quenched approximation, chiral extrapolations are particularly problematic, since the chiral limit is intrinsically pathological, due to the appearance of singularities in the quark mass dependence. This is illustrated by the NLO expression for the ratio $m_{\mathrm{PS}}^2/(m_1 + m_2)$, i.e.

$$\frac{m_{\mathrm{PS}}^2}{m_1 + m_2} = B_0\Big\{1 - \Big(\delta - \tfrac{2}{3}\alpha_\Phi y\Big)(\ln y + 1) + \Big[(2\alpha_8 - \alpha_5) - \tfrac{1}{3}\alpha_\Phi\Big]y\Big\}, \tag{5.82}$$

where B_0, α_5, α_8, δ and α_Φ are low-energy constants. For notational convenience we have introduced

$$y = \frac{M^2}{(4\pi F_0)^2}, \qquad M^2 = B_0(m_1 + m_2), \tag{5.83}$$

where F_0 denotes the pion decay constant in the chiral limit. The low-energy constants δ and α_Φ, which multiply the so-called "quenched chiral logarithms", have no counterpart in the unquenched case. Since δ has a non-zero value [37], the quenched chiral logarithm in Eq. (5.82) gives rise to a singularity in the chiral limit. For many applications, the singularity can be ignored, since its effect is numerically small even at the physical pion mass. However, it signals that the quenched approximation suffers from fundamental conceptual problems.

5.2.6 Simulations with Dynamical Quarks

Although one may argue that the quenched approximation describes hadronic properties fairly well, it is clearly unsatisfactory, both from a conceptual point of view, and also because it introduces an unknown systematic error. Below we shall discuss some general issues relating to simulations with dynamical quarks. It must be stressed that several different techniques how to treat the quark determinant of Eq. (5.57) efficiently are currently being explored. A preferred or clearly superior method has not emerged so far, and it is likely that some of the approaches presented below may become obsolete in the years to come.

In order to illustrate the main difficulties, we start by introducing the Hybrid Monte Carlo (HMC) algorithm [31], which has been the standard algorithm for simulations with dynamical quarks for many years. In order to produce one step in the Markov chain, the algorithm evolves the link variables according to the equations of motion of a classical Hamiltonian system. To this end one introduces a conjugate momentum variable $\Pi_\mu(x)$ for every link $U_\mu(x)$. The Hamiltonian is defined as

$$H[U,\Pi] = \tfrac{1}{2}\sum_{x\in\Lambda_{\mathrm{E}}}\sum_{\mu=0}^{3}\Pi_\mu(x)\Pi_\mu(x) + S_{\mathrm{G}}[U] + S_{\mathrm{F}}^{\mathrm{eff}}[U,\phi^*,\phi], \tag{5.84}$$

where $S_{\mathrm{G}}[U]$ is the lattice gauge action, and $S_{\mathrm{F}}^{\mathrm{eff}}[U,\phi^*,\phi]$ denotes an effective lattice fermion action, which is obtained by rewriting the quark determinant as a functional integral over complex *bosonic* fields $\phi(x)$ and $\phi^*(x)$. Explicitly, for $N_{\mathrm{f}} = 2$ one has

$$(\det D_{\mathrm{lat}})^2 = \int D[\phi^*,\phi]\,\exp\left\{-\sum_{x\in\Lambda_{\mathrm{E}}}\phi^*(x)\left[(D_{\mathrm{lat}}^\dagger D_{\mathrm{lat}})^{-1}\phi\right](x)\right\}. \tag{5.85}$$

For each step in the Markov chain, the conjugate momenta are drawn randomly from a Gaussian distribution ("momentum refreshment"). The Hamiltonian $H[U, \Pi]$ governs the dynamics of the variables $U_\mu(x)$ and $\Pi_\mu(x)$ with respect to "simulation time" τ, which parameterizes the evolution of $U_\mu(x)$ and $\Pi_\mu(x)$ as the simulation algorithm progresses. The evolution is described by Hamilton's equations, which read

$$\frac{\mathrm{d}}{\mathrm{d}\tau}U_\mu(x) = \Pi_\mu(x)U_\mu(x), \qquad \frac{\mathrm{d}}{\mathrm{d}\tau}\Pi_\mu(x) = -F_{\mathrm{G},\mu}(x) - F_{\mathrm{F},\mu}(x), \tag{5.86}$$

where

$$F_{\mathrm{G},\mu}(x) = \frac{\partial S_{\mathrm{G}}[U]}{\partial U_\mu(x)}, \qquad F_{\mathrm{F},\mu}(x) = \frac{\partial}{\partial U_\mu(x)} \sum_{x\in\Lambda_{\mathrm{E}}} \phi^*(x)\left[(D_{\mathrm{lat}}^\dagger D_{\mathrm{lat}})^{-1}\phi\right](x) \tag{5.87}$$

are the forces associated with the gluon and quark fields, respectively. The algorithm then proceeds by integrating the equations of motion numerically. As in any numerical integration scheme, the total time interval is divided into a number of sub-intervals of finite length $\Delta\tau$, which is called the step size. Starting from an initial gauge configuration $\{U_\mu(x)\}$ and a set of conjugate momenta $\{\Pi_\mu(x)\}$, one obtains new sets $\{U'_\mu(x)\}$, $\{\Pi'_\mu(x)\}$ after the integration. In the language of classical mechanics, the variables $U_\mu(x)$ and $\Pi_\mu(x)$ evolve along a trajectory in phase space which connects the initial and final configurations. However, since numerical integration is not exact, owing to the finite step size, the energy is not conserved. In the HMC algorithm this is rectified by introducing a global accept/reject step: if ΔH denotes the energy difference between the initial and final configurations, i.e.

$$\Delta H \equiv H[U', \Pi'] - H[U, \Pi], \tag{5.88}$$

then the new configuration $\{U'_\mu(x)\}$ is accepted with probability[7]

$$P\{U \to U'\} = \min(1, \mathrm{e}^{-\Delta H}). \tag{5.89}$$

In other words, a configuration $\{U'_\mu(x)\}$ associated with a large value for the energy violation ΔH is less likely to be accepted. This final step completes the Monte Carlo update. The name "Hybrid Monte Carlo" reflects the fact that one combines a deterministic classical dynamics procedure with a pseudo-random accept/reject step.

One major problem which has plagued simulations with dynamical quarks over many years is the fact that the efficiency of the conventional HMC algorithm

---

[7]In practice this is achieved by drawing a random number r, with $0 < r \leq 1$. If $r < \mathrm{e}^{-\Delta H}$ the new configuration is rejected.

deteriorates sharply when the lattice spacing is decreased and the masses of the light (up and down) quarks are tuned to their physical values. The poor scaling behaviour is driven by the condition number of the lattice Dirac operator $D_{\rm lat}$, i.e. the ratio of the largest to the smallest eigenvalue. This quantity is known to grow inversely proportional to the lattice spacing and the quark mass. In particular, the HMC algorithm scales with the second, perhaps the third power of the light quark mass. Thus, simulations based on the Wilson-Dirac operator were found to be unpractical for lattice spacings below 0.1 fm and quark masses significantly smaller than half of the strange quark mass.[8] This is related to the afore-mentioned fact that even the massive Wilson-Dirac operator is not protected against arbitrarily small eigenvalues. Its condition number may thus fluctuate strongly in the course of the simulation, leading not only to numerical instabilities, but also to large fluctuations in the quark force term $F_{{\rm F},\mu}(x)$, and, in turn, ΔH. In order to keep a reasonably large acceptance rate of well over 75%, one must reduce the step size $\Delta\tau$ accordingly, and thus the numerical effort to integrate the equations of motion for an interval τ of fixed length, increases.

Two basic strategies to address this problems have been followed: the first is based on using fermionic discretizations that avoid the problem of arbitrarily small eigenvalues, while the aim of the second approach is to improve the simulation algorithms.

Staggered fermions have been advocated as a numerically more efficient alternative to the Wilson-Dirac formulation: since the staggered Dirac operator couples one-component Grassmann fields rather than four-component spinors, fewer floating point operations are required for one application of the operator. Moreover, the residual $\mathrm{U}(1)\otimes\mathrm{U}(1)$ symmetry protects the quark mass against additive renormalization and thus prevents the occurrence of very small eigenvalues. However, the fact that the staggered formulation describes four "tastes" per quark flavour makes a physical interpretation difficult. Technically, the degeneracy implies that the statistical weight of the quark determinant is too large compared with that of one physical flavour. An *ad hoc* method to compensate for this is to take fractional powers of the staggered quark determinant. For instance, to simulate QCD with a doublet of degenerate up and down quarks with mass $\hat{m}$, and a single heavier (strange) quark with mass m_s, the probability measure is taken as

$$P=\frac{1}{Z}\left\{\det\left(D_{\rm stagg}+\hat{m}\right)\right\}^{1/2}\left\{\det\left(D_{\rm stagg}+m_s\right)\right\}^{1/4}{\rm e}^{-S_{\rm G}[U]}, \tag{5.90}$$

where $D_{\rm stagg}$ is the massless staggered Dirac operator. This procedure is known as the "fourth root trick". The main question, which has been hotly debated, is whether or not the rooted staggered operator corresponds to a local field theory, or whether it induces spurious interactions among the fermionic degrees of freedom, which might lead to a violation of the universality of the continuum limit. A thorough analysis

[8]This should be compared to the physical mass ratio of $\hat{m}\approx m_s/24$ [38].

of this problem was given in [39], but so far no firm conclusion has been reached. Nevertheless, the probability measure Eq. (5.90) and the "rooting trick" it is based on, have been employed in large-scale simulations (see, e.g. Ref. [40]).

Discretizations based on twisted mass QCD have also been proposed as a numerically more efficient quark action. Here, the twisted mass parameter μ_q protects the operator against arbitrarily small eigenvalues. The smallest mass in the pion channel that has been reached with this formulation was as low as 300 MeV [41]. This corresponds to a physical quark mass of about $m_s/5$, which may be sufficient to enter the regime where the quark mass behaviour of observables can be described analytically using Chiral Perturbation Theory.

Owing to several major algorithmic improvements, simulations based on the Wilson-Dirac operator can now be performed much more efficiently. Without going into much detail, we simply state that most of the gain is due to the use of suitably chosen factorizations of the Wilson-Dirac operator into its low- and high-frequency parts. The various factors are then "better conditioned". In particular, fluctuations in the condition number can be controlled via a separate and optimized treatment of the low-energy part. In this way the step size $\Delta\tau$ can be increased whilst keeping a reasonably high acceptance rate for fixed total trajectory length τ. Algorithmic implementations of factorization range from Hasenbusch's "mass preconditioning" [42, 43], Lüscher's domain decomposition technique based on the Schwarz Alternating Procedure (DD-HMC algorithm) [44], to factorizations based on mass preconditioning combined with rational approximations of the contributions from multiple pseudo-fermion fields [45]. Thanks to these developments, it appears that the spectral properties of the Wilson-Dirac operator are no longer an obstacle to the efficient simulation of lattice QCD with light dynamical quarks. At the same time, large-scale simulations employing the recent algorithmic improvements are only just starting.

5.3 Hadron Spectroscopy

The determination of the spectrum of hadrons, i.e. mesons, baryons, glueballs, and possibly "exotic" hadronic states, starting from the underlying gauge theory of quarks and gluons has traditionally been one of the main applications of lattice QCD. The rôle of lattice calculations in this context is twofold: first, the determination of the experimentally known values of hadron masses from first principles represents a stringent test of QCD. Second, lattice calculations can make predictions for the masses of undiscovered or poorly established states. For instance, lattice results have been instrumental in the search for glueball candidates, and have also contributed significantly to the debate on the existence of pentaquarks.

The principles of hadronic mass calculations have already been outlined at the end of Sect. 5.2.3: After defining a suitable interpolating operator with the quantum numbers of the desired hadronic channel, one computes its Euclidean two-point function. The mass (energy) of the ground state in that channel is then extracted from the exponential fall-off of the correlation function at large Euclidean times. The detailed functional form of the asymptotic behaviour depends on the choice of boundary conditions. Thus, it is not always described by a cosh function, as in the example of a pseudoscalar meson on a lattice with periodic boundary conditions in time, c.f. Eq. (5.64). In the limit of infinite temporal lattice size T, the effect of the boundary conditions is sufficiently weak, so that one may approximate the functional form of the correlation function for a generic interpolating operator $\phi_{\rm had}(x)$ by

$$C_{\rm had}(x_0;\vec{p}) = \sum_{\vec{x}} \mathrm{e}^{\mathrm{i}\vec{p}\cdot\vec{x}} \left\langle \phi_{\rm had}(x)\phi_{\rm had}^\dagger(0)\right\rangle \overset{T\to\infty}{=} \sum_{\alpha} w_\alpha(\vec{p})\mathrm{e}^{-\epsilon_\alpha(\vec{p})x_0}. \tag{5.91}$$

Here, the quantity $w_\alpha(\vec{p})$ is referred to as the spectral weight of the state $|\alpha\rangle$. A large value for the spectral weight of the ground state, $w_1(\vec{p})$, will lead to an early domination of the correlation function by the ground state energy. The choice of $\phi_{\rm had}$ in a given channel can be optimized such that

$$w_1(\vec{p}) \gg w_i(\vec{p}), \qquad i = 2, 3, \ldots. \tag{5.92}$$

An optimal choice of interpolating operator is not only important to ensure a reliable determination of the ground state energy: In order to determine the energies in the excitation spectrum, the associated spectral weights must be maximized by specifying appropriate operators.

Below we provide examples for interpolating operators in several mesonic and baryonic channels:

$$\begin{aligned}
K\text{-meson}: &\quad \phi_{\rm K} = (s\gamma_5\bar{u}), \quad (s\gamma_0\gamma_5\bar{u}) \\
K^*\text{-meson}: &\quad \phi_{\rm K^*} = (s\gamma_j\bar{u}), \qquad j = 1,2,3 \\
\text{nucleon}: &\quad \phi_{\rm N} = \varepsilon_{abc}\{u^{a\,T}\, C\gamma_5\, d^b\}u^c \\
\Delta: &\quad \phi_\Delta = \varepsilon_{abc}\{u^{a\,T}\, C\gamma_\mu\, d^b\}u^c
\end{aligned} \tag{5.93}$$

Here, parentheses indicate summation over spinor and colour indices, while curly brackets denote that only spinor indices are summed over.

5.3.1 *Light Hadron Spectrum*

The determination of the spectrum of light hadrons was historically one of the first attempts to compute hadronic properties on the lattice. Since the masses of the low-lying hadrons are known from experiment, such calculations serve as benchmarks to test the intrinsic accuracy of the lattice approach.

The quenched approximation has been widely used to compute a number of quantities that are of great phenomenological interest. However, these results are of limited value, since the inherent quenching error is left undetermined. A precise calculation of the masses of the lowest lying hadrons in quenched QCD will expose the typical magnitude of the systematic error incurred by neglecting dynamical quark effects. To this end, several calculations of the quenched light hadron spectrum, using different lattice actions, have been performed [46–51].

In Ref. [47], the CP-PACS Collaboration presented a comprehensive study of the masses of the lowest pseudoscalar and vector mesons, as well as octet and decuplet baryons. The Wilson fermion action without O(a) improvement was used at four different values of the lattice spacing, and a continuum extrapolation linear in a has been performed for all quantities. CP-PACS adopted a hadronic renormalization scheme in which the lattice scale was fixed using the mass of the ρ-meson. The average up and down quark mass was set using m_π. In order to fix m_s, either the kaon mass ("K"-input) or the mass of the ϕ-meson ("ϕ"-input) was used. Chiral extrapolations were either based on the form expected from quenched Chiral Perturbation Theory at NLO (see Eq. (5.82)), or on the leading-order formula supplemented by a quadratic term in the quark mass. The resulting (small) differences in the extrapolated values were added as systematic errors in the final results, which are summarily displayed in Fig. 5.5. Although the lattice results are in remarkable overall agreement with the experimentally observed spectrum, one finds significant deviations. For instance, the ratio of the nucleon and the ρ-meson masses is determined as

$$\frac{m_{\mathrm{N}}}{m_\rho} = 1.143 \pm 0.033 \pm 0.018, \tag{5.94}$$

where the first error is statistical, and the second is an estimate of systematic uncertainties other than quenching. The above value is 6.7% (2.5 standard deviations) below the experimental value of 1.218. Similarly, vector-pseudoscalar mass

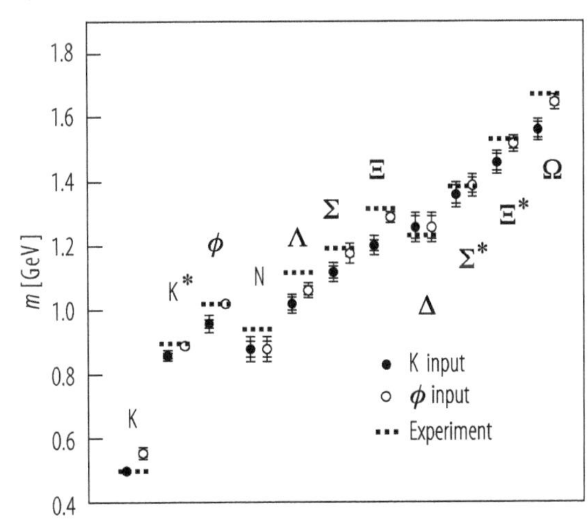

Fig. 5.5 Quenched light hadron spectrum computed in [47], compared with experiment. The statistical error and the sum of the statistical and systematic errors are indicated

splittings, such as $m_{K^*} - m_K$, are underestimated by 10–15% (4–6σ), depending on whether m_K or m_ϕ was used to fix the strange quark mass.

The findings reported by CP-PACS, which were based on unimproved Wilson fermions, have been broadly confirmed by other collaborations employing different lattice actions [48–51]. Thereby, the universality of the continuum limit of quenched QCD has been established: although different discretizations may yield statistically inconsistent results at non-zero lattice spacing, they converge to a common continuum limit, provided that the same hadronic renormalization scheme has been employed. The latter requirement is important, as there is considerable freedom in choosing a particular scheme. This leads to ambiguities in the quenched approximation, since different quantities are affected in different way by quark loops. In Ref. [51] it was found that, by using only stable or narrow states to define the hadronic renormalization scheme, the discrepancies between the quenched and experimental spectra could be shifted to the broad resonances, ρ, Δ, N^*, while the agreement for states like K, ϕ, N, Ω could be improved. Yet this observation does not alter the conclusion that the quenched approximation is unable to reproduce the spectrum of light hadrons with an accuracy better than 10%.

The obvious question is whether sea quark effects can account for the observed deviation between the quenched and experimental spectra. Owing to the larger numerical effort required to simulate QCD with dynamical quarks, unquenched studies have not yet reached the same level of control over systematic effects—notably lattice artefacts and chiral extrapolations—compared with the quenched benchmark [47]. Thus, a "definitive" unquenched calculation of the light hadron spectrum is still lacking, and thus we refrain from presenting an overview of recent results.

Nevertheless, the observed tendency in all simulations performed to date is that dynamical quarks "do the right thing", i.e. the deviation from experiment is decreased. An example is shown in Fig. 5.6, where continuum extrapolations of meson masses in the quenched and unquenched theories are compared. The plot

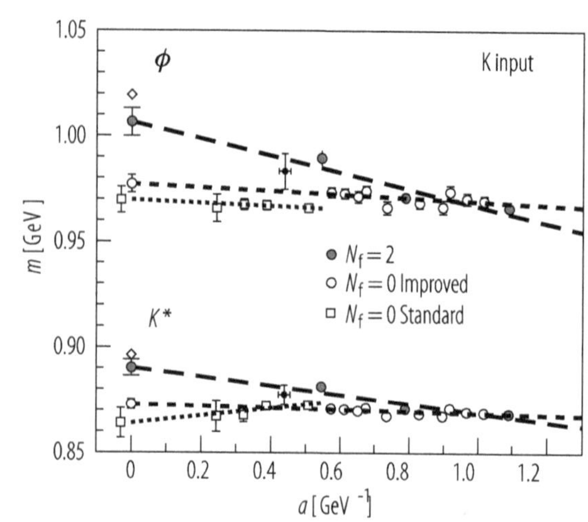

Fig. 5.6 Continuum extrapolations of the masses of the K^* and ϕ mesons in full ($N_f = 2$, full symbols) and quenched QCD (open symbols), compared with experiment (diamonds) [52]

shows that the data obtained for $N_f = 2$ are closer to the experimental results in the continuum limit in comparison with their quenched counterparts. However, the figure also shows that the extrapolation of unquenched data is not well constrained, since only three data points are available. Clearly, additional simulations at smaller lattice spacings and quark masses are required for a solid determination of the total error in unquenched calculations of the light hadron spectrum.

It should also be noted that the various discretizations of the quark action have complementary advantages and shortcomings. While simulations with Wilson quarks have in the past been restricted to quark masses not much smaller than half the strange quark mass for algorithmic reasons, the use of staggered fermions in conjunction with the rooting procedure may be afflicted with conceptual problems (see the discussion in Sect. 5.2.6). Domain wall and overlap fermions are *per se* more expensive to simulate. In simulations based on tmQCD the incorporation of a third, heavier quark flavour is quite complicated. Thus, progress in this area is likely to be made through the combined information from different discretizations.

5.3.2 *Glueballs*

In addition to bound states composed of a quark-antiquark pair or, alternatively, three quarks, QCD is also widely believed to support the existence of glueballs, i.e. bound states consisting mainly of gluonic degrees of freedom. Although several candidates for such states have been proposed (e.g. the $f_0(1370)$, $f_0(1500)$ and $f_0(1710)$), the experimental difficulty consists in their unambiguous identification as glueballs. To this end, they need to be distinguished from "conventional" flavour-singlet meson resonances in the scalar channel. Predictions for the masses and widths of glueballs from lattice QCD provide crucial input for this task.

The basic principles of mass calculations for glueballs in lattice QCD are the same as for bound states composed of quark degrees of freedom: first one must define an interpolating operator with the appropriate quantum numbers of the glueball state in question. That is, the operator must transform correctly under spin, parity and charge conjugation. At this point a complication arises: the lattice breaks all continuous space-time symmetries, such that Lorentz-invariance or—in the language of Euclidean field theory—rotational invariance is only recovered in the continuum limit. At non-zero lattice spacing the spin assignment is therefore ambiguous. Since the gluon field is represented by link variables, any glueball operator must be constructed from particular combinations of Wilson loops, i.e. products of link variables along closed paths on a hypercubic lattice (see Fig. 5.7).

Operators constructed in this way transform under irreducible representations (IRs) of the octahedral group O_h, which are conventionally labelled A_1, A_2, E, T_1 and T_2. By computing the relations between the IRs of O_h and SU(2) one finds that each IR in the set $\{A_1, A_2, E, T_1, T_2\}$ corresponds to infinitely many spins in the continuum. For instance, A_1 transforms not only like a scalar (spin 0) state, but also contributes to spin 4 and yet higher spin states. Similarly, the lowest states to

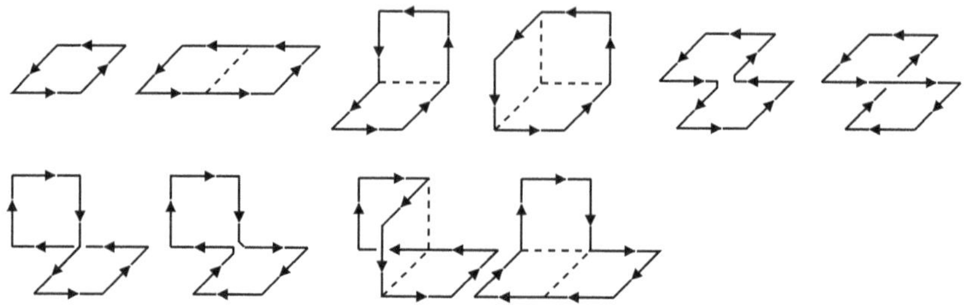

Fig. 5.7 Wilson loops used in the construction of glueball operators (from Ref. [53])

which T_1 makes a contribution are spin 1 and spin 3, while E corresponds to spins 2, 4, 5,.... In order to fully classify lattice glueball operators, the representations of O_h are supplemented by the transformation properties under parity and charge conjugation, in full analogy with the usual J^{PC}-assignment in the continuum. For example, an operator labelled A_1^{++} corresponds to the scalar channel 0^{++} in the continuum.

The above discussion implies that the two-point correlation function of an operator transforming under A_1^{++}, which is used to describe the scalar glueball, will be contaminated by contributions from a spin 4 state. However, in accordance with Regge theory one may expect that the latter dies out quickly, since higher spin states are more massive.

Another technical complication arises from the empirical observation that the spectral weight, $w_1(\vec{p})$, of the ground state in Eq. (5.91) is usually quite small. This implies that the asymptotic behaviour of the two-point correlation function is only isolated at large Euclidean times. However, the statistical accuracy deteriorates quickly as x_0 is increased, and in the asymptotic regime the correlation function is numerically comparable to the statistical noise. This precludes a precise determination of the mass of the ground state. A heuristic explanation for the small spectral weight can be given by noting that the operators constructed from the usual link variables are point-like and thus have little projection onto an extended object such as a glueball. The situation can be much improved if the links in the Wilson loops of Fig. 5.7 are replaced by so-called "smeared" or "fuzzed" links [54, 55]. For instance, the approach of [54] replaces the spatial link $U_j(x)$ by the combination

$$U_j(x) \equiv U_j^0(x) \longrightarrow \mathcal{P}\left\{U_j^0(x) + \alpha \sum_{\pm k=1, k\neq j}^{3} U_k(x)U_j(x+a\hat{k})U_k(x+a\hat{\jmath})^{-1}\right\}, \quad j = 1, 2, 3, \tag{5.95}$$

where α is a real, tunable parameter, and the symbol $\mathcal{P}$ denotes the projection back into the group manifold of SU(3). The procedure can be iterated, so that links at smearing level s, i.e. $U_j^s(x)$, are constructed from those at level $s-1$ via Eq. (5.95). One may say that smearing reduces the UV fluctuations of the gauge field, so that the smeared, extended link variables are better suited to project onto the IR regime, i.e. the long-distance properties. It should be stressed that the links in the temporal direction do *not* undergo the fuzzing procedure: Fuzzed temporal links will alter the transfer matrix and the spectral information it contains.

In order to obtain detailed information on the glueball spectrum one also seeks to determine the masses of the excited states in a given channel. This requires another level of refinement, since one normally hopes that excited state contributions die out quickly, while they now become the very focus of interest. A widely used method to gain information on the higher excitations is to construct a whole set of interpolating operators $\{O_1, \ldots, O_r\}$ in a given channel, say, A_1^{++}. This is achieved either by considering different shapes of Wilson loops that share the same transformation properties, or by applying several different smearing levels to one particular Wilson loop. Thus, each individual member of the set $\{O_1, \ldots, O_r\}$ is a perfectly valid operator in a given channel, but the projection properties, i.e. the associated spectral weights $w_\alpha^{(i)}$ for a particular state α in the spectral sum will in general be different for each member $i = 1, \ldots, r$. One then computes the matrix

$$C_{ij}(x_0) := \sum_{\vec{x}} \left\langle O_i(x) O_j^\dagger(0) \right\rangle, \qquad i, j = 1, \ldots, r, \tag{5.96}$$

whose elements consist of the correlations of all combinations of operators in the set. The diagonalization of the matrix correlator then yields the appropriate linear combination of operators which correspond to the states $\alpha = 1, 2, \ldots$ in the spectral decomposition. Diagonalization is achieved by solving the generalized eigenvalue problem

$$C_{ij}(x_0)\phi_j = \lambda_i(x_0, x_0')C_{ik}(x_0')\phi_k, \quad x_0' < x_0, \tag{5.97}$$

where ϕ denotes a vector, x_0' is fixed, and $C(x_0)$, $C(x_0')$ denote the matrix correlators taken at Euclidean times x_0 and x_0', respectively. As shown in [56], the set of eigenvalues $\lambda(x_0, x_0')$ converges rapidly towards

$$\lambda_\alpha(x_0, x_0') = \mathrm{e}^{-(x_0 - x_0')\epsilon_\alpha}, \qquad \alpha = 1, \ldots, r, \tag{5.98}$$

where ϵ_α is the mass (energy) of the state α in the spectral sum.

After all these technicalities, we now report on the status of glueball calculations. Recent results obtained in the quenched approximation were published in [53, 57–60]. In Fig. 5.8 we show the results from Ref. [57]. The three lowest-lying states are the scalar (0^{++}), tensor (2^{++}) and the 0^{-+} glueballs, whose masses are determined as

$$\begin{aligned} m_{0^{++}} &= 1710(50)(80)\,\mathrm{MeV}, \quad m_{2^{++}} = 2390(30)(120)\,\mathrm{MeV}, \\ m_{0^{-+}} &= 2560(35)(120)\,\mathrm{MeV}. \end{aligned} \tag{5.99}$$

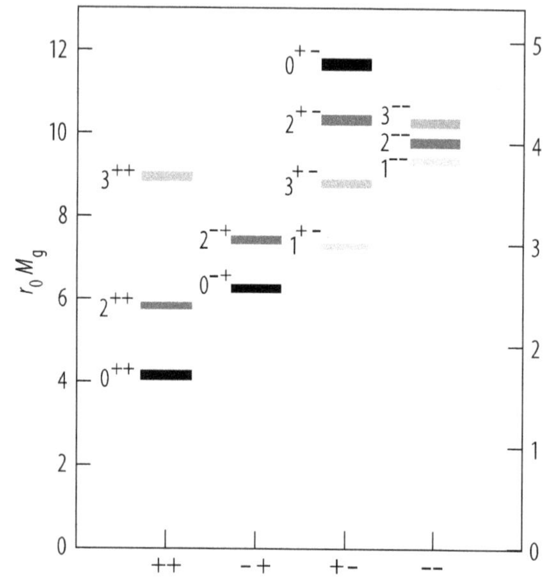

Fig. 5.8 Glueball spectrum in quenched QCD (from Ref. [57])

Here, the first error is statistical, while the second is an estimate of systematic uncertainties, which is dominated by the ambiguity in the scale setting in the quenched approximation.

While it is tempting to identify the experimentally established resonance $f_0(1710)$ as a scalar glueball in the light of the above results, the situation is more complicated. Since lattice predictions for the mass of the lightest glueballs fall into the mass range of conventional scalar mesons, mixing of glueballs with conventional $q\bar{q}$ states in conjunction with the observed decay patterns must be considered before drawing any definite conclusions. More details on the current phenomenological and experimental situation can be found in [61, 62]. So far, there have been only exploratory attempts to study glueball-meson mixing directly on the lattice. Any meaningful investigation must inevitably include dynamical quark effects, whose influence on the glueball spectrum have so far only been poorly understood.

5.4 Confinement and String Breaking

The empirical fact that quarks and gluons are not observed as free particles is commonly referred to as confinement. Since all experimentally observed states are singlets under $SU(3)_{\text{colour}}$, confinement is tantamount to saying that isolated colour charges are not allowed. A theoretical understanding of this phenomenon must inevitably go beyond the perturbative level, since QCD is a strongly coupled theory.

In Ref. [6], Wilson formulated a criterion for the confinement of colour charges known as the "area law". Let $U(\mathcal{C})$ denote the product of link variables around a

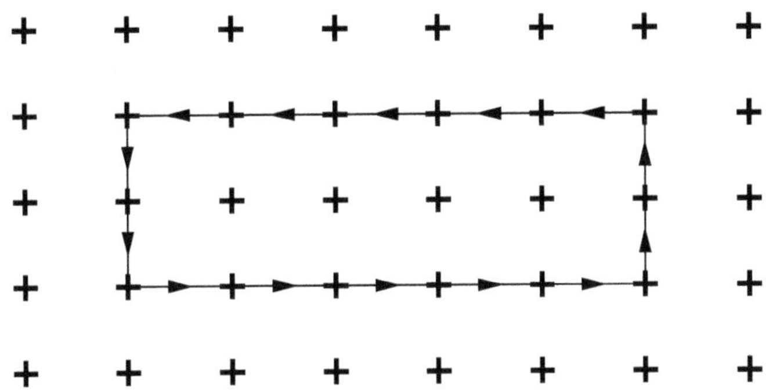

Fig. 5.9 Oriented product of link variables around a rectangle of area $r \cdot t$

closed loop $\mathcal{C}$ on a hyper-cubic lattice. The trace over colour indices is called the "Wilson loop", i.e.

$$W(\mathcal{C}) = \mathrm{tr}\,\{U(\mathcal{C})\}. \tag{5.100}$$

The area law then states that colour charges are confined if the expectation value of $W(\mathcal{C})$ decays exponentially with a rate proportional to the area $A(\mathcal{C})$ enclosed by the curve $\mathcal{C}$, i.e.

$$\langle W(\mathcal{C})\rangle \equiv \langle \mathrm{tr}\,\{U(\mathcal{C})\}\rangle \propto \mathrm{e}^{-\sigma A(\mathcal{C})}, \tag{5.101}$$

where σ is a constant. An example for a rectangular Wilson loop is shown in Fig. 5.9.

The interpretation of the area law rests on the observation that a Wilson loop of area $r{\cdot}t$ is equal to the Euclidean correlator which describes the propagation of a static, i.e. infinitely heavy, quark-antiquark pair separated by a distance r over a Euclidean time interval t. If t is taken to infinity at fixed r, the correlator yields the energy of the quark-antiquark pair:

$$\langle W(\mathcal{C})\rangle \overset{t\gg 0}{\sim} \mathrm{e}^{-V(r)t}. \tag{5.102}$$

The area law then implies $\sigma A(\mathcal{C}) = V(r)t$, and for a rectangular loop one obtains

$$V(r) \sim \sigma r. \tag{5.103}$$

Hence the energy of a static quark-antiquark pair increases linearly with the distance r. To achieve a full separation of static colour sources would therefore require an infinite amount of energy.

It has long been believed that SU(3) gauge theory is related to some kind of string theory. Heuristically, confinement may be viewed as due to the formation of a narrow tube of chromo-electric and -magnetic flux between static colour charges, the dynamics of which can be described by a string theory. The bosonic string model yields an asymptotic expansion for the static quark potential

$$V(r) = \sigma r + V_0 + \frac{c}{r} + \mathrm{O}(1/r^2), \tag{5.104}$$

where $V_0 = \text{const}$, and the universal coefficient c has been computed as [63]

$$c = -\frac{\pi}{12} \tag{5.105}$$

in the four-dimensional theory. The proportionality factor σ is called the "string tension". Instead of the potential one often considers the force, $F(r) \equiv \mathrm{d}V(r)/\mathrm{d}r$. The *ansatz* Eq. (5.104) yields

$$F(r) = \sigma - \frac{c}{r^2} + \mathrm{O}(1/r^3), \tag{5.106}$$

so that the string tension is obtained as the limiting value of the force, as $r \to \infty$,

$$\sigma = \lim_{r\to\infty} F(r). \tag{5.107}$$

String models of hadrons have been known since the late 1960s, and a phenomenological value for σ has been determined from Regge theory, $\sqrt{\sigma} = 440\,\text{MeV}$.

In QCD with light sea quarks the linear rise of the potential cannot persist for arbitrarily large distances. Instead, the creation of a light quark-antiquark pair from the vacuum will cause the hadronization of the static colour charges, leading to the formation of two static-light mesonic states. Thus, the string or flux-tube is expected to "break" when the two-meson state is energetically favoured over the linearly rising potential. The breaking of the string should set in at a characteristic value for the separation distance, r_b, causing the potential to flatten off for $r \gtrsim r_b$, since the energy of a state of two mesons is independent of their separation.

Lattice simulations have been instrumental for establishing that the area law, the string picture of confinement, as well as string breaking (i.e. hadronization) are indeed properties of SU(3) gauge theory and/or QCD. However, computations of large Wilson loops in lattice simulations suffer from the same problem encountered in glueball mass calculations: due to the strong exponential fall-off, the correlator in the asymptotic region, $r, t \to \infty$, is of the same order of magnitude than the statistical noise. Consequently, the same techniques have been applied, namely the smearing of link variables and the variational approach, which is based on the diagonalization of a matrix correlator. By combining these techniques with procedures designed to reduce statistical fluctuations [64] in the computation of large Wilson loops, one could verify the linear rise of the potential up to distances of $r \lesssim 1.5\,\text{fm}$ [65, 66] (See Fig. 5.10).

Since a phenomenological value for $\sqrt{\sigma}$ could be inferred from Regge theory, the string tension used to be a popular quantity to set the lattice scale. However, as lattice calculations became increasingly precise, it was realized that the extrapolation

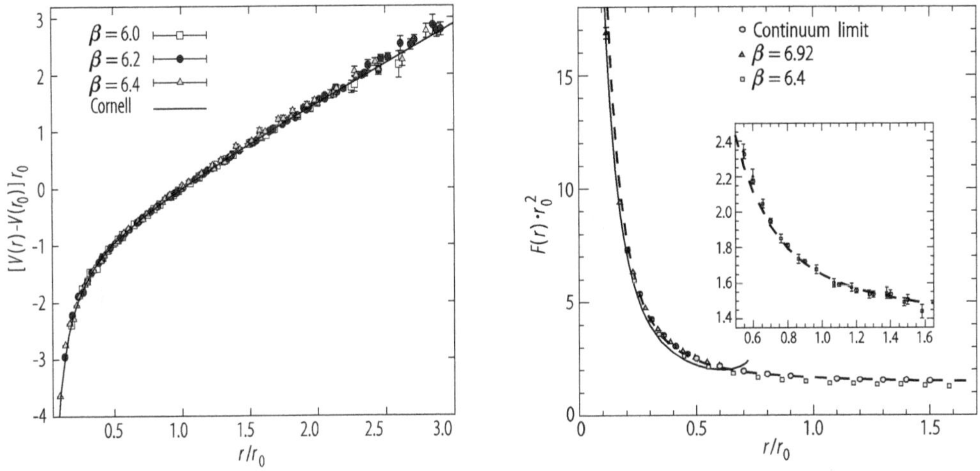

Fig. 5.10 Left panel: static quark potential in SU(3) gauge theory (from Ref. [66]). Right panel: force (from Ref. [65]) compared to the bosonic string model (dashed curve) and perturbation theory (solid curve). To compare results at different lattice spacings, all dimensionful quantities have been expressed in units of the hadronic radius $r_0 = 0.5\,\text{fm}$ (see text)

$r \to \infty$ is not easy to perform on the basis of lattice data restricted to $r \lesssim 1.5\,\text{fm}$. An alternative, conceptually much more reliable scale is obtained from the force between static colour charges [33]. The hadronic radius r_0 is defined by requiring that the force $F(r)$ evaluated at $r = r_0$ assumes a given reference value. The latter is fixed by matching $F(r)$ to phenomenological, non-relativistic potential models for heavy quarkonia. The scale r_0 is defined as the solution of

$$F(r)r^2\Big|_{r=r_0} = 1.65, \tag{5.108}$$

where the constant on the right-hand side is chosen such that r_0 has a value of $r = 0.5\,\text{fm}$ in QCD. Choosing r_0 to set the scale avoids the systematic uncertainty associated with the extrapolation of the force to infinite distance. Furthermore, r_0 remains well-defined in QCD with dynamical quarks, where string breaking must occur and the concept of a string tension as the limiting value of the force is intrinsically flawed. The quantity r_0/a has been determined numerically with good statistical accuracy over a wide range of bare couplings, corresponding to lattice spacings between $0.026 - 0.17\,\text{fm}$ [34, 65].

To test whether the bosonic string model for confinement is consistent with lattice data, one must confront the value of the Coulombic coefficient c in Eq. (5.104) with the predicted value of $c = -\pi/12$. As in the case for the string tension, such a comparison is difficult to perform reliably, since $-\pi/12$ represents the asymptotic value at infinite distance, which must be determined from data computed over a narrow range of accessible distances. Using highly accurate data for the potential $V(r)$, generated by an algorithm which allows for an exponential suppression of statistical fluctuations at large r and t, it could be shown [67] that the quantity

$$c_{\text{eff}}(r) = \frac{1}{2} r^3 \frac{\mathrm{d}^2 V(r)}{\mathrm{d}r^2} \tag{5.109}$$

indeed converges towards the predicted value of $-\pi/12$. This result confirms the string picture of confinement and suggests that string-like behaviour already sets in at rather small distances of $r \gtrsim 0.5\,\text{fm}$.

The incorporation of dynamical quarks should drastically change the string picture beyond a characteristic scale r_b, where due to $q\bar{q}$ pair creation string breaking occurs, since a two-meson state is energetically favoured over the flux-tube. However, the static quark potential determined from Wilson loops on dynamical configurations typically does not show any clear signs of flattening off, even at distances as large as 1 fm, where one expects hadronization to set in. This is attributed to the Wilson loop having little overlap onto the state of a broken string, such that the spectral weight associated with the broken string is extremely small. Therefore, extracting its energy reliably would require large Euclidean time separations, for which the statistical signal is usually lost.

It was thus proposed to address this problem by constructing a matrix correlator of Wilson loops supplemented by operators that directly project onto a two-meson state, and to consider their cross-correlations with the unbroken flux-tube. This strategy was first applied to Higgs models, i.e. non-Abelian gauge theory coupled to bosonic matter fields ("scalar QCD"), which are computationally much more efficient, whilst preserving the mechanism for string breaking to occur [68, 69]. The method was later extended to QCD with two flavours of dynamical quarks [70]. The plots in Fig. 5.11 clearly show that the ground state energy at short distances is linearly rising, while the first excited state (i.e. the two-meson state) is constant in r.

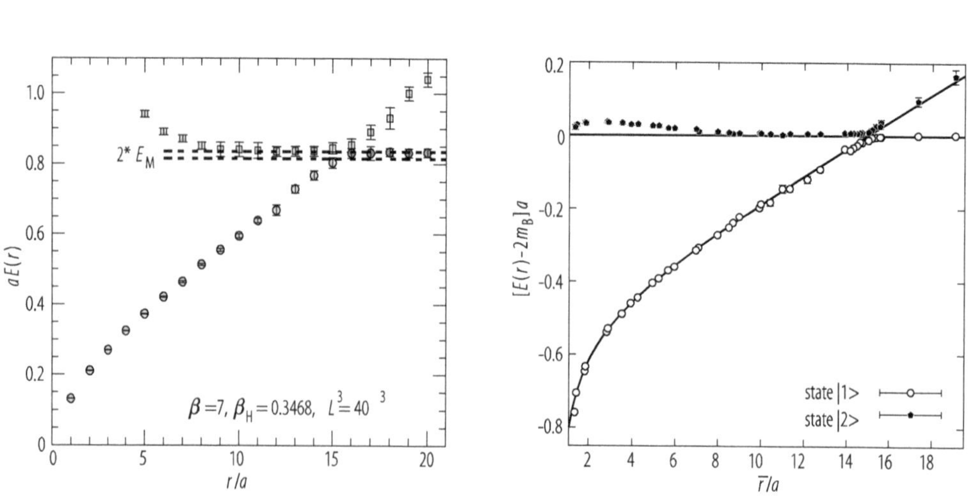

Fig. 5.11 Ground state and first excited state of the static quark potential computed using matrix correlators in the SU(2) Higgs model [68] (left panel) and QCD with $N_f = 2$ flavours of dynamical quarks [70] (right panel)

At a certain separation r_b one observes a crossing of energy levels and a continuing flat behaviour of the ground state energy. Near the crossing point one actually observes a repulsion of the energy levels, which is characteristic for the breaking phenomenon. The diagonalization of the matrix correlator also yields information on the composition of the states in the spectral decomposition. Indeed, for distances $r < r_b$ the combination of operators describing the ground state is dominated by Wilson loops, whereas for $r > r_b$, two-meson operators are the most relevant.

5.5 Fundamental Parameters of QCD

We have noted already that QCD is parameterized in terms of the gauge coupling and the masses of the quarks. In order to make predictions for cross sections, decay rates and other observables, their values must be fixed from experiment. As was discussed in detail in Sect. 4.3 , the renormalization of QCD leads to the concept of a "running" coupling constant, which depends on some momentum (energy) scale μ, and the same applies to the quark masses[9]:

$$\alpha_s(\mu) \equiv \frac{\bar{g}^2(\mu)}{4\pi}, \quad \bar{m}_u(\mu),\ \bar{m}_d(\mu),\ \bar{m}_s(\mu),\ \bar{m}_c(\mu),\ \bar{m}_b(\mu),\ \bar{m}_t(\mu). \tag{5.110}$$

The property of asymptotic freedom implies that the coupling becomes weaker as the energy scale μ is increased. This explains why the perturbative expansion of cross sections in the high-energy domain allows for an accurate determination of α_s from experimental data.

The scale dependence of the coupling and the quark masses is encoded in the renormalization group (RG) equations, which are formulated in terms of the β-function and the anomalous dimension τ,

$$\mu\frac{\partial\bar{g}(\mu)}{\partial\mu} = \beta(\bar{g}), \qquad \mu\frac{\partial\bar{m}(\mu)}{\partial\mu} = \bar{m}\tau(\bar{g}). \tag{5.111}$$

At high enough energy the RG functions β and τ admit perturbative expansions according to

$$\beta(\bar{g}) = -b_0\bar{g}^3 - b_1\bar{g}^5 + \ldots, \qquad \tau(\bar{g}) = -d_0\bar{g}^2 - d_1\bar{g}^4 + \ldots. \tag{5.112}$$

Here, b_0, b_1 and $d_0 = 8/(4\pi)^2$ are universal, while the higher coefficients depend on the adopted renormalization scheme.

[9] As usual we denote the running parameters by a bar across the symbol.

From the asymptotic scaling behaviour at high energies one can extract the fundamental scale parameter of QCD via

$$\Lambda = \lim_{\mu\to\infty}\left\{\mu(b_0\bar{g}^2)^{-b_1/2b_0^2}\,\mathrm{e}^{-1/2b_0\bar{g}^2}\right\}, \qquad \bar{g}\equiv\bar{g}(\mu). \tag{5.113}$$

Like the running coupling itself, the Λ-parameter depends on the chosen renormalization scheme.[10] A related, but less commonly used variable is the renormalization group invariant (RGI) quark mass

$$M_\mathrm{f} = \lim_{\mu\to\infty}\left\{\bar{m}_\mathrm{f}(2b_0\bar{g}^2)^{-d_0/2b_0}\right\}, \quad f=u,d,s,\ldots, \qquad \bar{m}\equiv\bar{m}(\mu). \tag{5.114}$$

Unlike Λ, the RGI quark masses are scheme-independent quantities. Instead of using the running coupling and quark masses of Eq. (5.110), one can parameterize QCD in an entirely equivalent way through the set

$$\Lambda,\ M_u,\ M_d,\ M_s,\ M_c,\ M_b,\ M_t. \tag{5.115}$$

At the non-perturbative level these quantities represent the most appropriate parameterization of QCD, since their values are defined without any truncation of perturbation theory.

The perturbative renormalization of QCD is accomplished by replacing the bare parameters with renormalized ones, whose values are fixed by considering the high-energy behaviour of Green's functions, usually computed in the $\overline{\mathrm{MS}}$-scheme of dimensional regularization. However, at low energies it is convenient to adopt a hadronic renormalization scheme, in which the bare parameters are eliminated in favour of quantities such as hadron masses and decay constants (see Sect. 5.2.4). Since QCD is expected to describe both the low- and high-energy regimes of the strong interaction, one should be able to express the quantities of Eq. (5.115), which are determined from the high-energy behaviour, in terms of hadronic quantities. In other words, by matching a hadronic renormalization scheme to a perturbative scheme like $\overline{\mathrm{MS}}$ one achieves the non-perturbative renormalization of QCD at all scales. In particular, one can express the fundamental parameters of QCD (running coupling and masses, or, equivalently, the Λ-parameter and RGI quark masses) in terms of low-energy, hadronic quantities. This amounts to predicting the values of these fundamental parameters from first principles.

[10]The expressions for b_0 and b_1, as well as the Λ-parameter have already been shown in Sect. 5.2.4.

5.5.1 Non-perturbative Renormalization

To illustrate the problem of matching hadronic and perturbative schemes like $\overline{\text{MS}}$, it is instructive to discuss the determination of the light quark masses. A convenient starting point is the PCAC relation, which for a charged kaon can be written as

$$f_{\text{K}} m_{\text{K}}^2 = (\bar{m}_u + \bar{m}_s) \langle 0|(\bar{u}\gamma_5 s)|K^+\rangle . \tag{5.116}$$

In order to determine the sum of quark masses $(\bar{m}_u + \bar{m}_s)$, using the experimentally determined values of f_{K} and m_{K}, it suffices to compute the matrix element $\langle 0|\bar{u}\gamma_5 s|K^+\rangle$ in a lattice simulation, as outlined in Sect. 5.2.3 (see Eq. (5.64)). The dependence on the renormalization scale and scheme cancels in Eq. (5.116), since the quantities on the left hand side are physical observables. Thus, in order to determine the combination $(\bar{m}_u + \bar{m}_s)$ in the $\overline{\text{MS}}$-scheme, one must compute the relation between the bare matrix element of the pseudoscalar density evaluated on the lattice and its counterpart in the $\overline{\text{MS}}$-scheme:

$$(\bar{u}\gamma_5 s)_{\overline{\text{MS}}} = Z_{\text{P}}(g_0, a\mu)(\bar{u}\gamma_5 s)_{\text{lat}}. \tag{5.117}$$

Here, μ is the subtraction point (renormalization scale) in the $\overline{\text{MS}}$-scheme. Provided that Z_{P} and the matrix element of $(\bar{u}\gamma_5 s)_{\text{lat}}$ are known, one can use Eq. (5.116) to compute $(\bar{m}_u + \bar{m}_s)/f_{\text{K}}$, which is just the ratio of a renormalized fundamental parameter expressed in terms of a hadronic quantity, up to lattice artefacts. In Fig. 5.4 we have already shown the continuum extrapolation of this ratio.[11]

The factor Z_{P} is obtained by imposing a suitable renormalization condition involving Green's functions of the pseudoscalar densities in the $\overline{\text{MS}}$ as well as the hadronic scheme. Since the $\overline{\text{MS}}$-scheme is intrinsically perturbative, in the sense that masses and couplings are only defined at a given order in the perturbative expansion, it is actually impossible to formulate such a condition at the non-perturbative level. In perturbation theory at one loop one finds

$$Z_{\text{P}}(g_0, a\mu) = 1 + \frac{g_0^2}{4\pi}\left\{\frac{2}{\pi}\ln(a\mu) + C\right\} + O(g_0^4), \tag{5.118}$$

where C is a constant that depends on the chosen discretization of the QCD action. Expressions like these are actually not very useful, since perturbation theory formulated in terms of the bare coupling g_0 converges rather slowly, so that reliable estimates of renormalization factors at one- or even two-loop order in the expansion cannot be obtained. Thus it seems that the problem of non-perturbative renormalization is severely hampered by the intrinsically perturbative

[11] The figure actually shows the ratio for the RGI quark masses, instead of those renormalized in the $\overline{\text{MS}}$-scheme.

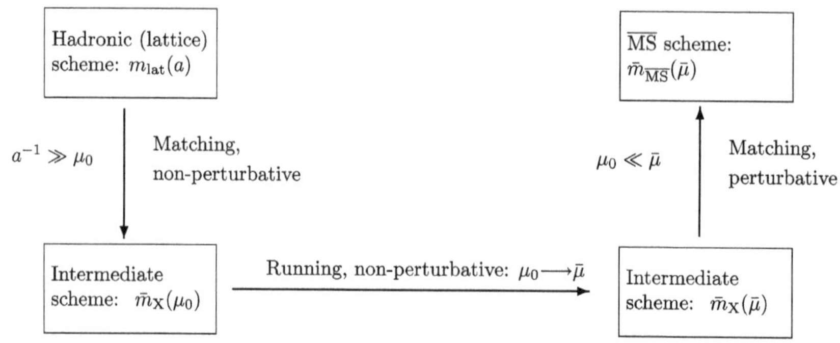

Fig. 5.12 Sketch of the matching of quark masses computed in lattice regularization and the $\overline{\text{MS}}$-scheme, via an intermediate renormalization scheme X

nature of the $\overline{\text{MS}}$ scheme in conjunction with the bad convergence properties of lattice perturbation theory.

This problem can, in fact, be resolved by introducing an intermediate renormalization scheme. Schematically, the matching procedure for the pseudoscalar density (or, equivalently, the quark mass) via such a scheme is sketched in Fig. 5.12. At low energies, corresponding to typical hadronic scales, it involves computing a non-perturbative matching relation between the hadronic and the intermediate scheme X at some scale μ_0. This matching step can be performed reliably if μ_0 is much smaller than the regularization scale a^{-1}. In the following step one computes the scale dependence within the intermediate scheme non-perturbatively from μ_0 up to a scale $\bar{\mu} \gg \mu_0$, which is large enough so that perturbation theory can be safely applied. At that point one may then determine the matching relation to the $\overline{\text{MS}}$-scheme perturbatively. Alternatively, one can continue to compute the scale dependence within the intermediate scheme to infinite energy via a numerical integration of the perturbative RG functions. According to Eq. (5.114) this yields the relation to the RGI quark mass. Since the latter is scale- and scheme-independent, one can use directly the perturbative RG functions, which in the $\overline{\text{MS}}$-scheme are known to four-loop order [71], to compute the relation to $\bar{m}_{\overline{\text{MS}}}$ at some chosen reference scale. By applying this procedure, the direct perturbative matching between between the hadronic and $\overline{\text{MS}}$-schemes (upper two boxes in Fig. 5.12), using the expression in Eq. (5.118) is thus completely avoided.

Decay constants of pseudoscalar mesons provide another example for which the renormalization of local operators is a relevant issue. For instance, the kaon decay constant is defined by the matrix element of the axial current, i.e.

$$f_{\text{K}} m_{\text{K}} = \left\langle 0 \left| (\bar{u}\gamma_0\gamma_5 s)(0) \right| K^+ \right\rangle. \tag{5.119}$$

If the matrix element on the right hand side is evaluated in a lattice simulation, then the axial current in the discretized theory must be related to its counterpart in the continuum via a renormalization factor Z_{A}:

$$(\bar{u}\gamma_0\gamma_5 s) = Z_{\text{A}}(g_0)(\bar{u}\gamma_0\gamma_5 s)_{\text{lat}}. \tag{5.120}$$

Normally one would expect that the chiral Ward identities ensure that the axial current does not get renormalized. However, this no longer applies if the discretization conflicts with the symmetries of the classical action. This is clearly the case for Wilson fermions, which break chiral symmetry, such that the resulting short-distance corrections must be absorbed into a renormalization factor Z_A. Similar considerations apply to the vector current: if the discretization does not preserve chiral symmetry, current conservation is only guaranteed if the vector current is suitably renormalized by a factor Z_V, which must be considered even in the massless theory. Unlike the case of the renormalization factor of the pseudoscalar density, Z_A and Z_V are scale-independent, i.e. they only depend on the bare coupling g_0. From the above discussion it is obvious that perturbative estimates of Z_A and Z_V are inadequate in order to compute hadronic matrix elements of the axial and vectors currents with controlled errors. A non-perturbative determination of Z_A and Z_V can be achieved by imposing the chiral Ward identities as a renormalization condition.

Two widely used intermediate schemes, namely the *Schrödinger functional* (SF) and the *Regularization independent momentum subtraction* (RI/MOM) schemes are briefly reviewed in the following. We strongly recommend that the reader consult the original articles (Refs. [72–75] for the SF, and [76] for RI/MOM) for further details.

5.5.2 Finite Volume Scheme: The Schrödinger Functional

The Schrödinger functional is based on the formulation of QCD in a finite volume of size $L^3 \cdot T$—regardless of whether space-time is discretized or not—with suitable boundary conditions. Assuming that lattice regularization is employed, one imposes periodic boundary conditions on the fields in all spatial directions, while Dirichlet boundary conditions are imposed at Euclidean times $x_0 = 0$ and $x_0 = T$. In order to make this more precise, let C and C' denote classical configurations of the gauge potential. For the link variables at the temporal boundaries one then imposes

$$U_k(x)|_{x_0=0} = \mathrm{e}^{aC}, \qquad U_k(x)|_{x_0=T} = \mathrm{e}^{aC'}. \tag{5.121}$$

In other words, the links assume prescribed values at the temporal boundaries, but remain unconstrained in the bulk (see Fig. 5.13).

Quark fields are easily incorporated into the formalism. Since the Dirac equation is first order, only two components of a full Dirac spinor can be fixed at the boundaries. By defining the projection operator $P_\pm = \frac{1}{2}(1 \pm \gamma_0)$, one requires that the quark fields at the boundaries satisfy

$$\begin{aligned} P_+\psi(x)|_{x_0=0} &= \rho(\vec{x}), \quad & P_-\psi(x)|_{x_0=T} &= \rho'(\vec{x}), \\ \bar{\psi}(x)P_-\big|_{x_0=0} &= \bar{\rho}(\vec{x}), \quad & \bar{\psi}(x)P_+\big|_{x_0=T} &= \bar{\rho}'(\vec{x}), \end{aligned} \tag{5.122}$$

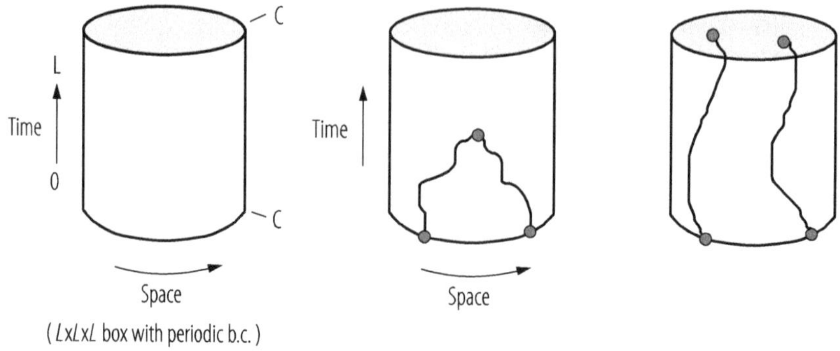

Fig. 5.13 Left panel: sketch of the SF geometry, indicating the classical gauge potentials at the temporal boundaries. Middle panel: correlation function of boundary quark fields $\zeta, \bar{\zeta}$ with a fermionic bilinear operator in the bulk. Right panel: boundary-to-boundary correlation function

where $\rho, \ldots, \bar{\rho}'$ denote prescribed values of the fields. The functional integral over all dynamical fields in a finite volume with the above boundary conditions is called the Schrödinger functional of QCD:

$$\mathcal{Z}[C', \rho', \bar{\rho}'; C, \rho, \bar{\rho}] = \int D[U]D[\bar{\psi}, \psi]\, \mathrm{e}^{-S}. \tag{5.123}$$

The classical field configurations at the boundaries are not integrated over. Using the transfer matrix formalism, one can show that this expression is the quantum mechanical amplitude for going from the classical field configuration $\{C, \rho, \bar{\rho}\}$ at $x_0 = 0$ to $\{C', \rho', \bar{\rho}'\}$ at $x_0 = T$.

Functional derivatives with respect to $\rho, \ldots, \bar{\rho}'$ behave like quark fields located at the temporal boundaries, and hence one may identify

$$\zeta(\vec{x}) = \frac{\delta}{\delta\bar{\rho}(\vec{x})}, \quad \bar{\zeta}(\vec{x}) = -\frac{\delta}{\delta\rho(\vec{x})}, \quad \zeta'(\vec{x}) = \frac{\delta}{\delta\bar{\rho}'(\vec{x})}, \quad \bar{\zeta}'(\vec{x}) = -\frac{\delta}{\delta\rho'(\vec{x})}. \tag{5.124}$$

The boundary fields $\zeta, \bar{\zeta}, \ldots$ can be combined with local composite operators (such as the axial current or the pseudoscalar density) of fields in the bulk to define correlation functions. Particular examples are the correlation function of the pseudoscalar density, f_{P} and the boundary-to-boundary correlation f_1

$$f_{\mathrm{P}}(x_0) = -\frac{a^6}{3}\sum_{\vec{y},\vec{z}} \left\langle \bar{\psi}(x)\gamma_5 \tfrac{1}{2}\tau^a \psi(x)\bar{\zeta}(\vec{y})\gamma_5 \tfrac{1}{2}\tau^a \zeta(\vec{z})\right\rangle,$$

$$f_1 = -\frac{a^{12}}{3L^6}\sum_{\vec{u},\vec{v},\vec{y},\vec{z}} \left\langle \bar{\zeta}'(\vec{u})\gamma_5 \tfrac{1}{2}\tau^a \zeta'(\vec{v})\bar{\zeta}(\vec{y})\gamma_5 \tfrac{1}{2}\tau^a \zeta(\vec{z})\right\rangle, \tag{5.125}$$

which are shown schematically in the middle and right panels of Fig. 5.13. In the above expressions, the Pauli matrices act on the first two flavour components of the fields.

The specific boundary conditions of the Schrödinger functional ensure that the Dirac operator has a minimum eigenvalue proportional to $1/T$ in the massless case [73]. As a consequence, renormalization conditions can be imposed at vanishing quark mass. If the aspect ratio T/L is set to some fixed value, the spatial length L is the only scale in the theory, and thus the masses and couplings in the SF scheme run with the box size. The recursive finite-size scaling study described below can then be used to map out the scale dependence of running quantities non-perturbatively from low to high energies. It is important to realize that in this way the relevant scale for the RG running (the box size L) is decoupled from the regularization scale (the lattice cutoff a). It is this features which ensures that the running of masses and couplings can be obtained in the continuum limit.

Let us now return to our earlier example of the renormalization of quark masses. The transition from lattice regularization and the associated hadronic scheme to the SF scheme is achieved by computing the scale-dependent renormalization factor which links the pseudoscalar density in the intermediate scheme to the bare one, i.e.

$$(\bar{s}\gamma_5 u)_{\rm SF}(\mu_0) = Z_{\rm P}(g_0, a\mu_0)\,(\bar{s}\gamma_5 u)_{\rm lat}(a). \tag{5.126}$$

A renormalization condition that defines $Z_{\rm P}$ can be formulated in terms of SF correlation functions:

$$Z_{\rm P}(g_0, a\mu_0) = c\,\left.\frac{\sqrt{f_1}}{f_{\rm P}(x_0)}\right|_{x_0=T/2}, \qquad \mu_0 = 1/L_{\rm max}, \tag{5.127}$$

where the constant c must be chosen such that $Z_{\rm P} = 1$ in the free theory. In order to determine the RG running of the quark mass non-perturbatively one can perform a sequence of finite-size scaling steps, as illustrated in Fig. 5.14. To this end one simulates pairs of lattices with box lengths L and $2L$, at fixed lattice spacing a. The ratio of $Z_{\rm P}$ evaluated for each box size yields the ratio $\bar{m}_{\rm SF}(L)/\bar{m}_{\rm SF}(2L)$ (upper horizontal step in Fig. 5.14), which amounts to the change in the quark mass when the volume is scaled by a factor 2. In a subsequent step, the physical volume can be doubled once more, which gives $\bar{m}_{\rm SF}(2L)/\bar{m}_{\rm SF}(4L)$. The important point to realize is that the lattice spacing can be adjusted for a given physical box size. In this way the number of lattice sites can be kept at a manageable level, while the physical volume is gradually scaled over several orders of magnitude, as indicated by the zig-zag pattern in Fig. 5.14. Furthermore, each horizontal step can be performed for several lattice resolutions, so that the continuum limit can be taken. By contrast, if one attempted to scale the physical volume for fixed lattice spacing, one would, after only a few iterations, end up with systems so large that they would not fit into any computer's memory.

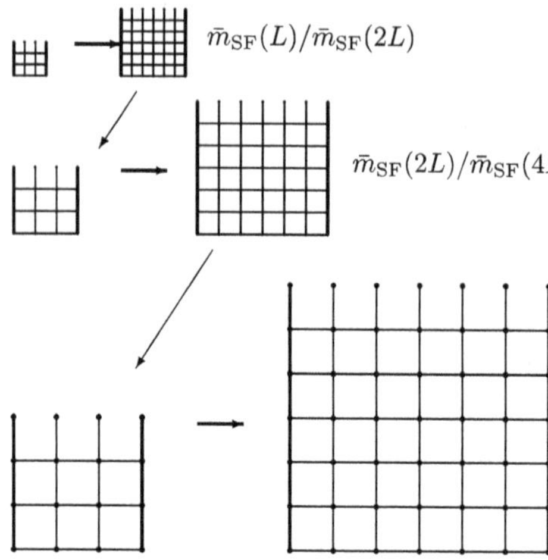

Fig. 5.14 Illustration of the recursive finite-size scaling procedure to determine the running of $\bar{m}(L)$ for $L \to 2L \to 4L \to 8L$. In any horizontal step L is scaled by a factor 2 for fixed lattice spacing a. In every diagonal shift one keeps the physical box size L fixed and increases a by an appropriate tuning of the bare coupling g_0

In an entirely analogous fashion one can set up the finite-size scaling procedure for the running coupling constant in the SF scheme, $\bar{g}_{\rm SF}(L)$.[12] Setting a value for the coupling actually corresponds to fixing the box size L, since the renormalization scale and the coupling in a particular scheme are in one-to-one correspondence. The sequence of scaling steps begins at the matching scale $\mu_0 = 1/L_{\rm max}$ between the hadronic and SF schemes, and in order to express the scale evolution in physical units, the maximum box size $L_{\rm max}$ must be determined in terms of some hadronic quantity, such as f_π or r_0. In typical applications of the method, $L_{\rm max}$ corresponds to an energy scale of about 250 MeV. After n steps, the box size has decreased by a factor 2^n (typically $n = 7-9$), and at this point one is surely in the regime where the perturbative approximations to the RG functions are reliable enough to extract the Λ-parameter (in the SF scheme) and the RGI quark masses according to Eqs. (5.113) and (5.114). The transition to the $\overline{\rm MS}$-scheme is easily performed, since the ratios $\Lambda_{\rm SF}/\Lambda_{\overline{\rm MS}}$, as well as $\bar{m}_{\overline{\rm MS}}/M$ are computable in perturbation theory. At that point one has completed the steps in Fig. 5.12, and all reference to the intermediate SF scheme has dropped out in the final result.

As examples we show the running coupling and quark mass in the SF scheme from actual simulations of lattice QCD for $N_{\rm f} = 2$ flavours of dynamical quarks in Fig. 5.15. The numerical data points in these plots originate from simulations with two flavours of O(a)-improved Wilson fermions and have been extrapolated to the continuum limit.

[12]The precise definition of $\bar{g}_{\rm SF}$ is specified in Sect. 5.5.5 below.

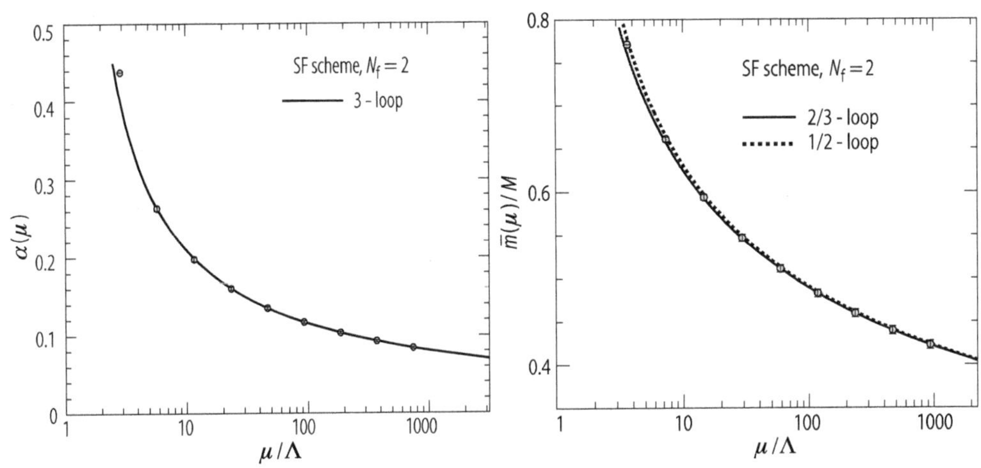

Fig. 5.15 Running of α_s (left panel) [77] and quark mass in units of the RGI mass M (right panel) [78] in the SF scheme. The results from simulations (full circles) are compared to the integration of the perturbative RG equations

5.5.3 Regularization-Independent Momentum Subtraction Scheme

An alternative choice of intermediate renormalization scheme is based on imposing renormalization conditions in terms of Green's functions of external quark states in momentum space, evaluated in a fixed gauge (e.g. Landau gauge) [76]. The external quark fields are off-shell, and their virtualities are identified with the momentum scale. Here we summarize the basic steps in this procedure by considering a quark bilinear non-singlet operator $O_\Gamma = \bar{\psi}_1\Gamma\psi_2$, where Γ denotes a generic Dirac structure, e.g. $\Gamma = \gamma_5$ in the case of the pseudoscalar density. The corresponding renormalization factor Z_Γ is fixed by requiring that a suitably chosen renormalized vertex function $\Lambda_{\Gamma,\mathrm{R}}(p)$ be equal to its tree-level counterpart:

$$\Lambda_{\Gamma,\mathrm{R}}(p)\big|_{p^2=\mu^2} = Z_\Gamma Z_\psi^{-1}\Lambda_\Gamma(p)\Big|_{p^2=\mu^2} = \Lambda_{\Gamma,0}(p). \tag{5.128}$$

This condition defines Z_Γ up to quark field renormalization. Such a prescription can be formulated in any chosen regularization, which is why the method is said to define a regularization-independent momentum subtraction (RI/MOM) scheme. However, Z_Γ does depend on the external states and the gauge.

In order to connect to our previous example of the renormalization of quark fields, we consider the pseudoscalar density for concreteness: $\Gamma = \gamma_5 =$ "P". In this case, $\Lambda_{\mathrm{P},0} = \gamma_5 \otimes \mathbb{1}_{\mathrm{colour}}$, and Eq. (5.128) can be cast into the form

$$Z_\mathrm{P}^{\mathrm{MOM}}(g_0, a\mu)\, Z_\psi^{-1}(g_0, ap)\frac{1}{12}\mathrm{Tr}\,\{\Lambda_\mathrm{P}(p)\gamma_5\}\Big|_{p^2=\mu^2} = 1, \tag{5.129}$$

where the trace is taken over Dirac and colour indices.

In practice, the unrenormalized vertex function $\Lambda_{\rm P}(p)$ is obtained by computing the quark propagator in a fixed gauge in momentum space and using it to amputate the external legs of the Green's function of the operator in question, evaluated between quark states, i.e.

$$\Lambda_{\rm P}(p) = S(p)^{-1}\, G_{\rm P}(p)\, S(p)^{-1}, \quad S(p) = \int {\rm d}^4x\, {\rm e}^{-{\rm i}px}\, \langle S(x,0)\rangle\,,$$

$$G_{\rm P}(p) = \int {\rm d}^4x\, {\rm d}^4y {\rm e}^{-{\rm i}p(x-y)} \left\langle \psi_1(x)\left(\bar\psi_1(0)\gamma_5\psi_2(0)\right)\bar\psi_2(y)\right\rangle. \qquad (5.130)$$

The quark field renormalization constant $Z_\psi^{1/2}$ can be fixed, e.g. via the vertex function of the vector current[13]:

$$Z_\psi = \frac{1}{48}{\rm Tr}\left\{\Lambda_{V_\mu^{\rm C}}(p)\gamma_\mu\right\}\Big|_{p^2=\mu^2}. \qquad (5.131)$$

The numerical evaluation of the Green's function and quark propagators in momentum space is performed on a finite lattice with periodic boundary conditions. Unlike the situation encountered in the Schrödinger functional, there is thus no additional infrared scale, so that the renormalization conditions cannot be evaluated directly at vanishing bare quark mass. A chiral extrapolation is then required to determine mass-independent renormalization factors.

Equation (5.128) is also imposed to define the subsequent matching of the RI/MOM and $\overline{\rm MS}$ schemes. In this case, the unrenormalized vertex function on the left-hand side is evaluated to a given oder in perturbation theory, using the $\overline{\rm MS}$-scheme of dimensional regularization. For a generic quark bilinear this yields the factor $Z_\Gamma^{\overline{\rm MS}}(\bar g_{\overline{\rm MS}}(\mu))$. In our specific example of the pseudoscalar density operator in the PCAC relation, Eq. (5.116), the transition between the RI/MOM and $\overline{\rm MS}$ schemes is provided by

$$(\bar u\gamma_5 s)_{\overline{\rm MS}}(\bar\mu) = R_{\rm P}(\bar\mu/\mu) Z_{\rm P}^{\rm MOM}(g_0, a\mu)(\bar u\gamma_5 s)_{\rm lat}(a). \qquad (5.132)$$

The ratio $R_{\rm P}$ admits a perturbative expansion in terms of the coupling in the $\overline{\rm MS}$-scheme, i.e.

$$R_{\rm P}(\bar\mu/\mu) \equiv \frac{Z_{\rm P}^{\overline{\rm MS}}(\bar g_{\overline{\rm MS}}(\bar\mu))}{Z_{\rm P}^{\rm MOM}(g_0, a\mu)} = 1 + R_{\rm P}^{(1)}\bar g_{\overline{\rm MS}}^2 + {\rm O}(\bar g_{\overline{\rm MS}}^4), \qquad (5.133)$$

[13]In this expression $V_\mu^{\rm C}$ denotes the conserved lattice vector current, which involves quark fields at neighbouring lattice sites, and which is known not to undergo any finite renormalization, such that $Z_{\rm V} \equiv 1$.

which is not afflicted with the bad convergence properties encountered in the direct matching of hadronic and $\overline{\text{MS}}$-schemes. Finally, for the whole method to work, one must be able to fix the virtualities μ of the external fields such that

$$\Lambda_{\text{QCD}} \ll \mu \ll 1/a. \tag{5.134}$$

In other words, the method relies on the existence of a "window" of scales in which lattice artefacts in the numerical evaluation are controlled, $\mu \ll 1/a$, and where μ is also large enough such that the perturbative matching to the $\overline{\text{MS}}$ scheme can be performed reliably. In the ideal situation one expects that the dependence of $Z_\Gamma^{\text{MOM}}(g_0, a\mu)$ on the virtuality μ inside the "window" is well described by the perturbative RG function.

The RI/MOM prescription is a flexible method to introduce an intermediate renormalization scheme and can easily be adapted to a range of operators and lattice actions. In particular, the extension to discretizations of the quark action based on the Ginsparg-Wilson relation is straightforward. This contrasts with the situation encountered in the Schrödinger functional, where extra care must be taken to ensure that imposing Schrödinger functional boundary conditions is compatible with the Ginsparg-Wilson relation [79–81]. On the other hand, the non-perturbative scale evolution, for which the Schrödinger functional is tailored, is not so easy to incorporate into the RI/MOM framework. Hence, the matching between RI/MOM and $\overline{\text{MS}}$ schemes is usually performed at fairly low scales, i.e. $\bar{\mu} = \mu_0$ in the notation of Fig. 5.12. Furthermore, the accessible momentum scales in the matching of hadronic and RI/MOM schemes are typically quite narrow, i.e. $a\mu_0 \approx 1$. Special care must also be taken when one considers operators that couple to the pion, such as the pseudoscalar density. In this case the vertex function receives a contribution from the Goldstone pole, which for $p \equiv \mu = 0$ diverges in the limit of vanishing quark mass. The fact that the chiral limit is ill-defined may spoil a reliable determination of the renormalization factor, in particular when the accessible "window" is narrow such that μ cannot be set to large values.

5.5.4 *Mean-Field Improved Perturbation Theory*

Another widely used strategy is to avoid the introduction of an intermediate renormalization scheme altogether and attempt the direct, perturbative matching between hadronic and $\overline{\text{MS}}$ schemes via an effective resummation of higher orders in the expansion. In this sense one regards the bare coupling and masses as parameters that run with the cutoff scale a^{-1}.

The bad convergence properties of perturbative expansions such as Eq. (5.118) has been attributed to the presence of large gluonic tadpole contributions in the relation between the link variable $U_\mu(x)$ and the continuum gauge potential $A_\mu(x)$. It was already suggested by Parisi [82] that the convergence of lattice perturbation theory could be accelerated by replacing the bare coupling g_0^2 by an "improved" coupling $\tilde{g}^2 \equiv g_0^2/u_0^4$, where u_0^4 denotes the average plaquette:

$$u_0^4 = \tfrac{1}{3}\mathrm{Re}\,\langle \mathrm{tr}\, P\rangle, \qquad P \equiv \frac{1}{6}\sum_{\mu,\nu,\nu<\mu} P_{\mu\nu}. \tag{5.135}$$

A more systematic extension of the idea of setting up such a "tadpole" or "mean-field" improved version of lattice perturbation theory was presented in Ref. [83]. The main strategy is to factor out tadpole contributions through a redefinition of the link variable:

$$U_\mu(x) \to \tilde{U}_\mu(x) \equiv U_\mu(x)/u_0, \tag{5.136}$$

where u_0 is the average link, defined e.g. via the average plaquette. A factor of u_0 is then absorbed into the normalization of the quark fields. According to [83], the mismatch between non-perturbative estimates for u_0 and its expression in lattice perturbation theory can be used to improve the convergence properties of lattice perturbation theory via a relative rescaling of quark fields in the continuum and lattice formulations. To make this more explicit, we consider Wilson fermions (see Sect. 5.2.2). Factoring out the average link u_0 modifies the quark field normalization of Eq. (5.36) according to

$$\psi^{\mathrm{cont}}(x) = \sqrt{2\kappa u_0}\,\psi(x), \qquad \bar{\psi}^{\mathrm{cont}}(x) = \bar{\psi}(x)\sqrt{2\kappa u_0}. \tag{5.137}$$

The general expression for the perturbative expansion of Z_P in powers of the bare coupling reads

$$Z_\mathrm{P}(g_0, a\mu) = 1 + g_0^2 Z_\mathrm{P}^{(1)}(a\mu) + \mathrm{O}(g_0^4), \tag{5.138}$$

where $Z_\mathrm{P}^{(1)}(a\mu)$ denotes the one-loop expansion coefficient. The convergence of Eq. (5.138) can be accelerated by dividing out u_0 in the rescaling factors of the quark and antiquark fields using its perturbative expansion and replacing it by its non-perturbative estimate computed in simulations. In other words, the rescaling of the quark fields is exploited to divide out the relative mismatch between the perturbative and non-perturbative estimates for the average link in expressions like Eq. (5.138):

$$1 = u_0(u_0)^{-1} \simeq u_0\left\{1 - u_0^{(1)} g_0^2 + \mathrm{O}(g_0^4)\right\}, \tag{5.139}$$

where the one-loop coefficient $u_0^{(1)} = -1/12$ for the average plaquette. In this way, i.e. by combining non-perturbatively determined values for u_0 with its perturbative expansion, and after replacing the bare coupling by $\tilde{g}^2$, one arrives at the mean-field improved version of Eq. (5.138), viz.

$$Z_\mathrm{P}^{\mathrm{mf}} = u_0\left\{1 + \left[Z_\mathrm{P}^{(1)}(a\mu) - u_0^{(1)}\right]\tilde{g}^2\right\}. \tag{5.140}$$

Instead of Parisi's "boosted" coupling $\tilde{g}$ other expansion parameters have been suggested, which are expected to accelerate the convergence of the perturbative series [83]. While mean-field improvement is a general procedure, which is easily adapted to a wide range of actions and operators, it is difficult to estimate the effectiveness of the resummation and, in turn, the size of higher-order corrections. Also, a principal problem is the identification of the running scale with the cutoff, since it is difficult to separate renormalization effects from lattice artefacts.

5.5.5 The Running Coupling from the Lattice

Having discussed the non-perturbative renormalization of QCD in detail, we shall now present results for the running coupling constant, α_s, from two different approaches. This complements the discussion in Sect. 4.6, where determination of α_s from experimental data has been described in detail. Any lattice calculation of α_s proceeds along the following steps:

1. A non-perturbative definition of the coupling must be provided in terms of some quantity which can be evaluated in lattice simulations with high precision. This amounts to specifying the running coupling in a particular renormalization scheme, $\alpha_{\mathrm{X}}(a\mu_0)$, which can be related to the $\overline{\mathrm{MS}}$ scheme of dimensional regularization.
2. *Scale setting:* the matching to a hadronic scheme is performed via the calibration of the lattice spacing, which yields the scale μ_0 at which α_{X} is evaluated in units of some physical quantity Q:

$$\mu_0\,[\mathrm{MeV}] = (a\mu_0)\cdot a^{-1}\,[\mathrm{MeV}] = (a\mu_0)\cdot\frac{Q\,[\mathrm{MeV}]}{(aQ)}. \tag{5.141}$$

3. *Running and matching:* provided that the energy scale at which α_{X} has been determined is large enough, one can use perturbation theory to relate α_{X} to the coupling in the $\overline{\mathrm{MS}}$ scheme, e.g.

$$\alpha_{\overline{\mathrm{MS}}}(\bar{\mu}) = \alpha_{\mathrm{X}}(\mu) + c_{\mathrm{X}}^{(1)}(\bar{\mu}/\mu)\alpha_{\mathrm{X}}(\mu)^2 + \dots. \tag{5.142}$$

4. The Λ-parameter can be determined from the asymptotic behaviour of α_{X} via Eq. (5.113).

The attentive reader has surely noticed that the above steps follow closely the general strategy for non-perturbative renormalization via an intermediate renormalization scheme outlined in Sect. 5.5.1 and Fig. 5.12.

First we discuss the determination of α_s from the Schrödinger functional. The definition of the running coupling is somewhat technical in this case. The starting point is the effective action of Eq. (5.123); the classical field configurations at the boundaries at $x_0 = 0,\ T$ can be parameterized in terms of a real variable η:

$$C = C(\eta), \qquad C' = C'(\eta). \tag{5.143}$$

For explicit expressions we refer the reader to the original article [84]. The associated effective action is defined by

$$\Gamma(\eta) = -\ln \mathcal{Z}[C'(\eta), 0, 0;\ C(\eta), 0, 0] \tag{5.144}$$

and admits a perturbative expansion in terms of the bare coupling g_0, viz.

$$\Gamma(\eta) = \frac{1}{g_0^2}\Gamma_0 + \Gamma_1 + g_0^2\Gamma_2 + \ldots. \tag{5.145}$$

A renormalized coupling can then be defined in terms of the effective action via

$$\frac{1}{\bar{g}_{\rm SF}^2(L)} = \left\{\frac{\partial}{\partial\eta}\Gamma(\eta) \Big/ \frac{\partial}{\partial\eta}\Gamma_0(\eta)\right\}_{\eta=0,\, m=0}. \tag{5.146}$$

This definition is imposed at vanishing quark mass, $m = 0$, and provided that the aspect ratio T/L has been fixed, the spatial dimension is the only scale in the theory, such that $\bar{g}_{\rm SF}(L)$ runs with the box size L. From the perturbative expansion of $\Gamma(\eta)$ one easily infers that $\bar{g}_{\rm SF}^2(L) = g_0^2$ at tree level. The quantity on the right-hand side is given in terms of plaquettes attached to the SF boundaries and can be computed with good statistical precision.

If $L_{\rm max}$ denotes the largest box size for which $\bar{g}_{\rm SF}$ is computed, then the scale is set by expressing $L_{\rm max}$ in terms of some known dimensionful quantity, for instance, by computing the combination $L_{\rm max}/r_0$ in the continuum limit and using $r_0 = 0.5\,$fm.

The finite-size scaling procedure described earlier in Sect. 5.5.1 allows to compute the scale evolution of $\bar{g}_{\rm SF}$ over several orders of magnitude. In particular, each of the horizontal steps in Fig. 5.14 can be repeated for several values of the lattice spacing, so that the continuum limit is reached by taking $a/L \to 0$ for fixed physical box size L. The resulting scale evolution of $\alpha_{\rm SF} \equiv \bar{g}_{\rm SF}^2/4\pi$ is shown in Fig. 5.15 and compared to the perturbative evolution. Although the non-perturbatively determined points are described very well by perturbation theory, using the three-loop expression for the RG function, one should realize that this behaviour may be specific to the SF scheme and should not be generalized to other schemes.

Starting from $\mu_0 = 1/L_{\text{max}}$ one obtains the coupling at $\mu = 2^9/L_{\text{max}}$ after nine steps in the scaling procedure. At that point one can extract the Λ-parameter by evaluating the exact expression

$$\Lambda_{\text{SF}} = \mu \left(b_0 \bar{g}^2(\mu)\right)^{-b_1/(2b_0^2)} \mathrm{e}^{-1/(2b_0\bar{g}^2(\mu))} \exp\left\{-\int_0^{\bar{g}(\mu)} dx \left[\frac{1}{\beta(x)} + \frac{1}{b_0 x^3} - \frac{b_1}{b_0^2 x}\right]\right\}, \tag{5.147}$$

where $\mu = 2^9/L_{\text{max}}$. The integral can be computed using the three-loop approximation to the RG β-function in the SF scheme. Equation (5.147) yields the combination $\Lambda_{\text{SF}} L_{\text{max}}$, and knowledge of L_{max} in physical units allows to express the Λ-parameter in MeV. Conversion to the $\overline{\text{MS}}$ scheme is easily achieved, since the ratio of Λ-parameters in two different schemes is computable via a one-loop calculation in which $\bar{g}^2_{\overline{\text{MS}}}$ is expanded in powers of $\bar{g}^2_{\text{SF}}$. This gives

$$\Lambda_{\overline{\text{MS}}} = \Lambda_{\text{SF}} \cdot c_\Lambda. \tag{5.148}$$

The entire procedure of determining the Λ-parameter via the Schrödinger functional has so far been carried out for the pure SU(3) gauge theory ($N_{\text{f}} = 0$) and for QCD with two flavours of dynamical quarks. The values of the coefficient c_Λ are 2.04872(4) for $N_{\text{f}} = 0$ [84] and $c_\Lambda = 2.382035(3)$ for $N_{\text{f}} = 2$ [85], and the resulting values for $\Lambda_{\overline{\text{MS}}}$ are [75, 77]

$$\begin{aligned} \Lambda^{(0)}_{\overline{\text{MS}}} r_0 &= 0.602 \pm 0.048 &&\Leftrightarrow \Lambda^{(0)}_{\overline{\text{MS}}} = 238 \pm 19\,\text{MeV} \\ \Lambda^{(2)}_{\overline{\text{MS}}} r_0 &= 0.62 \pm 0.04 \pm 0.04 &&\Leftrightarrow \Lambda^{(2)}_{\overline{\text{MS}}} = 245 \pm 16 \pm 16\,\text{MeV}, \end{aligned} \tag{5.149}$$

where $r_0 = 0.5\,\text{fm}$ is used to convert into physical units. There is room for improvement in several respects: for $N_{\text{f}} = 2$ the extrapolation to the continuum limit can be made more reliable by including simulations at smaller lattice spacings, which should reduce the first of the two quoted errors. Also, the conversion into physical units should be performed in terms of a quantity such as f_π, which is directly accessible in experiment. Finally, the calculation must be repeated with more dynamical quark flavours, in order to allow for a direct comparison with phenomenology, since all experimental determinations yield the Λ-parameter for $N_{\text{f}} = 4$ or 5 quark flavours.

The determination of α_s and $\Lambda_{\overline{\text{MS}}}$ via the Schrödinger functional is quite involved. However, it is the only method so far, which allows to map out the running of α_s in a completely non-perturbative manner, including the systematic elimination of lattice artefacts. In particular, perturbation theory is used only for energy scales well above 50 GeV.

The second method that we will discussed here in some detail is the determination of α_s via heavy quarkonia. Below we present an account of the calculation published in [86]. Here, the dynamical quark effects of the light (u, d, s) quarks have been accounted for in simulations with improved staggered quarks employing the fourth-root trick (see Sect. 5.2.6). In this approach, the coupling constant is defined in the so-called "V-scheme" via the heavy quark potential in momentum space:

$$V(q) = -C_{\mathrm{F}}\frac{4\pi}{q^2}\,\alpha_{\mathrm{V}}(q). \tag{5.150}$$

Small Wilson loops such as the plaquette can be expanded in powers of α_{V}

$$-\ln\tfrac{1}{3}\,\langle \mathrm{Re\,tr}\,P\rangle = c_{\mathrm{P}}^{(1)}\alpha_{\mathrm{V}}(s_{\mathrm{P}}/a) + c_{\mathrm{P}}^{(2)}\,[\alpha_{\mathrm{V}}(s_{\mathrm{P}}/a)]^2 + \ldots, \tag{5.151}$$

where s_{P} is a real dimensionless variable which can be chosen to optimize the convergence properties of the expansion [83]. Equation (5.151) thus provides the link between the coupling and a quantity that is easily computed in lattice simulations. The above expression can be generalized to (small) rectangular Wilson loops W_{rt} with area $r \cdot t$:

$$-\ln\tfrac{1}{3}\,\langle W_{rt}\rangle = \sum_{k=0}^{\infty} c_{rt}^{(k)}\,[\alpha_{\mathrm{V}}(s_{\mathrm{rt}}/a)]^k\,. \tag{5.152}$$

Knowledge of the expansion coefficients in conjunction with lattice data for the quantity on the left hand side allows for the determination of α_{V}.

The second step, namely the calibration of the momentum scale which appears in the argument of α_{V}, is done by determining the lattice spacing from mass splittings in the bottomonium system. Here one typically considers the mass differences between the Υ and Υ', or alternatively, between the χ_b and Υ states. Of course, any other low-energy quantity like f_π or r_0 could be used. It can be argued, however, that mass splittings in heavy quarkonia are a natural choice for setting the scale in this particular approach, chiefly because of their relative insensitivity to the exact value of the heavy quark mass. Since the b-quark mass of $m_b \approx 4\,\mathrm{GeV}$ is greater than typical values of the inverse lattice spacing, a^{-1} one must employ special techniques to deal with heavy quarks on the lattice. In [86] this is done via an approach based on non-relativistic QCD. A detailed discussion of the specific treatment of heavy quarks in lattice simulations is deferred to Sect. 5.7.2.

After setting the scale, the Wilson loops $\langle W_{rt}\rangle$ computed on ensembles with $N_{\mathrm{f}} = 3$ flavours of rooted staggered quarks are used to determine α_{V} via a global fit involving data at three different values of the lattice spacing. This yields

$$\alpha_{\mathrm{V}}^{(3)}(7.5\,\mathrm{GeV}) = 0.2082 \pm 0.0040, \tag{5.153}$$

where the superscript on the coupling reminds us that the result is valid in the three-flavour theory. The relation to the coupling in the $\overline{\mathrm{MS}}$-scheme at the Z-pole is determined in perturbation theory, by employing the third-order expansion of $\alpha_{\overline{\mathrm{MS}}}$

in terms of α_V [87]:

$$\alpha_{\overline{\rm MS}}^{(3)}(\mathrm{e}^{-5/6}q) = \alpha_{\rm V}^{(3)}(q) + \frac{2}{\pi}\left[\alpha_{\rm V}^{(3)}(q)\right]^2 - (0.3111\ldots)\left[\alpha_{\rm V}^{(3)}(q)\right]^3, \tag{5.154}$$

which yields $\alpha_{\overline{\rm MS}}^{(3)}(3.26\,\mathrm{GeV})$. This coupling is then translated to $\alpha_{\overline{\rm MS}}^{(5)}(M_Z)$ via the numerical integration of the four-loop RG β-function, including the effects from quark mass thresholds at m_c and m_b, which finally yields

$$\alpha_{\overline{\rm MS}}^{(5)}(M_Z) = 0.1170 \pm 0.0012. \tag{5.155}$$

This result is included in the world average of $\alpha_{\overline{\rm MS}}^{(5)}(M_Z) = 0.1176 \pm 0.002$ in Ref. [61]. It is also in very good agreement with the non-lattice global estimate of $\alpha_{\overline{\rm MS}}^{(5)}(M_Z) = 0.1182 \pm 0.0027$ [88].

The running and matching in this approach is done perturbatively, involving energy scales from M_Z down to m_c. In this sense the method may be regarded as similar in spirit to, say, the determination of α_s from the semi-leptonic branching ratio of τ decays, as in both cases the coupling is extracted from the perturbative expansion of a particular observable. While for τ-lepton decays an experimentally measured quantity is considered, it is the non-perturbatively computed data for the Wilson loops in the lattice approach which are expressed in terms of the running coupling. This contrasts with the Schrödinger functional approach, where also the running is computed non-perturbatively, albeit with considerable numerical effort.

The error on the result in Eq. (5.155) is rather small. It is left for future studies to confirm this level of precision, which must entail further investigations into the influence of lattice artefacts, as well as the validity of the fourth root trick.

5.5.6 *Light Quark Masses*

We shall now apply the general framework of non-perturbative renormalization to the determination of quark masses. Typically one distinguishes the "light" u, d, s quarks from the "heavy" c, b, t quarks. At first, this distinction may seem rather arbitrary. It is actually based on the relative magnitude of the quark masses compared with the chiral symmetry breaking scale Λ_χ, which separates "soft" from "hard" momentum scales. Masses and momenta well below Λ_χ break chiral symmetry only softly, so that spontaneous chiral symmetry breaking still dominates over the explicit breaking generated by non-zero values of the quark masses. Gasser and Leutwyler [89, 90] have demonstrated that QCD with u, d, s flavours can be studied via an "effective" theory of Goldstone boson fields. This approach, called Chiral Perturbation Theory (ChPT), has an $\mathrm{SU}(3)_L \otimes SU(3)_R$ chiral symmetry, which is spontaneously broken to the SU(3) vector subgroup. The associated Goldstone bosons are then identified with the pions, kaons and η-mesons, whose masses are indeed small compared to typical hadronic scales, such as the mass of the nucleon, for instance. Thus, the magnitude of Λ_χ is identified with a value close to 1 GeV. In ChPT, quantities like hadron masses, decay rates or cross sections are computed through an expansion in powers of quark masses (and 4-momenta) about the chiral limit. The inclusion of the charm quark into the formalism is rather useless, since the masses if the lightest charmed pseudoscalar mesons are far greater than $\Lambda_\chi \approx 1\,\text{GeV}$.

The top quark can be safely ignored in this context, since its lifetime is an order of magnitude shorter than typical QCD processes. As a consequence, the top quark does not undergo any hadronization effects (for instance, "toponium", i.e. $t\bar{t}$ bound states have never been observed), but rather decays weakly into a W-boson and a b-quark.

The mass of the b-quark is rather large (and to some extent this is also true for the charm quark), so that one may attempt to determine their values from perturbative expansions in α_s of some mass-dependent quantity. By contrast, in the light quark sector non-perturbative effects such as spontaneous chiral symmetry breaking dominate. As far as the determination of the masses of the u, d, s quarks is concerned, ChPT is of limited value, since only ratios of quark masses can be predicted, but not their absolute values. The reason is that although the light quark masses appear as parameters of ChPT, their values cannot be fixed by chiral symmetry (see Sect. 5.6.1 for more details). The absolute normalization must therefore be provided by non-perturbative methods such as lattice simulations or QCD sum rules.

Below we will focus on attempts to compute the values of the light quark masses in units of some hadronic quantity. As indicated in Sect. 5.5.1, this entails the knowledge of the renormalization factor that links lattice regularization to the chosen continuum scheme. Lattice simulations have maximum impact in the light quark sector, owing to the dominance of non-perturbative effects, which is in fact signified by the large uncertainties quoted for the values of the u, d and s quark masses in the particle data book [61].

The general procedure for the determination of light quark masses in lattice QCD starts from the PCAC relation, Eq. (5.116). Assuming exact isospin symmetry, $m_u = m_d$, one can consider a generic light flavour ℓ with mass $m_\ell \equiv \hat{m} = \frac{1}{2}(m_u + m_d)$. In order to determine, say, the combination $\hat{m} + m_s$, one must define a particular hadronic renormalization scheme, by specifying the lattice scale and the hadronic quantity that fixes the value of $\hat{m} + m_s$. Furthermore, the renormalization factor which connects hadronic and continuum schemes must be known. Equation (5.116) can then be rewritten such that it yields the sum of RG-invariant quark masses $\hat{M} + M_s$ in units of the quantity Q which sets the lattice spacing:

$$\frac{\hat{M} + M_s}{Q} = Z_{\mathrm{M}} \times \left(\frac{f_{\mathrm{PS}}^{\mathrm{bare}}\, Q}{G_{\mathrm{PS}}^{\mathrm{bare}}} \right) \Bigg|_{m_{\mathrm{PS}}=m_{\mathrm{K}}} \times \left(\frac{m_{\mathrm{K}}^2}{Q^2} \right) \Bigg|_{\mathrm{exp}} + \mathrm{O}(a^p). \tag{5.156}$$

In this expression, the subscript "exp" denotes the experimental values for the respective quantities, while the matrix element $G_{\mathrm{PS}}^{\mathrm{bare}}$ is given by

$$G_{\mathrm{PS}}^{\mathrm{bare}}\Big|_{m_{\mathrm{PS}}=m_{\mathrm{K}}} \equiv G_{\mathrm{K}}^{\mathrm{bare}} = \big\langle 0\,\big|(\bar{\ell}\gamma_5 s)_{\mathrm{lat}}\big|\,K\big\rangle\,. \tag{5.157}$$

The pseudoscalar decay constant $f_{\mathrm{PS}}^{\mathrm{bare}}$ parameterizes the matrix element of the unrenormalized axial current, i.e.

$$f_{\mathrm{PS}}^{\mathrm{bare}} m_{\mathrm{PS}}\Big|_{m_{\mathrm{PS}}=m_{\mathrm{K}}} \equiv f_{\mathrm{K}}^{\mathrm{bare}} m_{\mathrm{K}} = \big\langle 0\,\big|(\bar{\ell}\gamma_0\gamma_5 s)_{\mathrm{lat}}\big|\,K\big\rangle\,. \tag{5.158}$$

The renormalization factor Z_{M} relates the bare current quark mass to the RG-invariant mass. Thus, the task for lattice calculations is to compute the ratio $f_{\mathrm{PS}}^{\mathrm{bare}} Q/G_{\mathrm{PS}}^{\mathrm{bare}}$ for a generic pseudoscalar state and tune the bare quark mass such that $m_{\mathrm{PS}} = m_{\mathrm{K}}$. By combining the result with the renormalization factor Z_{M} and the experimental value of m_{K}^2/Q^2, the RGI quark masses in units of Q are obtained up to lattice artefacts of order a^p, where p is characteristic of the details of the discretization. Since the RGI quark masses are scale- and scheme-independent quantities, the factor Z_{M} depends only on the bare coupling g_0. Using the Schrödinger functional as the intermediate renormalization scheme, non-perturbative estimates of Z_{M} computed for $\mathrm{O}(a)$ improved Wilson fermions within a wide range of bare couplings, have been published in Refs. [75] and [78]. In this case, Z_{M} is given by

$$Z_{\mathrm{M}}(g_0) = \frac{M}{\bar{m}_{\mathrm{SF}}(\mu_0)}\,\frac{Z_{\mathrm{A}}(g_0)}{Z_{\mathrm{P}}(g_0, a\mu_0)}, \tag{5.159}$$

where the ratio $M/\bar{m}_{\mathrm{SF}}(\mu_0)$ is computed via the finite-size scaling procedure. The transition between lattice regularization and the SF-scheme is accomplished by determining Z_{P} and the renormalization factor Z_{A} of the axial current.[14] Note that the dependence on the intermediate matching scale μ_0 drops out completely in this expression. Finally, the conversion to the $\overline{\mathrm{MS}}$-scheme is performed by considering

$$Z_{\mathrm{m}}(g_0, a\mu) \equiv \frac{\bar{m}_{\overline{\mathrm{MS}}}(\mu)}{M}\, Z_{\mathrm{M}}(g_0), \tag{5.160}$$

where the ratio $\bar{m}_{\overline{\mathrm{MS}}}(\mu)/M$ can be computed through the numerical integration of the perturbative approximation of the anomalous dimension τ and the β-function at four loops. This yields [35, 78]

$$\frac{\bar{m}_{\overline{\mathrm{MS}}}(2\,\mathrm{GeV})}{M} = \begin{cases} 0.7208, & N_{\mathrm{f}} = 0 \\ 0.7013, & N_{\mathrm{f}} = 2\,. \end{cases} \tag{5.161}$$

[14] If the fermionic discretization preserves chiral symmetry $Z_{\mathrm{A}} = 1$, while for Wilson fermions Z_{A} should be computed non-perturbatively. For the SF this was performed in Refs. [91, 92].

Table 5.2 Results for the strange quark mass in the $\overline{\text{MS}}$-scheme at $\mu = 2\,\text{GeV}$ and for the ratio $m_s/\hat{m}$, in the continuum limit of the quenched approximation

Collaboration	Action	Q	Renorm.	$m_s^{(Q)}$ [MeV]	$m_s/\hat{m}$	$m_s^{(r_0)}$ [MeV]
SPQ$_{\text{cd}}R$ [93]	Wilson	m_{K^*}	RI	105(9)(6)	24.3(2)(6)	95(9)(5)
CP-PACS [47]	Wilson	m_ρ	pert.	$114(2)(^{6}_{3})$	$26.5(^{5.1}_{3.4})$	$98(2)(^{6}_{3})$
ALPHA/UKQCD [35]	Clover	f_K	SF	97(4)		99(4)
JLQCD [94]	Stagg.	m_ρ	RI	106(7)	25.1(2.4)	95(6)

The table includes information on the fermionic action employed in the simulations, the quantity Q that sets the scale, and the type of renormalization (RI/MOM, SF or tadpole improved perturbation theory. The right-most column contains the results for the strange quark mass when converted into a common hadronic scheme, in which the scale is set by $Q' = 1/r_0$, assuming that $r_0 = 0.5\,\text{fm}$

Estimates for the strange quark mass itself can be obtained in two ways: first, one combines $\hat{M} + M_s$ with the ratio $M_s/\hat{M} = 24.4 \pm 1.4$ estimated in ChPT [38]. Alternatively, one might attempt to compute $\hat{M}$ directly from lattice data, by considering Eq. (5.116) for a pion. In this case, however, one relies on chiral extrapolations, because of the difficulties involved when tuning the masses of the light quarks towards the values of the physical up- and down-quark masses.

In Table 5.2 we present a selection of results for the mass of the strange quark in the quenched approximation, normalized in the $\overline{\text{MS}}$-scheme at $\mu = 2\,\text{GeV}$, as well as the ratio $M_s/\hat{M}$. Two observations are worth mentioning: first, direct determinations of $M_s/\hat{M}$ via chiral extrapolations agree well with the estimate from ChPT, even though the chiral limit is ill-defined in the quenched approximation. Second, the different systematics in the simulations (lattice actions, renormalization of local operators) generate a spread of seemingly incompatible results for the mass of the strange quark. However, the spread can be traced to the particular choice of hadronic renormalization scheme. To this end one can compute the relation between quark masses computed for two different lattice scales, Q and Q'. From Eq. (5.156) one easily infers that the strange quark mass $m_s^{(Q')}$ estimated using Q', is related to its counterpart $m_s^{(Q)}$ via [37]

$$m_s^{(Q')}\,[\text{MeV}] = \left(\frac{Q'}{Q}\right)_{\text{lat}} \left(\frac{Q}{Q'}\right)_{\text{exp}} m_s^{(Q)}\,[\text{MeV}]. \tag{5.162}$$

Here, the subscripts "lat" and "exp" refer to lattice and experimental estimates of the scale ratios. The ratio $(Q'/Q)_{\text{lat}}$ can be determined in the continuum limit using published lattice data, and the deviation of the proportionality factor from unity is a measure of the relative quenching effects, when either Q or Q' is chosen to set the scale. Once the results have been converted to the common scale r_0, the estimates for m_s in the continuum limit show remarkable consistency, despite the very different systematic effects among the simulations included in this analysis (c.f. Table 5.2). This demonstrates that lattice artefacts and renormalization effects can be controlled at the level of a few percent with the available techniques.

Table 5.3 Selection of recent unquenched results for the light quark masses

Collaboration	N_f	Action	Q	Ren.	m_s [MeV]	$m_s/\hat{m}$	$\hat{m}$ [MeV]
CPPACS/JLQCD [95]	2+1	Clover	m_ρ	pert.	$91.1(^{14.6}_{6.2})$		$3.54(^{0.64}_{0.35})$
HPQCD [96]	2+1	Stagg.	$\Delta_{\Upsilon'-\Upsilon}$	pert.	87(8)	27.4(4)	3.2(3)
QCDSF/UKQCD [97]	2	Clover	m_N, r_0	RI	111(9)	27(3)	4.1(4)
$\mathrm{SPQ_{cd}}R$ [98]	2	Wilson	m_{K^*}	RI	$101(^{26}_{8})$		4.3(4)
ALPHA [78]	2	Clover	r_0	SF	97(22)		
CPPACS [99]	2	Clover	m_ρ	pert.	$88(^{4}_{6})$	26(2)	$3.44(^{14}_{22})$
ETM [100]	2	tmQCD	f_π	RI	105(3)(9)	27.3(3)(1.2)	3.85(12)(40)

The challenge for current and future simulations is to eliminate the remaining uncertainty due to quenching. Several simulations with $N_\mathrm{f} = 2$ or $2+1$ flavours of dynamical quarks[15] based on different fermionic discretizations have produced results for the light quark masses, which are shown in Table 5.3. Despite the enormous progress that has been made in simulating light dynamical quarks, it is important to realize that systematic effects such as lattice artefacts and/or renormalization effects are currently not as well controlled as in the quenched theory. The fact that affordable lattice spacings are still relatively large implies that extrapolations to the continuum limit are in general longer than in the quenched approximation, thereby leading to larger errors. In some cases it is not even clear whether the leading lattice artefacts in dynamical simulations have been isolated. Also, the quantity Q that sets the scale must be known at least as accurately as the quark mass itself, and hence the determination of these observables may prove just as costly. Finally, dynamical quark masses are still fairly large, especially in many simulations using Wilson fermions, and thus the long and potentially uncontrolled chiral extrapolations significantly affect estimates for the isospin-averaged light quark mass $\hat{m}$.

5.6 Spontaneous Chiral Symmetry Breaking

Chiral symmetry has already been mentioned in connection with the masses of the light quarks. Here we will extend the general framework and elaborate on effective descriptions of QCD at low energies, which can be treated analytically. As we shall see, much can be learnt via the interplay of such effective theories and lattice simulations of QCD.

[15] $N_\mathrm{f} = 2$ usually denotes a degenerate doublet of light (u, d) quarks, while $N_\mathrm{f} = 2 + 1$ denotes a degenerate doublet together with a heavier third flavour, i.e. the strange quark.

Massless QCD with N_f flavours is invariant under independent rotations of the left- and right-handed components of the quarks fields. If one defines the field Ψ as the vector of N_f Dirac spinors ψ_i via

$$\Psi = \left(\psi_1, \ldots, \psi_{N_f}\right)^T, \tag{5.163}$$

its left- and right-handed components are given by

$$\Psi_L := \left(\mathbb{1}_{N_f} \otimes P_-\right)\Psi, \quad \Psi_R := \left(\mathbb{1}_{N_f} \otimes P_+\right)\Psi, \quad P_\pm = \tfrac{1}{2}\left(1 \pm \gamma_5\right). \tag{5.164}$$

The action of the massless theory is then invariant under transformations like

$$\Psi \to \Psi' = \exp\left\{\mathrm{i}P_-(\boldsymbol{\omega_L}\cdot\boldsymbol{T}) + \mathrm{i}P_+(\boldsymbol{\omega_R}\cdot\boldsymbol{T})\right\}\Psi, \tag{5.165}$$

where $\boldsymbol{\omega_L}$, $\boldsymbol{\omega_R}$ are real vectors, and $\boldsymbol{T}$ denotes the generators of SU(N_f), which satisfy

$$\left[T^a, T^b\right] = \mathrm{i}f^{abc}T^c, \qquad \mathrm{Tr}\left(T^a T^b\right) = \tfrac{1}{2}\delta^{ab}. \tag{5.166}$$

The above transformation can be rewritten in terms of vector and axial rotations, i.e.

$$\Psi \to \Psi' = \exp\left\{\mathrm{i}\boldsymbol{\alpha_V}\cdot\boldsymbol{T} + \mathrm{i}\boldsymbol{\alpha_A}\cdot\boldsymbol{T}\gamma_5\right\}\Psi, \tag{5.167}$$

where $\boldsymbol{\alpha_V} \equiv \frac{1}{2}(\boldsymbol{\omega_R} + \boldsymbol{\omega_L})$ and $\boldsymbol{\alpha_A} \equiv \frac{1}{2}(\boldsymbol{\omega_R} - \boldsymbol{\omega_L})$. Invariance under these transformation laws is what one usually means when one says that (massless) QCD is invariant under a global $\mathrm{SU}(N_f)_L \otimes \mathrm{SU}(N_f)_R$ symmetry.

Actually, QCD has even more global symmetries, namely a $\mathrm{U}(1)_V$ symmetry, which corresponds to a common rotation of all quark flavours. The conserved charge derived from the Noether current, which is associated with this unbroken symmetry, is the quark number. The conservation of the axial current associated with the remaining axial U(1) symmetry is, however, severely broken by an anomalous term, which gives rise to strong non-perturbative effects generated by instantons. Without going into further detail here, we refer to common textbooks.

Returning now to $\mathrm{SU}(N_f)_L \otimes \mathrm{SU}(N_f)_R$, we note that symmetries in sub-nuclear physics are usually deduced from the particle spectrum. That is, symmetries manifest themselves through the occurrence of mass-degenerate (or nearly degenerate) particle multiplets that can be grouped according to the irreducible representations of the symmetry group. Indeed, for $N_f = 3$ one finds that the light pseudoscalar mesons, i.e. the pions, kaons and η-mesons form an octet. The mass splittings among the members of the octet are small when viewed on typical hadronic scales, and arise due to the unequal, non-zero masses of the light quarks. However, if the pseudoscalar octet were interpreted as a manifestation of an (approximate) $\mathrm{SU}(3)_L \otimes SU(3)_R$ chiral symmetry, one would expect that each member of the octet is accompanied by a parity partner, i.e. a scalar meson, whose mass is of the same order of magnitude. This is not observed in experiment, where the lightest scalar mesons are found to lie 600–700 MeV above the pseudoscalar octet. One therefore concludes that the symmetry must be spontaneously broken. The term "spontaneous breaking" refers to the fact that theories like QCD possess more internal symmetries than those that can be inferred from the particle spectrum. In general, spontaneously broken symmetries are not realized as symmetry transformations involving the physical states of the theory. In particular, the ground state, i.e. the vacuum, is not invariant under the transformation. As discussed in many textbooks, it is precisely the invariance of the vacuum under the symmetry transformation that is required to ensure the degeneracy of the particle spectrum. If the vacuum is not invariant, certain operators may acquire a non-vanishing expectation value. In fact, a sufficient condition for the spontaneous breaking of the physical $\mathrm{SU}(3)_L \otimes SU(3)_R$ chiral symmetry is fulfilled if the expectation value of the scalar density, $\bar{\Psi}\Psi$, is non-zero, i.e.

$$\langle\bar{\Psi}\Psi\rangle \equiv \langle\bar{u}u + \bar{d}d + \bar{s}s\rangle \neq 0. \tag{5.168}$$

Furthermore, according to Goldstone's theorem [101], the generator of each broken symmetry is associated with a massless particle. Since the masses of the members of the pseudoscalar octet are rather small in comparison with the proton mass, they are identified as the Goldstone bosons of the spontaneously broken chiral symmetry.

Spontaneous chiral symmetry breaking is an entirely non-perturbative phenomenon. The task is then to explore the breaking mechanism and compute the value of the quark condensate $\langle\bar{\Psi}\Psi\rangle$. As shall be outlined below, this can be achieved through the interplay of lattice simulations and effective low-energy descriptions of QCD.

5.6.1 Chiral Perturbation Theory

Chiral Perturbation Theory (ChPT) has already been mentioned in connection with extrapolations of lattice data to the physical values of the up- and down-quark masses, and also in the context of lattice determinations of the strange quark mass. Here we present a brief introduction into the general formalism. More thorough reviews can be found in Refs. [102, 103].

Chiral Perturbation Theory is an effective theory, based on a systematic expansion of the low-energy dynamics of QCD in powers of the 4-momentum and the quark mass about the chiral limit [89, 90], i.e.

$$\mathcal{L}_{\mathrm{eff}} = \mathcal{L}_{\mathrm{eff}}^{(2)} + \mathcal{L}_{\mathrm{eff}}^{(4)} + \dots, \tag{5.169}$$

where the superscripts label the order of the expansion in powers of p. In contrast to QCD, the basic degrees of freedom which appear in $\mathcal{L}_{\mathrm{eff}}$ are the Goldstone bosons, rather than the fundamental quarks and gluons. ChPT is parameterized in terms of a set of empirical couplings, usually called "low-energy constants" (LECs). At lowest order, the effective chiral Lagrangian (in Euclidean space-time) reads

$$\mathcal{L}_{\text{eff}}^{(2)} = \tfrac{1}{2}F_0^2\left\{\tfrac{1}{2}\text{Tr}\left(\partial_\mu U^\dagger \partial_\mu U\right) - B_0\text{Tr}\left(\mathcal{M}(U+U^\dagger)\right)\right\}, \tag{5.170}$$

where $\mathcal{M} = \text{diag}(m_u, m_d, m_s)$ is the quark mass matrix, and $U(x)$ collects the Goldstone boson fields, i.e.

$$U(x) = \exp\left(\frac{\text{i}}{F_0}\boldsymbol{\lambda}\cdot\boldsymbol{\phi}(x)\right); \quad \boldsymbol{\lambda}\cdot\boldsymbol{\phi} \equiv \sum_{a=1}^{8}\lambda^a\phi_a = \begin{pmatrix} \pi^0 + \frac{1}{\sqrt{3}}\eta & \sqrt{2}\pi^+ & \sqrt{2}K^+ \\ \sqrt{2}\pi^- & -\pi^0 + \frac{1}{\sqrt{3}}\eta & \sqrt{2}K^0 \\ \sqrt{2}K^- & \sqrt{2}\bar{K}^0 & -\frac{2}{\sqrt{3}}\eta \end{pmatrix}. \tag{5.171}$$

The λ^a's denote the Gell-Mann matrices which are normalized as $\text{Tr}\,(\lambda^a\lambda^b) = 2\delta^{ab}$. The LECs at leading order are B_0 and F_0, where the latter corresponds to the pion decay constant in the chiral limit.[16] The expression for $\mathcal{L}_{\text{eff}}^{(4)}$, i.e. the interaction terms at next-to-leading order in the chiral expansion, contains 12 additional interaction terms, multiplied by the LECs $L_1, \ldots, L_{10}, H_1, H_2$. The values of the LECs are usually determined by matching the expressions of ChPT for physical observables to experimental data. However, it turns out that the complete set of LECs cannot be obtained in this way. Rather, in order to fix the values of some LECs, one must resort to additional theoretical assumptions. One particular example is the value of B_0, which appears in the chiral expansion of the pion mass at lowest order (see also Eq. (5.80)):

$$m_\pi^2 = B_0(m_u + m_d). \tag{5.172}$$

From this expressions it is clear that B_0 can only be determined using m_π as input if the physical values of the quark masses are known in the first place. By the same token, the value of $\hat{m} = \frac{1}{2}(m_u + m_d)$ can only be inferred if an estimate for B_0 is available. However, the a priori unknown parameter B_0 drops out in suitably chosen ratios of $m_\pi^2, m_{\text{K}}^2, \ldots$. This explains why ChPT can be used to predict the ratios of the light quark masses but fails to provide an absolute mass scale. Another reason why the complete set LECs cannot be determined from chiral symmetry considerations alone is the fact that the effective Lagrangian beyond leading order is invariant under a symmetry transformation which involves the LECs and the mass matrix $\mathcal{M}$, but which is absent in QCD. This is the so-called "Kaplan-Manohar ambiguity" [104]. At this point it is clear that lattice simulations of QCD can provide valuable input for the determination of LECs. For instance, since the values of the

---

[16] We use capital symbols for decay constants whenever we refer to a normalization in which $F_\pi \simeq 93\,\text{MeV}$.

quark masses are input parameters in the simulations, lattice QCD allows to map out the quark mass dependence of the masses of Goldstone bosons and thus determine the LEC B_0. We shall see below that B_0 is related to the quark condensate $\langle\bar{\Psi}\Psi\rangle$ which can be considered as the order parameter for spontaneous chiral symmetry breaking. Furthermore, as we have already discussed in Sect. 5.5.6, absolute values of quark masses are accessible via lattice QCD.

We end our brief introduction to ChPT with the derivation of a few relations which will be useful for our discussion of chiral symmetry breaking below. In particular, we shall derive the leading-order mass formulae such as Eq. (5.80) and establish a link between the quark condensate and B_0. To this end we expand the field U in the chiral Lagrangian $\mathcal{L}_{\text{eff}}^{(2)}$ in powers of the Goldstone boson fields. Assuming exact isospin symmetry, $m_u = m_d$, one finds at lowest order in ϕ_a:

$$\begin{aligned}\mathcal{L}_{\text{eff}}^{(2)} &= \frac{1}{2}\sum_{a=1}^{8}\partial_\mu\phi_a\partial_\mu\phi_a + \ldots \\ &+\frac{1}{2}(m_u+m_d)B_0\sum_{a=1}^{3}\phi_a^2+\frac{1}{2}(\hat{m}+m_s)B_0\sum_{a=4}^{7}\phi_a^2 \\ &+\frac{1}{3}(\hat{m}+2m_s)B_0\phi_8^2+\ldots. \end{aligned} \tag{5.173}$$

After identifying ϕ_1, ϕ_2, ϕ_3 with the pions, $\phi_4, \ldots, \phi_7$ with the kaons, and $\phi_8 \equiv \eta$, one derives the leading-order relations between the quark masses and the masses of the Goldstone bosons, viz.

$$m_\pi^2 = 2B_0\hat{m}, \quad m_{\mathrm{K}}^2 = B_0(\hat{m}+m_s), \quad m_\eta^2 = \tfrac{2}{3}B_0(\hat{m}+2m_s). \tag{5.174}$$

Thus, the relation for a generic pseudoscalar Goldstone boson made up of quarks with masses m_1 and m_2 is precisely what was already shown in Eq. (5.80). We note that from Eq. (5.174) one easily derives the Gell-Mann–Okubo mass relation, i.e.

$$3m_\eta^2 + m_\pi^2 - 4m_{\mathrm{K}}^2 = 0, \tag{5.175}$$

which is satisfied experimentally within a few percent. Furthermore, Eq. (5.174) yields the ratio $m_s/\hat{m}$ at lowest order, viz.

$$\frac{m_s}{\hat{m}} = \frac{2m_{\mathrm{K}}^2 - m_\pi^2}{m_\pi^2} \simeq 24, \tag{5.176}$$

which is already close to the estimate at next-to-leading order of $m_s/\hat{m} = 24.4 \pm 1.5$ [38], quoted in Sect. 5.5.6.

For the discussion of spontaneous symmetry breaking, it is useful to establish a connection between the quark condensate in QCD, $\langle\bar{\Psi}\Psi\rangle$, and the LECs which parameterize the effective chiral Lagrangian. This link is provided by the so-called Gell-Mann–Oakes–Renner relation [105], which we are going to derive below. To this end we consider the QCD Lagrangian in the continuum:

$$\mathcal{L}_{\mathrm{QCD}} = -\frac{1}{4}F^a_{\mu\nu}(x)F^a_{\mu\nu}(x) + \sum_f \bar{\psi}_f(x)\left(\gamma_\mu D_\mu + m_f\right)\psi_f(x). \tag{5.177}$$

The path integral is defined as

$$Z_{\mathrm{QCD}} = \int D[A_\mu]D[\bar{\psi},\psi]\exp\left\{-\int \mathrm{d}^4x\,\mathcal{L}_{\mathrm{QCD}}\right\}, \tag{5.178}$$

and the expression for the quark condensate can be formally derived by taking derivatives with respect to the light quark masses, i.e.

$$\sum_{f=u,d,s} \left.\frac{\partial \ln Z_{\mathrm{QCD}}}{\partial m_f}\right|_{m_f=0} = -\left.\left\langle \bar{u}u + \bar{d}d + \bar{s}s\right\rangle\right|_{m_f=0} \equiv -\left\langle \bar{\Psi}\Psi\right\rangle. \tag{5.179}$$

What is the analogue of this expression in the effective chiral theory? To answer this question one takes the lowest-order chiral Lagrangian of Eq. (5.170) and defines the corresponding path integral[17]

$$Z_{\mathrm{ChPT}} = \int D[U]\exp\left\{-\int \mathrm{d}^4x\,\mathcal{L}^{(2)}_{\mathrm{eff}}\right\}. \tag{5.180}$$

Since $\mathcal{L}^{(2)}_{\mathrm{eff}}$ contains the quark mass matrix one can consider similar derivatives, i.e.

$$\sum_{f=u,d,s} \left.\frac{\partial \ln Z_{\mathrm{ChPT}}}{\partial m_f}\right|_{m_f=0} = \frac{F_0^2 B_0}{2}\sum_{f=u,d,s}\frac{\partial}{\partial m_f}\left.\left\langle \mathrm{Tr}\,\mathcal{M}(U+U^\dagger)\right\rangle\right|_{m_f=0} = 3\cdot F_0^2 B_0 + \ldots, \tag{5.181}$$

and comparison with Eq. (5.179) yields

$$-\frac{1}{3}\left\langle \bar{u}u + \bar{d}d + \bar{s}s\right\rangle \equiv \Sigma = F_0^2 B_0. \tag{5.182}$$

In other words, the quark condensate is related to the slope parameter in the lowest-order mass formulae and the pion decay constant in the chiral limit, F_0. This result is known as the Gell-Mann–Oakes–Renner relation.

[17] It should be obvious that the field U must not be confused with the link variable considered in previous sections.

5.6.2 Lattice Calculations of the Quark Condensate

The Gell-Mann–Oakes–Renner relation is the starting point for many lattice determinations of the quark condensate. For a generic pseudoscalar meson consisting of a mass-degenerate quark and antiquark, i.e. $m_1 = m_2 \equiv m$, the LEC Σ is given by

$$\Sigma = \lim_{m \to 0} \left(\frac{m_{\text{PS}}^2 F_{\text{PS}}^2}{2m} \right). \tag{5.183}$$

The technical drawback of this straightforward approach is that the chiral limit in the above expression is difficult to take in practice, as we have mentioned several times already. In the quenched approximation the situation is even worse: due to the appearance of quenched chiral logarithms (c.f. Eq. (5.82)) the ratio m_{PS}^2/m becomes singular at vanishing quark mass, and hence the chiral limit does not exist. Since the quenched approximation is being abandoned, this issue will gradually become irrelevant.

However, a more serious obstacle remains in the case of dynamical simulations with Wilson fermions: since this particular type of regularization breaks chiral symmetry explicitly, the matching of simulation data at non-zero lattice spacing to the expressions of ChPT is—strictly speaking—not permitted. Matching is certainly justified if a fermionic discretization is employed which preserves chiral symmetry, such as overlap or domain wall fermions, or if results obtained using Wilson fermions are extrapolated to the continuum limit before a comparison to ChPT is performed.

A complementary approach for determining the condensate on the lattice is based on the Banks–Casher relation [106]. It provides a link between the LEC Σ and the spectral properties of the Dirac operator, viz.

$$\Sigma = \lim_{\lambda \to 0} \lim_{m \to 0} \lim_{V \to \infty} \frac{\pi}{V} \rho(\lambda), \tag{5.184}$$

where V is the space-time volume. The spectral density $\rho(\lambda)$ is defined as follows: Let $\mathcal{D}$ denote the massless Dirac operator in the continuum, satisfying $\{\gamma_5, \mathcal{D}\} = 0$. Its eigenvalue equation reads

$$\mathcal{D}\psi_n = \mathrm{i}\lambda_n \psi_n, \qquad \lambda_n \in \mathbb{R}, \tag{5.185}$$

where the eigenvalues and eigenfunctions depend on the gauge field. A suitable definition of the spectral density is then represented by

$$\rho(\lambda) := \sum_n \langle \delta(\lambda - \lambda_n) \rangle, \tag{5.186}$$

where the expectation value is taken with respect to the QCD functional integral.[18] Note that in Eq. (5.184) the ordering of limits must be obeyed. In particular, since the spontaneous breaking of a continuous symmetry cannot occur in finite volume, the limit $V \to \infty$ must be taken before the chiral limit and the spectrum in the deep infrared are considered.

The Banks–Casher relation provides not only a method to determine the condensate, but also suggests a mechanism how spontaneous chiral symmetry breaking comes about. Indeed, Eq. (5.184) implies that a non-zero value of the quark condensate is generated through a non-vanishing value of the spectral density in the deep infrared. In other words, spontaneous chiral symmetry breaking is driven by an accumulation of small eigenvalues. An immediate consequence of the Banks–Casher relation is that the level spacing $\Delta\lambda$ between the small eigenvalues is given by

$$\Delta\lambda \equiv \frac{1}{\rho(\lambda)} = \frac{\pi}{V\Sigma}. \tag{5.187}$$

Hence, as $V \to \infty$ the level spacing becomes arbitrarily small. In the free theory, i.e. in the absence of a non-trivial gauge field one finds that $\rho(\lambda) \propto \lambda^3$, which vanishes as $\lambda \to 0$. The accumulation of eigenvalues near zero with a rate predicted by Eq. (5.187) must therefore arise through the interaction with the gauge field.

In order to test the Banks–Casher scenario, a possible strategy is to compute the spectral density and check whether it actually produces an arbitrarily dense spectrum near the origin. Analytic predictions for $\rho(\lambda)$ can be derived in the framework of effective theories of QCD at low energies, namely ChPT, as well as chiral Random Matrix Theory (RMT). The latter also yields predictions for the distributions of individual eigenvalues, in addition to the spectral density.

Chiral Random Matrix Theory goes back to an idea of Wigner who tried to utilize statistical properties for the theoretical description of systems with many degrees of freedom and complicated dynamics, such as nuclear resonances. Rather than trying to model the local interactions within such a system explicitly, all possible interactions that are consistent with the symmetries of the theory are equally likely. The Hamiltonian is then approximated by a matrix whose elements are uncorrelated but obey a particular probability distribution. The main guiding principle for the RMT description of QCD is the requirement that all global symmetries must be respected. The massless Dirac operator can then be represented by an $N \times N$ matrix $\hat{D}$ with an off-diagonal block structure which is characteristic for systems with chiral symmetry:

$$\hat{D} = \begin{pmatrix} 0 & W \\ -W^\dagger & 0 \end{pmatrix} \begin{matrix} \}N_+ \\ \}N_- \end{matrix}. \tag{5.188}$$

[18] A normalization factor of V^{-1} must be included in Eq. (5.184) since $\rho(\lambda)$ is proportional to the volume.

As illustrated by the above expression, the matrix W is, in general, rectangular with N_+ rows and N_- columns, such that $N = N_+ + N_-$. For $N_+ \neq N_-$ the matrix $\hat{D}$ has $|N_+ - N_-|$ zero modes, and the index $\nu \equiv N_+ - N_-$ may be identified with the topological charge in QCD. With this definition, $\hat{D}$ is anti-hermitian and has purely imaginary eigenvalues which come in complex conjugate pairs:

$$\hat{D}\phi_n = \mathrm{i}\mu_n\phi_n, \qquad \mu_n \in \mathbb{R}. \tag{5.189}$$

One can define the system's partition function in a sector of fixed topological charge ν via

$$\mathcal{Z}_\nu = \int D[W] \det\left(\hat{D} + m\right)^{N_\mathrm{f}} \mathrm{e}^{-\frac{1}{2}N \mathrm{Tr}\,(W^\dagger W)}, \tag{5.190}$$

where N_f is—as usual—the number of dynamical quark flavours. It makes sense to identify the matrix size N with the physical volume V of the theory (up to some proportionality constant).

In order to study the spectral properties of $\hat{D}$ in the deep infrared, it is useful to rescale the eigenvalues by the system size

$$z \equiv \mu_n N, \qquad N \propto V \tag{5.191}$$

since, according to Eq. (5.187), the level spacing of the scaled eigenvalues z is of order one. The so-called *microscopic spectral density* in the sector of topological charge ν is then defined as

$$\rho_s^{(\nu)}(z) := \lim_{N\to\infty} \sum_n \langle \delta(z - \mu_n N)\rangle_\nu\,, \tag{5.192}$$

where the expectation value $\langle\cdots\rangle_\nu$ is taken with respect to the partition function $\mathcal{Z}_\nu$. An explicit expression for $\rho_s^{(\nu)}(z)$ in terms of Bessel functions has been worked out by Verbaarschot and Zahed [107]

$$\rho_s^{(\nu)}(z) = \frac{z}{2}\left\{\left[J_{\nu+N_\mathrm{f}}(z)\right]^2 - J_{N_\mathrm{f}+\nu+1}(z)\,J_{N_\mathrm{f}+\nu-1}(z)\right\}. \tag{5.193}$$

The microscopic spectral density is the convolution of the distribution functions $p_k^{(\nu)}$ of the individual scaled eigenvalues, i.e.

$$\rho_s^{(\nu)}(z) = \sum_{k=1}^{\infty} p_k^{(\nu)}(z), \qquad \int_0^\infty \mathrm{d}z\, p_k^{(\nu)}(z) = 1. \tag{5.194}$$

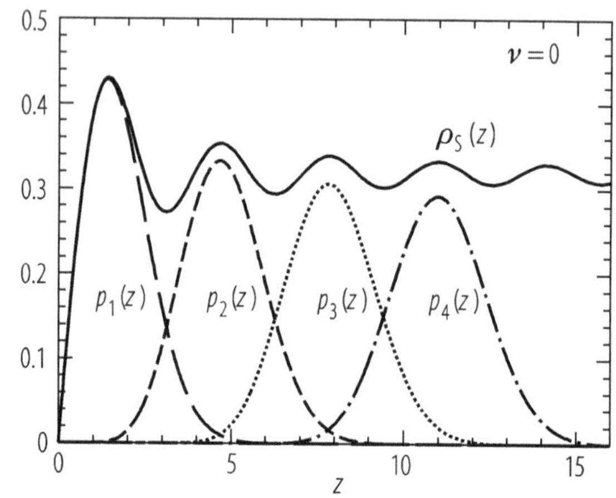

Fig. 5.16 RMT predictions for the microscopic spectral density and distributions for individual eigenvalues in the sector with topological charge $\nu = 0$

Chiral RMT yields predictions for these distributions. For instance, for the lowest eigenvalue in the sector with $\nu = 0$ one obtains for $N_{\mathrm{f}} = 0$

$$p_1^{(0)}(z) = \frac{1}{2} z\, \mathrm{e}^{-z^2/4}. \tag{5.195}$$

For further illustration the microscopic spectral density and the distribution functions for a few of the lowest eigenvalues are plotted in Fig. 5.16. The result for $\rho_s^{(\nu)}(z)$ indicates that an accumulation of small eigenvalues does indeed take place. Since one considers the simultaneous limits $\mu \to 0$ and $N \to \infty$ for fixed z, a non-zero value of $\rho_s^{(\nu)}(z)$ for finite z signals that the spectrum is packed more and more densely near the origin.

Can the predictions of RMT be verified from first principles in simulations of lattice QCD? The answer is 'yes', provided one considers a particular kinematical situation, commonly referred to as the "ϵ-regime" of QCD. It is based on the formulation of QCD in a large but finite volume of spatial size L and for arbitrarily small quark mass. The Compton wavelength of the pion then exceeds the spatial size, and thus the ϵ-regime is characterized by

$$m_\pi L \ll 1, \qquad F_\pi L \gg 1. \tag{5.196}$$

In this particular situation the path integral of the theory is dominated by zero momentum modes. In a symmetric finite box with volume $V = L^4$, the minimum non-zero momentum is given by $p_{\min} \propto 1/L$. Let us recall the expression for the lowest-order effective chiral Lagrangian, i.e.

$$\mathcal{L}_{\mathrm{eff}}^{(2)} = \tfrac{1}{2} F_0^2 \Big\{ \tfrac{1}{2} \mathrm{Tr}\, \left(\partial_\mu U^\dagger \partial_\mu U\right) - m\Sigma\, \mathrm{Tr}\, \left(\mathrm{e}^{\mathrm{i}\theta/N_{\mathrm{f}}} U + \mathrm{h.c.}\right) \Big\}, \tag{5.197}$$

where we have included the vacuum angle θ and assumed that $\mathcal{M} \equiv m\mathbb{1}$. If the quark mass m is tuned so that

$$m\Sigma \ll F_0^2 p_{\min}^2 \sim F_0^2/L^2, \tag{5.198}$$

the statistical weight of fields with $\partial_\mu U \neq 0$ will be strongly suppressed in the path integral. In other words, the mass term will dominate over the kinetic term, except for fields U with $\partial_\mu U = 0$. Since $2m\Sigma/F_0^2 = m_{\mathrm{PS}}^2$, the conditions in Eq. (5.196), which define the kinematical situation of the ϵ-regime, are equivalent to

$$m\Sigma V \ll 1. \tag{5.199}$$

The zero-momentum part can be represented by a constant SU(3) matrix U_0 such that

$$U(x) = U_0\,\mathrm{e}^{2\mathrm{i}\xi(x)/F_0}, \qquad U_0 \in \mathrm{SU}(3), \tag{5.200}$$

where the field ξ incorporates the fluctuations about the zero momentum mode. According to Leutwyler and Smilga [108], the path integral of the theory in topological sector ν can be written in the form

$$Z_\nu^{(0)} = \int D[U_0]\,(\det U_0)^\nu \exp\left(m\Sigma V\,\mathrm{ReTr}\,U_0\right). \tag{5.201}$$

After this somewhat lengthy preparatory discussion, the connection between QCD in the ϵ-regime and chiral RMT can finally be established. An important result derived by Shuryak and Verbaarschot [109] states that the path integral $Z_\nu^{(0)}$ can be mapped exactly onto the partition function $\mathcal{Z}_\nu$ of RMT. One therefore expects that the low-lying eigenvalues of QCD in the ϵ-regime are distributed in the same way as those in RMT. By computing the former in a lattice simulation and performing a comparison to the analytically known distributions in RMT, one may verify the Banks–Casher scenario of spontaneous chiral symmetry breaking.

The Neuberger-Dirac operator D_N of Eq. (5.47) is ideally suited for this task. Since it satisfies the Ginsparg-Wilson relation, chiral symmetry is preserved at the level of the discretized theory. Furthermore, D_N can be shown to satisfy an exact index theorem, so that it sustains $|\nu|$ exact zero modes on gauge configurations with topological charge ν. This allows for an unambiguous identification of topological sectors to which the path integral $Z_\nu^{(0)}$ is restricted [110]. Therefore, the investigation of spontaneous chiral symmetry breaking is a prime example where it is absolutely vital that the lattice-regularized theory obeys the same symmetries that are present in the continuum.

Before we proceed we must elucidate the relation of the spectra of the random matrix $\hat{D}$ and the Neuberger-Dirac operator. While the eigenvalues of $\hat{D}$ are purely imaginary, the operator D_N is unitary, and hence its eigenvalues lie on a circle with radius $1/\overline{a}$ in the complex plane, centered around the point $1/\overline{a}$ on the real axis.

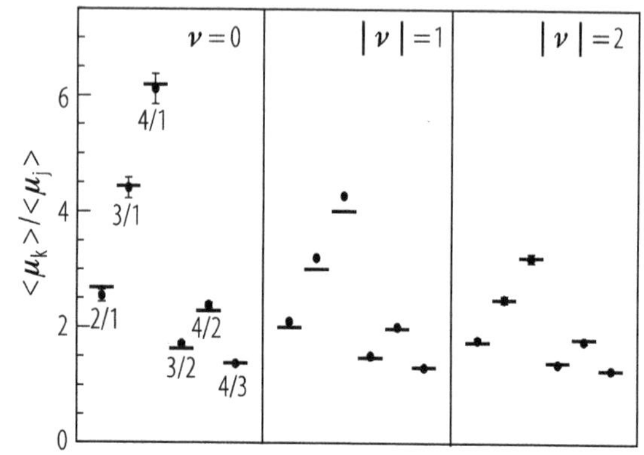

Fig. 5.17 Comparison of simulation results for ratios of eigenvalues with Random Matrix Theory (horizontal bars) in the sectors with topological charge $\nu = 0, 1, 2$ [111]

Thus, if γ denotes an eigenvalues of D_{N}, it can be parameterized as

$$\gamma = \frac{1}{\overline{a}}(1 - \mathrm{e}^{\mathrm{i}\phi}), \qquad \overline{a} = \frac{a}{1+s}. \tag{5.202}$$

Since the radius of the circle diverges in the continuum limit, the low-lying part of the spectrum satisfies $|\gamma| \ll 1/\overline{a}$, and hence $\mathrm{Re}\gamma \simeq 0$. One can then identify an eigenvalue μ of $\hat{D}$ with $\mathrm{Im}\gamma$, i.e.

$$\mu \quad \leftrightarrow \quad \mathrm{Im}\gamma \simeq |\gamma| = \frac{1}{\overline{a}}\,[2(1-\cos\phi)]\,. \tag{5.203}$$

A simple but effective check of the RMT description of the low-lying spectrum can be performed by comparing ratios of scaled eigenvalues. The combination $|\gamma_k|\Sigma V$ of the kth eigenvalue in QCD corresponds to $\mu_k N$ in RMT. If the low-lying spectra in the two theories indeed coincide one expects the following equalities in a given topological sector ν

$$\frac{\langle|\gamma_k|\rangle_\nu}{\langle|\gamma_j|\rangle_\nu} \stackrel{!}{=} \frac{\langle\mu_k\rangle_\nu}{\langle\mu_j\rangle_\nu} \equiv \int_0^\infty \mathrm{d}z\, z\, p_k^{(\nu)}(z) \bigg/ \int_0^\infty \mathrm{d}z\, z\, p_j^{(\nu)}(z). \tag{5.204}$$

While the ratio $\langle|\gamma_k|\rangle_\nu/\langle|\gamma_j|\rangle_\nu$ is determined in the simulation, the two integrals on the right-hand side can be evaluated analytically for the first few eigenvalues.[19]

In Refs. [111, 112] ratios for some of the lowest eigenvalues have been computed in the quenched approximation. The results from [111] are shown in Fig. 5.17 for a box size $L = 1.49$ fm. The agreement between lattice results and RMT is excellent. By contrast, a smaller box size of about 1 fm yields significant discrepancies between QCD and RMT, which can be as large as 10 standard deviations. This is a reflection of the fact that the large volume limit must be taken before the RMT

[19]The expressions for the distributions $p_k^{(\nu)}(z)$ become rapidly more complicated as k increases, so that one may have to resort to numerical evaluations of the integrals.

behaviour sets in. Similar findings have been reported for QCD with $N_{\mathrm{f}} = 2$ flavours of dynamical overlap quarks [113].

The confirmation of the RMT prediction for the distribution of the low-lying eigenvalues supports the Banks–Casher scenario of spontaneous chiral symmetry breaking. In a subsequent step one may therefore extract the LEC Σ via the relation

$$\langle|\gamma_k|\rangle_\nu \Sigma V = \langle\mu_k N\rangle_\nu \equiv \int_0^\infty \mathrm{d}z\, z\, p_k^{(\nu)}(z). \tag{5.205}$$

If Σ is identified with the expectation value of the scalar density, as suggested by the effective low-energy description of QCD, it must be related to a particular continuum scheme, like the $\overline{\mathrm{MS}}$-scheme of dimensional regularization. If the regularization prescription obeys chiral symmetry, the corresponding renormalization factor, Z_{S}, satisfies

$$Z_{\mathrm{S}} = Z_{\mathrm{P}} = 1/Z_{\mathrm{m}}. \tag{5.206}$$

where Z_{m} relates the bare quark mass to the chosen continuum scheme (for instance, $\overline{\mathrm{MS}}$). Provided that Z_{S}, or equivalently, Z_{m} has been computed for a range of bare couplings, the lattice estimates for Σ can be used to determine the renormalized condensate in units of some scale, e.g.

$$r_0^3 \Sigma_{\overline{\mathrm{MS}}}(\mu) = Z_{\mathrm{S}}(g_0, a\mu) r_0^3 \Sigma + \mathrm{O}(a^2). \tag{5.207}$$

For the Neuberger-Dirac operator, Z_{S} has been computed non-perturbatively in the quenched approximation [114], employing the technique outlined in Ref. [115]. The resulting values for Z_{S} could then be combined with the results for Σ extracted from the matching to RMT from [111]. A subsequent extrapolation to vanishing lattice spacing yields the results for the renormalized condensate in the continuum limit:

$$\Sigma_{\overline{\mathrm{MS}}}(2\,\mathrm{GeV}) = (285 \pm 9\,\mathrm{MeV})^3, \qquad \text{(scale set by } f_{\mathrm{K}}). \tag{5.208}$$

The quoted error represents the total uncertainty arising from statistics, the uncertainty in the renormalization factor, and the continuum extrapolation. If the nucleon mass is used to set the scale the central value drops to 261 MeV, as a consequence of the scale ambiguity encountered in the quenched approximation. We stress once more that the chiral condensate is ill-defined in the quenched theory, and thus great care must be taken when the results are interpreted in the context of the full theory. Nevertheless, it is encouraging that for $N_{\mathrm{f}} = 2$ flavours of dynamical quarks, a similar calculation [113] finds $\Sigma_{\overline{\mathrm{MS}}}(2\,\mathrm{GeV}) = (251 \pm 7 \pm 11\,\mathrm{MeV})^3$ at $a \simeq 0.11\,\mathrm{fm}$, in good agreement with the quenched result, given the inherent ambiguities and inconsistencies of the latter.

Lattice results for the condensate have been reported by many other authors (e.g. [116–125]), employing a variety of approaches. Although the various calculations are subject to different systematics, the overall picture is rather consistent, with values for the condensate centering around (250 MeV)^3. As for many other quantities, the influence of lattice artefacts and renormalization effects must be studied in more detail, especially in the case of fully dynamical calculations. It is also important to mention that analytic non-perturbative approaches to the strong interaction, such QCD sum rules, also give broadly consistent results with lattice simulations within the quoted uncertainties (see e.g. [126–128] and references therein). This completes the consistent picture of chiral symmetry and its spontaneous breaking in QCD.

5.7 Hadronic Weak Matrix Elements

The experimental programme at the B-factories BaBar and Belle, as well as many other experiments at high-energy colliders, such as the Tevatron and LEP, have greatly enhanced the accuracy of many observables related to flavour physics and the Cabibbo–Kobayashi–Maskawa (CKM) matrix. The main motivation for studying flavour physics is to gain a proper understanding of CP violation and, in turn, the matter-antimatter asymmetry which is apparently manifest in the universe. CP violation is incorporated into the Standard Model via a complex phase in the CKM matrix, and therefore a precise knowledge of its elements is required to decide whether or not additional sources of CP violation must be considered.

In order to make these statements more precise we recall some basic definitions. As is well known, the CKM matrix V_{CKM} relates flavour to mass eigenstates. For flavour-changing charged current transitions between up- and down-type quarks this implies that, in addition to the dominant transitions like $u \leftrightarrow d$, $c \leftrightarrow s$ and $t \leftrightarrow b$, there are further transitions of lesser strength. The CKM matrix is therefore expected to possess a hierarchical structure, with the diagonal elements V_{ud}, V_{cs} and V_{tb} being of order one. An approximate parameterization that takes this into account is due to Wolfenstein [129]. By expanding V_{CKM} in powers of the Cabibbo-angle $|V_{us}| \equiv \lambda \simeq 0.22$ one obtains

$$V_{\mathrm{CKM}} \equiv \begin{pmatrix} V_{ud} & V_{us} & V_{ub} \\ V_{cd} & V_{cs} & V_{cb} \\ V_{td} & V_{ts} & V_{tb} \end{pmatrix} \simeq \begin{pmatrix} 1-\lambda^2/2 & \lambda & A\lambda^3(\rho - i\eta) \\ -\lambda - iA^2\lambda^5\eta & 1-\lambda^2/2 & A\lambda^2 \\ A\lambda^3(1-\bar\rho - i\bar\eta) & -A\lambda^2 - iA\lambda^4\eta & 1 \end{pmatrix}, \tag{5.209}$$

with the remaining parameters A, $\bar\rho$ and $\bar\eta$ of order one.[20] In the standard model, V_{CKM} is unitary, and, provided that one can determine its elements with sufficient precision, any deviation from unitarity would be a signature of "new physics".

[20]The relation of rescaled parameter $\bar\rho$ to ρ is given by $\bar\rho = \rho(1 - \lambda^2/2 + \mathrm{O}(\lambda^4))$, and a similar relation holds for $\bar\eta$ and η.

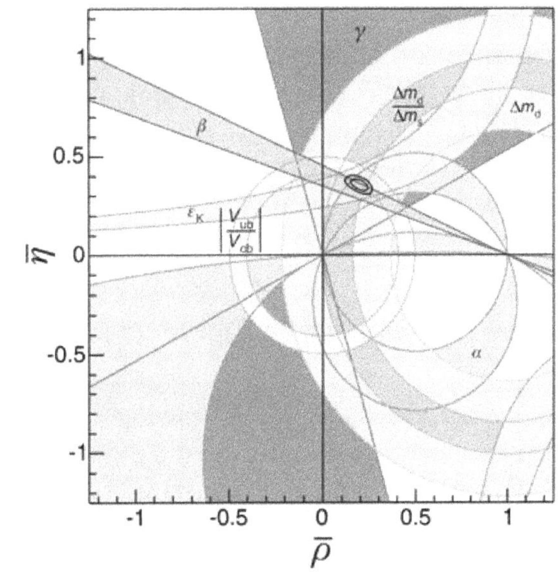

Fig. 5.18 Constraints on the apex of the unitarity triangle [130]

Unitarity gives rise to relations such as

$$V_{ud}V_{ub}^{*} + V_{cd}V_{cb}^{*} + V_{td}V_{tb}^{*} = 0, \tag{5.210}$$

which can be represented by a triangle. The strategy that has been adopted in order to search for hints of new physics, is to use experimental and theoretical input to over-constrain the unitarity relations like those in Eq. (5.210). The current status is depicted in Fig. 5.18, where the unitarity triangle is plotted in the $(\bar{\rho}, \bar{\eta})$-plane [130].

The experimentally measured quantities, i.e. the mass differences ΔM_s, ΔM_d and $\epsilon_{\rm K}$, the latter of which parameterizes indirect CP violation in the kaon system, serve to constrain the apex of the unitarity triangle. They are proportional to the relevant CKM matrix elements, i.e.

$$\Delta M_d = \frac{G_{\rm F}^2 M_{\rm W}^2}{6\pi^2}\eta_{\rm B} S\left(\frac{m_t}{M_{\rm W}}\right) f_{\rm B}^2 \hat{B}_{\rm B} \left|V_{td}V_{tb}^{*}\right|^2, \qquad \frac{\Delta M_s}{\Delta M_d} = \frac{f_{\rm B_s}^2 \hat{B}_{\rm B_s}}{f_{\rm B}^2 \hat{B}_{\rm B}}\frac{m_{\rm B_s}}{m_{\rm B}}\frac{|V_{ts}|^2}{|V_{td}|^2},$$

$$\epsilon_{\rm K} \propto \hat{B}_{\rm K}\,{\rm Im}(V_{td}V_{ts}^{*}), \tag{5.211}$$

where $G_{\rm F}$ is the Fermi constant, and $M_{\rm W}$, m_t denote the masses of the W-boson and top quark, respectively. The proportionality factors in the above expressions involve the leptonic B-meson decay constants $f_{\rm B}$ and $f_{\rm B_s}$, as well as the B-parameters $\hat{B}_{\rm B}$, $\hat{B}_{\rm B_s}$ and $\hat{B}_{\rm K}$, which in turn parameterize the transition amplitudes for $B^0 - \bar{B}^0$, $B_s^0 - \bar{B}_s^0$, and $K^0 - \bar{K}^0$ mixing. While the decay constants are difficult to measure with sufficient accuracy, due to the fact that the leptonic decay rates are suppressed, the B-parameters are not at all accessible in experiment. One must therefore resort to theoretical estimates of these quantities. Since non-perturbative effects must inevitably be included, lattice simulations of QCD are ideally suited for this task.

Lattice calculations of weak hadronic matrix elements is a major activity within the lattice community, and a thorough coverage of all aspects would easily fill an entire chapter. We shall therefore concentrate on some of the most important quantities, and point out the main conceptual issues. It is strongly recommended that the reader consult the regular reviews of the topic at the annual lattice conferences, e.g. [131–135].

5.7.1 Weak Matrix Elements in the Kaon Sector

In the kaon sector, $K^0 - \bar{K}^0$ mixing is one of the most important processes. The B-parameter $B_{\rm K}$ parameterizes the non-perturbative contribution to indirect CP violation. It is defined by the ratio of the relevant operator matrix element to its value in the so-called "vacuum saturation approximation":

$$B_{\rm K}(\mu) = \frac{\langle \bar{K}^0 | Q^{\Delta S=2}(\mu) | K^0 \rangle}{\frac{8}{3} f_{\rm K}^2 m_{\rm K}^2}. \tag{5.212}$$

Here, μ denotes the renormalization scale at which the $\Delta S = 2$ four-quark operator $Q^{\Delta S=2}$, defined by

$$Q^{\Delta S=2} = \left[\bar{s}\gamma_\mu(1-\gamma_5)d\right]\left[\bar{s}\gamma_\mu(1-\gamma_5)d\right] \equiv O_{\rm VV+AA} - O_{\rm VA+AV}, \tag{5.213}$$

is considered. The relation between $\epsilon_{\rm K}$ and the CKM matrix elements is provided by the RG-invariant B-parameter $\hat{B}_{\rm K}$. In NLO perturbation theory $\hat{B}_{\rm K}$ is related to $B_{\rm K}(\mu)$ via

$$\hat{B}_{\rm K} = \left(\frac{\bar{g}(\mu)^2}{4\pi}\right)^{\gamma_0/2b_0} \left\{1 + \bar{g}(\mu)^2 \left[\frac{b_0\gamma_1 - b_1\gamma_0}{2b_0^2}\right]\right\} B_{\rm K}(\mu), \tag{5.214}$$

where γ_0, γ_1 denote the coefficients in the perturbative expansion of the anomalous dimension of $Q^{\Delta S=2}$. Since QCD is parity-conserving, the physically relevant operator in the above expression is the parity-even combination $O_{\rm VV+AA}$. The typical left-handed chiral structure of this operator, which is characteristic for weak transitions, poses a problem for lattice calculations if Wilson fermions are employed. In this case the discretization breaks chiral symmetry explicitly, and thus $O_{\rm VV+AA}$ mixes under renormalization with operators involving the opposite chirality. Therefore, the general renormalization pattern is

$$O^{\rm R}_{\rm VV+AA}(\mu) = Z(g_0, a\mu)\left\{O^{\rm bare}_{\rm VV+AA} + \sum_{i=1}^{4} \Delta_i(g_0) O_i^{\rm bare}\right\} \tag{5.215}$$

Thus, in order to determine the physical matrix element, one must not only determine the overall renormalization factor Z, but also the mixing coefficients Δ_i. Several techniques have been developed [136–138] to address this problem, which is merely an inconvenience rather than a serious obstacle. In a formulation based on staggered fermions the problem is absent, since the remnant $\mathrm{U}(1) \otimes U(1)$ symmetry protects the operator from mixing with other chiralities. However, a drawback of the staggered formulation is the broken flavour ("taste") symmetry, which may lead to significant complications [139]. Fermionic discretizations based on the Ginsparg-Wilson relation, such as domain wall or overlap fermions do not suffer from the mixing problem, whilst preserving all flavour symmetries. Finally, the mixing problem can also be circumvented for Wilson-like discretizations in the context of twisted-mass QCD [140, 141]. With the help of a suitably chosen flavour rotation (see Eq. (5.51)), the matrix element of $O_{\mathrm{VV+AA}}$ in QCD can be mapped exactly onto that of the parity-odd operator $O_{\mathrm{VA+AV}}$ in the chirally twisted theory, viz.

$$\left\langle \bar{K}^0 \left| O_{\mathrm{VA+AV}}^{\mathrm{bare}} \right| K^0 \right\rangle_{\mathrm{tmQCD}} = \mathrm{i} \left\langle \bar{K}^0 \left| O_{\mathrm{VV+AA}}^{\mathrm{bare}} \right| K^0 \right\rangle_{\mathrm{QCD}} . \tag{5.216}$$

It has been shown that $O_{\mathrm{VA+AV}}$ renormalizes purely multiplicatively [142], i.e. all mixing coefficients vanish. The overall multiplicative, scale-dependent renormalization factor of $O_{\mathrm{VA+AV}}$ which yields the physical matrix element has been determined non-perturbatively [143], using the finite-size scaling procedure based on the Schrödinger functional formalism described in Sect. 5.5.2.

We now give a summary of the current status of B_{K}. Here, the calculation by the JLQCD Collaboration [154], based on staggered quarks in the quenched approximation, has served as a benchmark result for a long time. Their result, for which the perturbatively renormalized matrix element was extrapolated to the continuum limit, has since been confirmed by many other calculations employing different fermionic discretizations and different renormalization techniques. These include domain wall [148, 149] and overlap quarks [150, 151], as well as the Wilson formulation [153, 155]. Moreover, a calculation employing twisted mass QCD has been completed [152], which includes non-perturbative renormalization and a thorough investigation of the continuum limit.

Recently, results for B_{K} from simulations with dynamical quarks have become available, both for $N_{\mathrm{f}} = 2$ [146, 147] and $N_{\mathrm{f}} = 2 + 1$ flavours [144, 145]. A compilation of quenched and unquenched results is shown in Fig. 5.19. Although the figure suggests a trend in the data which points to slightly lower estimates for $\hat{B}_{\mathrm{K}}$ if dynamical quarks are switched on (see Fig. 5.19), the quoted uncertainties are still too large to point to a significant deviation. In particular, a systematic study of the continuum limit in the unquenched case is not yet available. It is interesting to compare the results for $\hat{B}_{\mathrm{K}}$ to the non-lattice determination in Ref. [130]. Here, the determinations of the angles of the unitarity triangle from experimental data in conjunction with direct measurements of ΔM_d, ΔM_s and ϵ_{K} allow to fit the values of several of the quantities in Eq. (5.211), which incorporate the hadronic

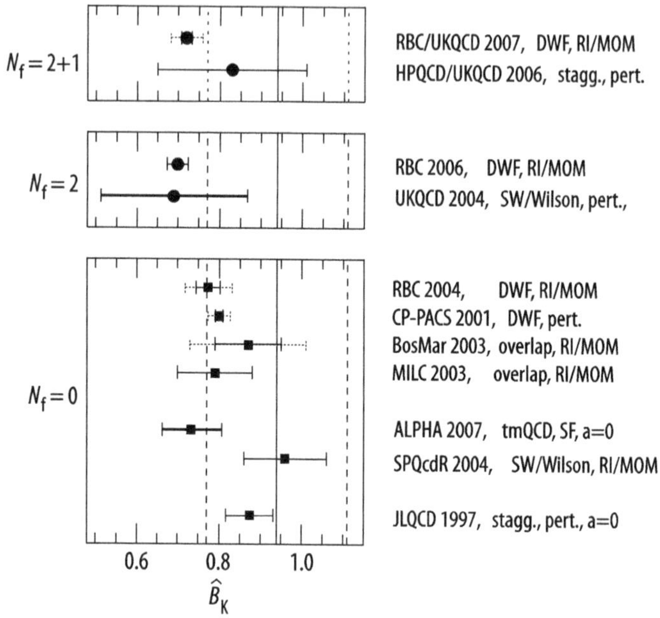

Fig. 5.19 Recent lattice results for the RGI kaon B-parameter $\hat{B}_K$. From top to bottom, the plotted values are taken from Refs. [144–154]. Dotted error bars (where shown) indicate the quoted systematic error. The labels include information on the fermionic discretization and the intermediate renormalization scheme, if non-perturbative renormalization was used. We also indicate whether or not the results have been extrapolated to the continuum limit. The vertical lines represent the (non-lattice) result from [130], with the quoted uncertainty (see text)

uncertainties. In this way one obtains $\hat{B}_K^{\text{non-lattice}} = 0.94 \pm 0.17$, which is shown as the vertical band in Fig. 5.19. Clearly, within the rather large error margins, this result is compatible with all lattice determinations, quenched or unquenched.

First Row Unitarity and the Value of $|V_{us}|$ In addition to Eq. (5.210), the unitarity of the CKM matrix implies many other constraints on its elements, such as those which appear in the first row:

$$|V_{ud}|^2 + |V_{us}|^2 + |V_{ub}|^2 = 1. \tag{5.217}$$

Owing to the smallness of $|V_{ub}|$, i.e. $|V_{ub}|^2 \simeq 2 \cdot 10^{-5}$, the direct verification of first row unitarity with the current experimental and theoretical accuracy rests on the precise knowledge of $|V_{ud}|$ and $|V_{us}|$. The value of $|V_{ud}|$ can be determined with high accuracy from super-allowed nuclear β-decays ($0^+ \to 0^+$ transitions), and in the current edition of the particle data book the best estimate is quoted as [61]

$$|V_{ud}| = 0.97377 \pm 0.00027. \tag{5.218}$$

The value of $|V_{us}|$ can be extracted from the decay rate of $K_{\ell 3}$ transitions, i.e.

$$\Gamma(K \to \pi \ell \nu_\ell) \propto \frac{G_F^2 m_K^5}{192\pi^3} |V_{us}|^2 \left| f_+^{K\pi}(0) \right|^2, \tag{5.219}$$

where $f_+^{K\pi}$ is one of the two form factors which parameterize the hadronic matrix element for semi-leptonic $K \to \pi \ell \nu_\ell$ transitions, i.e.

$$\begin{aligned} \left\langle \pi(\vec{p}_\pi) \left| (\bar{s}\gamma_\mu u)(0) \right| K(\vec{p}_K) \right\rangle &= f_+^{K\pi}(q^2)(p_K + p_\pi)_\mu + f_-^{K\pi}(q^2)(p_K - p_\pi)_\mu, \\ q_\mu &= (p_K - p_\pi)_\mu. \end{aligned} \tag{5.220}$$

In order to arrive at a precise estimate for $|V_{us}|$, $f_+^{K\pi}(q^2)$ must be determined with an accuracy at the level of 1%, since the decay rate and hence the combination $|V_{us}|^2 [f_+^{K\pi}]^2$ can be measured rather precisely. The form factor $f_+^{K\pi}$ admits a chiral expansion; At zero momentum transfer it reads

$$f_+^{K\pi}(0) = 1 + f_2 + f_4 + \ldots. \tag{5.221}$$

While the leading chiral correction, $f_2 = -0.023$, has been computed long ago [156], knowledge on f_4 and the higher corrections is still fairly limited. The strategy pursued in lattice calculations [157] is based on computing the quantity

$$\Delta f \equiv f_+^{K\pi}(0) - (1 + f_2), \tag{5.222}$$

which is a measure of the contributions beyond leading order. An old phenomenological estimate by Leutwyler and Roos [158] yields the value $\Delta f = -0.016(8)$. It is clearly desirable to check this result and ultimately replace it by a model-independent estimate based on QCD.

Semi-leptonic form factors can be determined in lattice simulations by computing suitable three-point correlation functions, in which the initial and final hadronic states are projected onto non-vanishing momentum. The main issues that must be addressed in order to judge the accuracy of the form factor determination are listed in the following:

- The dependence of the form factors on the momentum transfer q^2 must be modelled, in order to interpolate their values to $q^2 = 0$. Typical *ansätze* for the interpolation include linear or quadratic functions of q^2, as well as formulae based on pole dominance [159]. The freedom of choosing a particular *ansatz* introduces a certain ambiguity, since different model functions yield slightly different results. Via the introduction of so-called twisted boundary conditions [159–164], the q^2 resolution of form factors can be significantly improved;
- As for all quantities involving pions, a chiral extrapolation of lattice results must be performed. Clearly, in order to obtain $f_+^{K\pi}(0)$ and hence Δf with small controlled errors, a reliable chiral extrapolation is perhaps the single most

Table 5.4 Recently published lattice results for the quantity Δf

Collaboration	N_f	Action	$f_+^{K\pi}(0)$	Δf	$m_\pi^{\min}$ [MeV]
Bećirević et al. [157]	0	Clover	0.960(5)(6)	−0.017(5)(7)	490
RBC [165]	2	DWF	0.968(9)(6)	−0.009(9)(6)	490
UKQCD/RBC [166]	$2+1$	DWF	0.964(5)	−0.013(5)	330
Leutwyler and Roos [158]	./.	./.	0.961(8)	−0.016(8)	./.
Bijnens and Talavera [167]	./.	./.	0.976(10)	−0.001(10)	./.
Jamin et al. [168]	./.	./.	0.974(11)	−0.003(11)	./.
Cirigliano et al. [169]	./.	./.	0.984(12)	0.007(12)	./.

The minimum value of the pion mass used in the simulations is listed in the right-most column. The lower part of the table contains analytical estimates

important issue. Thus, the ability to simulate as deeply as possible in the chiral regime will be decisive for the final accuracy;

- Other systematic uncertainties include control over lattice artefacts, which is closely related to the renormalization of local operators, such as the vector current, which appears in Eq. (5.220). If chiral symmetry is broken explicitly, the (local) vector current is not conserved, and in order to guarantee a smooth approach to the continuum limit, its renormalization factor, Z_V, must be included. However, in all recent simulations the form factor has been extracted from suitably chosen ratios in which Z_V drops out.

A compilation of recent results for the form factor $f_+^{K\pi}(0)$ and the quantity Δf are presented in Table 5.4, where they are compared to analytical estimates. The agreement with the old result by Leutwyler and Roos is quite striking. Despite a tendency among the more recent analytical calculations to produce slightly larger estimates for Δf, all results are in good agreement within the quoted uncertainties.

An alternative method to determine $|V_{us}|$ from experimental data was proposed by Marciano [170]. Instead of considering semi-leptonic decays, it is based on the leptonic decay rates, i.e.

$$\frac{\Gamma(K \to \mu\bar{\nu}_\mu(\gamma))}{\Gamma(\pi \to e\bar{\nu}_e(\gamma))} \propto \frac{|V_{us}|^2}{|V_{ud}|^2}\,\frac{f_K^2 m_K}{f_\pi^2 m_\pi}. \tag{5.223}$$

Hence, the task is to provide an input value for the ratio of decay constants, f_K/f_π. This quantity is well-suited for lattice calculations in several respects: first, ratios of quantities can be computed with high statistical accuracy, owing to the fact that the fluctuations in the numerator and denominator are correlated. Second, the renormalization factor of the axial current, Z_A, drops out in the ratio f_K/f_π. However, since the quantity of interest involves a chiral extrapolation, the same caveats as in the case of the pion form factor, apply in this case. In particular, it is mandatory to go as close as possible to the physical mass of the pion. The quenched approximation is clearly of very limited value in this context, since the chiral behaviour and hence the actual value of f_K/f_π may strongly depend on the

Table 5.5 Recently published results for f_K/f_π in lattice QCD with dynamical quarks

Collaboration	N_f	Action	f_K/f_π	$m_\pi^{\min}$ [MeV]
CP-PACS [52]	2	Clover	1.19(3)	500
JLQCD [172]	2	Clover	$1.148(11)(^{12}_{5})(^{2}_{3})$	500
ETM [100]	2	tmQCD	1.227(9)(24)	290
MILC [173]	$2+1$	Stagg.	$1.208(2)(^{7}_{14})$	290
NPLQCD [174]	$2+1$	Stagg./DWF	$1.218(2)(^{11}_{24})$	290
RBC/UKQCD [175]	$2+1$	DWF	1.24(2)	330
HPQCD [176]	$2+1$	Stagg.	1.189(7)	250

number of active sea quarks. Furthermore, it is known that in the continuum limit of the quenched approximation the value f_K/f_π is underestimated by about 10% [171].

Recent results for f_K/f_π in lattice QCD with dynamical quarks are listed in Table 5.5. A caveat that applies to all such compilations is that systematic errors are not estimated in a uniform manner. For instance, none of the listed results (with the exception of [52]) is based on a systematic scaling study aimed at separating cutoff effects from the actual mass dependence, although the influence of lattice artefacts has been included in some error estimates by including cutoff effects into a generalized chiral fit. Moreover, not all of the listed values of f_K/f_π include finite-volume corrections, which can be computed in ChPT and incorporated into the *ansatz* for the chiral fit [177, 178]. Despite these caveats it appears, though, that the estimates for f_K/f_π based on fits including pion masses well below 500 MeV are compatible with each other.

5.7.2 *Weak Matrix Elements in the Heavy Quark Sector*

The main obstacle for calculations of weak matrix elements involving heavy quarks, and in particular the b-quark, is that one is faced with a multi-scale problem. In Sect. 5.2.5 we have already discussed systematic effects in lattice calculations that arise from finite-size effects and lattice artefacts. Translating the relations in (5.79) directly to the b-quark sector, one finds that the following inequalities cannot be satisfied simultaneously, at least not with the currently available computer power:

$$am_b \ll 1, \quad m_\pi L \gg 1, \quad L/a \lesssim 50. \qquad (5.224)$$

Violation of the first relation implies the presence of large lattice artefacts, the second inequality must be satisfied if one wants to avoid uncontrolled finite-volume effects, and the third is dictated by memory capacities of current computers. With a b-quark mass of $m_b \approx 4\,\text{GeV}$ and typical inverse lattice spacings of $a^{-1} \lesssim 4.5\,\text{GeV}$, it is evident that the b-quark cannot be studied directly, since its Compton wavelength is smaller or of the same order of magnitude than the lattice spacing itself.

Several strategies to deal with this problem have been applied over many years, among them the "static approximation" [179], the non-relativistic formulation (NRQCD) [180], the so-called "Fermilab-approach" [181] and finite-size scaling techniques [182, 183].

Since the charm quark is lighter than the b-quark by roughly a factor three, one may attempt to treat charm as a fully relativistic, propagating quark in simulations. Still, one can incur large lattice artefacts in this way, and a careful extrapolation to the continuum limit is then required. However, such an extrapolation may be spoilt if the leading lattice artefacts cannot be isolated in the results, due to the relatively large mass of the charm quark. Still, if one has reason to trust the results obtained for relativistic charm quarks, one may extrapolate them to the mass of b-quark, which is yet another way of circumventing the problem that the b-quark is too heavy to be treated relativistically. Typically, the *ansatz* for the extrapolation of a particular quantity to the mass of the b-quark is motivated by its expected quark mass dependence in Heavy Quark Effective Theory (HQET).

In the static approximation the b-quark is assumed to be infinitely heavy [179]. In this formalism it is convenient to represent the b-quark by a pair of spinors, $(\psi_h, \psi_{\bar{h}})$, which propagate forward and backward in time, respectively, and which satisfy

$$P_+\psi_h = \psi_h, \quad P_-\psi_{\bar{h}} = \psi_{\bar{h}}, \qquad P_\pm = \tfrac{1}{2}(1 \pm \gamma_0). \tag{5.225}$$

While the field ψ_h annihilates a heavy quark, $\psi_{\bar{h}}$ creates a heavy antiquark. The dynamics of these fields in the discretized version of the theory is described by the Eichten-Hill action [184]

$$S^{\rm stat} = a^4 \sum_x \left\{\mathcal{L}_h^{\rm stat} + \mathcal{L}_{\bar{h}}^{\rm stat}\right\}, \quad \mathcal{L}_h^{\rm stat} = \bar{\psi}_h(x)\nabla_0^*\psi_h(x), \quad \mathcal{L}_{\bar{h}}^{\rm stat} = -\bar{\psi}_{\bar{h}}(x)\nabla_0\psi_{\bar{h}}(x), \tag{5.226}$$

where ∇_0, ∇_0^* denote the forward and backward covariant lattice derivatives in the temporal direction. Although the numerical computation of the quark propagator based on the Eichten-Hill action is relatively "cheap", simulation results in the static approximation typically suffer from relatively large statistical noise. Without going into detail we note that the signal-to-noise ratio can be significantly improved if one replaces the temporal link variables in ∇_0 and ∇_0^* by suitably chosen generalized parallel transporters. A full account can be found in Ref. [185].

Obviously, the static approximation represents only the leading term in an expansion of the quark action in inverse powers of the heavy quark mass, and thus one expects corrections in powers of $1/m_h$. As described in Ref. [182], one can set up a formalism in which the leading corrections to physical observables can be systematically computed as operator insertions in correlation functions defined with respect to the static action $S^{\rm stat}$. Again, we refrain from describing any further details and refer the reader to the original literature [182, 186].

Higher-order corrections to the static approximation can also be incorporated into the theory by adding the appropriate $1/m_h$ terms to the action itself. In this way one obtains a non-relativistic version of QCD (NRQCD) [180], in which the mass of the heavy quark is imposed as a cutoff on relativistic momentum modes, i.e.

$$p \sim m_h v \ll m_h, \tag{5.227}$$

where v denotes the four-velocity of the heavy quark. Heuristically, the introduction of the cutoff is justified since the internal typical momentum modes of hadrons containing a heavy quark are much smaller than the mass of the latter. The loss of relativistic states can be compensated by adding new local interaction terms order by order in $p/m_h \sim v$ to $\mathcal{L}_h^{\text{stat}}$ and $\mathcal{L}_{\bar{h}}^{\text{stat}}$. In general, these additional interaction terms will generate mixing between quark and antiquark. However, by applying a Foldy-Wouthuysen transformation, the fields can be decoupled. At the level of the classical theory, the $1/m_h$ correction to the NRQCD Lagrangian for the forward propagating field reads

$$\mathcal{L}_h^{(1);\,\text{class}} = -\frac{1}{2m_h}\left\{\bar{\psi}_h \boldsymbol{D}\cdot\boldsymbol{D}\psi_h + \bar{\psi}_h\boldsymbol{\sigma}\cdot\boldsymbol{B}\psi_h\right\}, \tag{5.228}$$

and $\boldsymbol{D}$ is the vector of the covariant derivatives in the spatial directions.

In the quantized version of the theory, the coefficients which multiply the fields in the above expression become dependent on the gauge coupling and must be appropriately tuned to guarantee the correct matching of the non-relativistic theory to standard QCD at order in $1/m_h$. Thus, the lattice-regularized version of the $1/m_h$ correction reads

$$\mathcal{L}_h^{(1)} = -\left\{\omega_1\bar{\psi}_h\nabla\cdot\nabla\psi_h + \omega_2\bar{\psi}_h\boldsymbol{\sigma}\cdot\hat{\boldsymbol{B}}\psi_h\right\}, \tag{5.229}$$

where $\hat{\boldsymbol{B}}$ denotes a lattice representation of the magnetic field. The coefficients ω_1 and ω_2 are formally of order $1/m_h$ and are found to be linearly divergent in the lattice spacing a. Therefore, at a given order in the non-relativistic expansion of the action, a finite cutoff must be kept, and in this sense the effective theory is non-renormalizable. All this implies that in NRQCD the continuum limit, $a \to 0$, cannot be taken. Instead, one must argue that lattice artefacts are small in the range of lattice spacings where the calculations are performed.

Another approach can be based on the idea that the Wilson fermion action is suitably adapted for heavy quarks, such that the Wilson quark propagator docs not deviate from the continuum behaviour even for quark masses $am \gtrsim 1$, i.e. for quark masses near or above the cutoff [181]. According to Ref. [181] this can be achieved by modifying the normalization of the quark fields (see Eq. (5.36)) in the discretized lattice theory, i.e.

$$\psi(x) \to \sqrt{2\kappa}\,\mathrm{e}^{am_{\mathrm{P}}/2}\psi(x), \qquad \bar{\psi}(x) \to \bar{\psi}(x)\,\mathrm{e}^{am_{\mathrm{P}}/2}\sqrt{2\kappa}, \tag{5.230}$$

where the "pole mass" am_{P} of the Wilson propagator is given by

$$am_{\mathrm{P}} = \ln(1 + am), \tag{5.231}$$

and am denotes the bare subtracted quark mass in the Wilson theory (see Eq. (5.39)). The factor $\sqrt{2\kappa}\,\mathrm{e}^{am_{\mathrm{P}}/2}$ is designed to interpolate smoothly between the relativistic and non-relativistic regimes. As a consequence, in order to cancel the effects of large quark masses in hadronic matrix elements involving b-quarks, the normalization of quark fields is modified according to the above prescription. The so-called "Fermilab approach" to heavy quark physics on the lattice is based on the normalization in Eq. (5.230). Essentially it amounts to formulating an effective theory for quarks, whose spatial momenta are small, $|a\vec{p}| \ll 1$, with mass-dependent coefficients. Like in the case of the static approximation, the formalism allows to take the continuum limit. Related approaches to the Fermilab method have been presented in Refs. [187, 188].

Finally, we briefly introduce another strategy to deal with heavy quarks on the lattice and the related multi-scale problem [182, 183]. Here the condition $m_\pi L \gg 1$ in Eq. (5.224) is sacrificed in favour of $am_b \ll 1$. In this way one is able to accommodate a fully relativistic b-quark at the expense of having to deal with strong finite-volume effects. The key observation is that the "distortion" due to unphysically small volumes can be computed in a series of finite-size scaling steps, which relate the results obtained on a sequence of lattice sizes $L_0, L_1, \ldots$. Like in the case of the non-perturbative determination of the RG running of the coupling and the quark mass discussed in Sect. 5.5.2, one can set up a recursive finite-size scaling procedure, which traces the volume dependence of observables. Here it is mostly sufficient to apply two or three steps in the scaling sequence.

In the remainder of this section we shall discuss some selected results. Regarding the vast number of individual results, we do not attempt to provide a complete review of the current status of lattice calculations of weak matrix elements in the heavy quark sector. Regular appraisals of the progress made in studying these systems can be found in the rapporteur talks on the subject at the annual conferences on lattice field theory [132, 133, 135]. Instead we shall discuss the relation between CKM matrix elements and the quantities that must be computed in order to extract the former from experimental data without resorting to model assumptions.

Heavy-Light Decay Constants From Eq. (5.211) and Fig. 5.18 one infers that the ratio $\xi \equiv f_{\mathrm{B_s}}\sqrt{\hat{B}_{\mathrm{B_s}}}/f_{\mathrm{B}}\sqrt{\hat{B}_{\mathrm{B}}}$ of decay constants and B-parameters is a key quantity, since it links $\Delta M_s/\Delta M_d$ to the ratio $|V_{ts}|^2/|V_{td}|^2$ of CKM matrix elements. Typically, one determines decay constants and B-parameters separately, since the former can be easily extracted from hadronic two-point functions, while the latter may undergo complicated mixing patterns, depending on the fermionic discretization. The decay constant of, say, a B^+ meson, is defined via the matrix element of the heavy-light axial current, i.e.

$$f_{\mathrm{B}} m_{\mathrm{B}} = \left\langle 0\,|(\bar{u}\gamma_0\gamma_5 b)|\,B^+\right\rangle. \tag{5.232}$$

If the matrix element on the right-hand side is computed in a lattice simulation, then the axial current defined in the discretized theory must be matched to its counterpart in the continuum formulation. The details of the matching procedure depend on the type of fermionic discretization and the chosen treatment to represent the heavy-light axial current on the lattice (e.g. static approximation, NRQCD, etc.). If the b-quark is treated in the static approximation, the axial current has a non-vanishing anomalous dimension, and hence its running must be determined as well. Therefore, the various techniques which have been developed to compute the renormalization factors of local operators non-perturbatively, are of particular relevance also in the study of heavy-light decay constants [189]. In particular, non-perturbative estimates for the renormalization factor of the axial current, $Z_{\rm A}$, are required to ensure a smooth convergence towards the continuum limit.

We now present results for $f_{\rm B}$ and $f_{\rm B_s}$. From Chiral Perturbation Theory one expects that the bulk of the SU(3)-flavour breaking effect in ξ (i.e. the deviation of ξ from unity) is carried by the decay constants. The full expression at NLO for $f_{\rm B_s}/f_{\rm B}$ reads [190]

$$\frac{f_{\rm B_s}}{f_{\rm B}}-1=(m_{\rm K}^2-m_\pi^2)f_2(\mu)-\frac{1+3g^2}{(4\pi f_\pi)^2}\left[\frac{1}{2}I_{\rm P}(m_{\rm K})+\frac{1}{4}I_{\rm P}(m_\eta)-\frac{3}{4}I_{\rm P}(m_\pi)\right], \tag{5.233}$$

where $I_{\rm P}(m_{\rm PS})=m_{\rm PS}^2\ln(m_{\rm PS}^2/\mu^2)$ and f_2 is a low-energy constant, and g^2 is the strength of the $B^*B\pi$ vertex. As was pointed out by Kronfeld and Ryan [191], the contribution from the chiral logarithms can be sizeable, so that a naïve linear extrapolation of lattice data from the region of the strange quark mass tends to underestimate $f_{\rm B_s}/f_{\rm B}$. By contrast, the corresponding ratio $B_{\rm B_s}/B_{\rm B}$ is expected to be close to one, since the coefficient of the chiral logarithm nearly vanishes. Since $f_{\rm B_s}/f_{\rm B}$ enters directly into fits to the CKM parameters, many attempts were made to pin down its value precisely. As in the case of $f_{\rm K}/f_\pi$ discussed earlier, the main issue for lattice calculations is whether the quark masses employed in simulations are small enough to allow for a controlled chiral extrapolation based on the NLO formulae. The influence of the chiral logarithms has so far been detected only in simulations based on $N_{\rm f}=2{+}1$ flavours of rooted staggered quarks. Using NRQCD to treat the b-quark, the authors of [192] find

$$\frac{f_{\rm B_s}}{f_{\rm B}}=1.20\pm0.03\pm0.01,\qquad N_{\rm f}=2+1, \tag{5.234}$$

where the first error is statistical, while the second is an estimate of the systematic uncertainty. This result awaits confirmation from simulations with sea quark masses as small as those used in [192], but employing different fermionic discretizations, both in the sea and valence quark sectors. This is of particular relevance, since the typical spread among the recently published results is of the same order or even larger than the uncertainty quoted above. Further discussions and compilations of lattice data for f_{B_s}/f_B can be found in [133, 193].

Estimates for absolute values of heavy-light decay constants are also highly desirable, especially since f_B is hard to determine experimentally, even at the B-factories, since the $B \to \tau\nu_\tau$ decay rate is suppressed. For f_{B_s} the suppression is even stronger, and thus the prospects for an experimental determination of this quantity are extremely uncertain. The main issues facing lattice calculations are the influence of lattice artefacts in conjunction with the renormalization of the axial current, and the dependence of results on the number of dynamical quark flavours.

As an example for one of the most advanced quenched calculations for f_{B_s} we shall briefly discuss the result by the ALPHA collaboration [198], which also illustrates the interplay between various methods to treat the b-quark. In Ref. [198] the results obtained in the static approximation were combined with data computed around the mass of the charm quark. Provided that estimates for the decay constants in both datasets have been extrapolated to the continuum limit, a subsequent interpolation in the heavy quark mass yields the desired result for f_{B_s}. The *ansatz* for the interpolation is based on the expression

$$f_{PS}\sqrt{m_{PS}} = C_{PS}(M/\Lambda_{\overline{MS}})\,\gamma\left(1+\frac{\delta}{m_{PS}}\right), \tag{5.235}$$

where f_{PS} is a generic heavy-light decay constant, γ, δ are real constants, and the factor C_{PS} arises from the matching between the static approximation and QCD with fully relativistic quarks. Thus, using the static approximation as the limiting case removes the systematic error due to the uncontrolled extrapolation to the mass of the b-quark. The resulting estimate for f_{B_s} is [198]

$$f_{B_s} = 193 \pm 6\,\text{MeV}, \qquad N_f = 0. \tag{5.236}$$

Non-perturbative renormalization has been employed in both the static approximation and the relativistic formulation. Except for the unknown systematic error due to quenching, the quoted error contains all uncertainties. The above result has been confirmed by the approach based on the finite-size scaling method [199].

Turning now to unquenched simulations, we compare the above value to the result by the HPQCD Collaboration [192], which was obtained using NRQCD for the b-quark, while $N_f = 2+1$ rooted staggered quarks were used as sea quarks. Here, the estimate for f_{B_s} results from a combination of the value for f_B and the ratio f_{B_s}/f_B already quoted in Eq. (5.234). In this way one obtains

$$f_{B_s} = 259 \pm 32\,\text{MeV}. \tag{5.237}$$

Thus, in spite of the large error, it appears that the inclusion of dynamical quark effects increases the estimate for heavy-light decay constants. This is also supported by other simulations. For instance, using their simulation results in quenched QCD

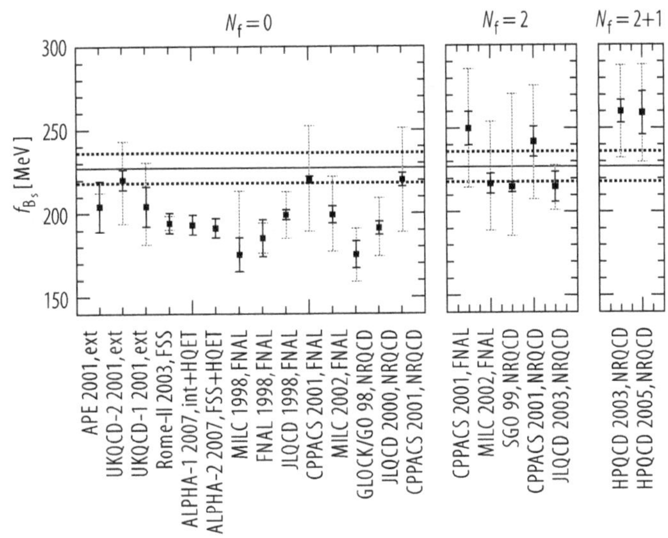

Fig. 5.20 Recent lattice results for f_{B_s}. From left to right the results are taken from the following papers. $N_f = 0$: [194–207]; $N_f = 2$: [203, 204, 207, 208], [209]; $N_f = 2 + 1$: [192, 210]. The labels indicate the method used to treat the b-quark in the simulation ("ext" and "int" stand for extrapolations and interpolations to the mass of the b-quark, respectively. The horizontal lines represent the (non-lattice) result from [130], with the quoted uncertainty

and with $N_f = 2$ flavour of dynamical Wilson quarks, the CP-PACS Collaboration find [203]

$$\frac{f_{B_s}^{N_f=2}}{f_{B_s}^{N_f=0}} = 1.14 \pm 0.05. \tag{5.238}$$

The "non-lattice" determination of f_{B_s} via fits using the experimental results for the angles of the unitarity triangle as input [130] also point to a larger value compared to the quenched theory, as can be seen from the horizontal band in the compilation in Fig. 5.20.

In current unquenched simulations, systematic effects such as lattice artefacts and the renormalization of local operators are not yet controlled at a similar level compared to the quenched approximation. Thus, despite the fact that these calculations are much more "realistic" in that they include sea quarks, the quoted overall uncertainties are still relatively large.

***B*-Parameters $\hat{B}_{B_d}$ and $\hat{B}_{B_s}$** Following the recent experimental determination of the mass difference ΔM_s at the Tevatron [211, 212], lattice determinations of the B-parameters $\hat{B}_{B_d}$ and $\hat{B}_{B_s}$ have received much attention. Although the first calculations date back to the 1990s, relatively few results are available, due to several specific technical difficulties. First, the complicated renormalization and mixing patterns of four-quark operators which afflict lattice calculations of the kaon B-parameter $\hat{B}_K$ are also encountered in the b-quark sector. Second, there is the

Table 5.6 Published lattice results for the B-parameter $B_{\mathrm{B_d}}(m_b)$ in the $\overline{\mathrm{MS}}$-scheme and the ratio $B_{\mathrm{B_s}}/B_{\mathrm{B_d}}$

Collaboration	N_f	Action	$B_{\mathrm{B_d}}(m_b)$	$B_{\mathrm{B_s}}/B_{\mathrm{B_d}}$	
Gadiyak et al. [213]	2	DWF		1.06(6)(3)	static
JLQCD [209]	2	Clover	$0.836(27)(^{56}_{62})$	$1.017(16)(^{56}_{17})$	NRQCD
Gimenez et al. [214, 215]	0	Clover	0.81(5)(4)	1.01(1)	Static
UKQCD [215, 216]	0	Clover	0.79(4)(4)	1.02(2)	Static
Christensen et al. [217]	0	Wilson	0.98(4)	0.99(1)(1)	Static
SPQ$_{\mathrm{cd}}$R [218, 219]	0	Clover	$0.87(4)(^{5}_{4})$	0.99(2)	rel./ext.
UKQCD [195]	0	Clover		$0.98(2)(^{0}_{2})$	rel./ext.

The method to treat the heavy quark is specified in the last column

added complication which arises from the fact that the b-quark cannot be simulated directly.

In Table 5.6 we list published results for $B_{\mathrm{B_d}}(m_b)$ and the ratio $B_{\mathrm{B_s}}/B_{\mathrm{B_d}}$ from a variety of methods to treat the heavy quark. The table shows that all results are broadly consistent with each other at the level of 10%, despite the different systematics. Moreover, none of the listed estimates is based on non-perturbative renormalization factors, and furthermore all entries have been computed for a fixed value of the lattice spacing, i.e. a systematic study of the continuum limit is lacking even in the quenched approximation. As for the ratio $B_{\mathrm{B_s}}/B_{\mathrm{B_d}}$, it should be mentioned that the quark masses in the simulations correspond to pion masses not much smaller than 500 MeV. However, in view of the fact that the bulk of the relevant SU(3)-flavour breaking effect in $\Delta M_s/\Delta M_d$ is expected to come from the ratio of decay constants, $f_{\mathrm{B_s}}/f_{\mathrm{B_d}}$, this may not be such a serious limitation. Results for $B_{\mathrm{B_d}}$ and $B_{\mathrm{B_d}}$ computed on dynamical gauge configurations with rooted staggered quarks should be published soon.

Another recent development is the implementation of non-perturbative renormalization for heavy-light four-quark operators in the static approximation [220, 221]. If the b-quark is treated in the static approximation, the $\Delta B = 2$ four-quark operator must be matched to its counterpart in the static theory, i.e.

$$Q^{\Delta B=2}(m_b) = C_\mathrm{L}(m_b, \mu)\widetilde{Q}_1(\mu) + C_\mathrm{S}(m_b, \mu)\widetilde{Q}_2(\mu), \tag{5.239}$$

where

$$\begin{aligned}\widetilde{Q}_1 &= \left(\bar{\psi}_h\gamma_\mu(1-\gamma_5)\ell\right)\left(\bar{\psi}_{\bar{h}}\gamma_\mu(1-\gamma_5)\ell\right) \equiv \widetilde{O}_{\mathrm{VV+AA}} + \widetilde{O}_{\mathrm{VA+AV}} \\ \widetilde{Q}_2 &= \left(\bar{\psi}_h(1-\gamma_5)\ell\right)\left(\bar{\psi}_{\bar{h}}(1-\gamma_5)\ell\right) \equiv \widetilde{O}_{\mathrm{SS+PP}} + \widetilde{O}_{\mathrm{SP+PS}},\end{aligned} \tag{5.240}$$

with ℓ denoting the light (d or s) flavour. For the physical matrix element only the parity-even operators $\widetilde{O}_{\mathrm{VV+AA}}$ and $\widetilde{O}_{\mathrm{SS+PP}}$ are relevant. If chiral symmetry is not preserved by the discretization, four-quark operators such as $\widetilde{O}_{\mathrm{VV+AA}}$ undergo complicated mixing patterns under renormalization, which necessitate finite subtractions similar to those required for the operator $O_{\mathrm{VV+AA}}$ in Eq. (5.215). However, just as in the case of $K^0 - \bar{K}^0$ mixing, the parity-even operators $\widetilde{O}_{\mathrm{VV+AA}}$ and $\widetilde{O}_{\mathrm{SS+PP}}$ can be mapped onto their parity-odd counterparts $\widetilde{O}_{\mathrm{VA+AV}}$ and $\widetilde{O}_{\mathrm{SP+PS}}$ by a flavour rotation, which realizes the transition to tmQCD at maximal twist angle. Moreover, it can be shown [220] that the combinations

$$\widetilde{O}_1' \equiv \widetilde{O}_{\mathrm{VA+AV}}, \qquad \widetilde{O}_2' \equiv \widetilde{O}_{\mathrm{VA+AV}} + 4\widetilde{O}_{\mathrm{SP+PS}} \tag{5.241}$$

renormalize purely multiplicatively. The RG running of these operators, as well as the matching to hadronic schemes based on tmQCD have been determined non-perturbatively in the SF scheme for $N_\mathrm{f} = 0$ [221] and $N_\mathrm{f} = 2$ [222], which will eventually allow for a determination of $\hat{B}_{\mathrm{B_s}}$ and $\hat{B}_\mathrm{B}$ with full control over renormalization and discretization effects. Corrections of order $1/m_b$ can be taken into account through an interpolation between the results obtained in the static approximation and for relativistic heavy quarks with masses in the region of that of the charm quark.

Semi-Leptonic *B*-Decays The CKM elements $|V_{ub}|$ and $|V_{cb}|$, which appear in the unitarity triangle relation equation (5.210), can be extracted from both inclusive and exclusive B-meson decays. However, $|V_{ub}|$ is still one of the most poorly constrained CKM elements. Its value can be determined by combining lattice calculations of semi-leptonic form factors for exclusive decays such as $\bar{B}^0 \to \pi^+\ell^-\bar{\nu}_\ell$ with the experimentally measured decay rate. If the leptons are assumed to be massless, the latter yields the combination $[|V_{ub}|\, f_+(q^2)]^2$, while the form factor $f_+(q^2)$ can be extracted from the matrix element

$$\langle \pi(\vec{p}_\pi)\, |(\bar{b}\gamma_\mu u)(0)|\, B(\vec{p}_\mathrm{B})\rangle = \left[(p_\mathrm{B} + p_\pi)_\mu - q_\mu \frac{m_\mathrm{B}^2 - m_\pi^2}{q^2}\right] f_+(q^2) + q_\mu \frac{m_\mathrm{B}^2 - m_\pi^2}{q^2} f_0(q^2). \tag{5.242}$$

Here, $q_\mu \equiv (p_\mathrm{B} - p_\pi)_\mu$ denotes the momentum transfer. For a B-meson at rest one has

$$q^2 = m_\mathrm{B}^2 + m_\pi^2 - 2m_\mathrm{B}\sqrt{m_\pi^2 + \vec{p}_\pi^{\,2}}. \tag{5.243}$$

In order to avoid large lattice artefacts, typical values of the pion momentum in simulations are restricted to

$$|\vec{p}_\pi| \lesssim 1\,\mathrm{GeV}. \tag{5.244}$$

Therefore, lattice calculations typically yield the form factors f_+ and f_0 near $q^2 = q^2_{\mathrm{max}}$. By contrast, the bulk of the experimental data is recorded in bins with small values of q^2, since the decay rate is suppressed near q^2_{max}. Therefore, an extrapolation to small values of q^2 must be performed, which requires an

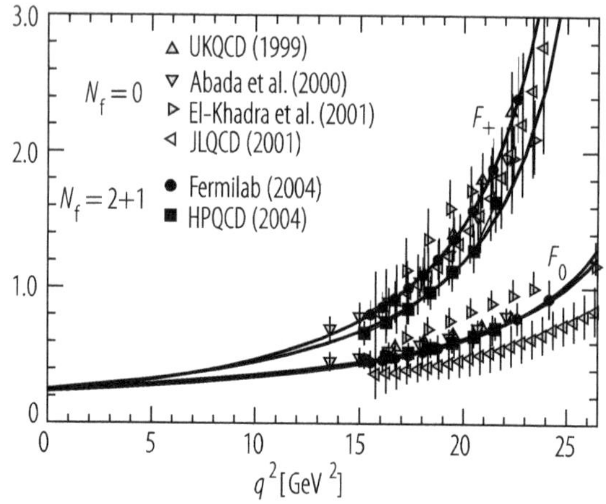

Fig. 5.21 Form factors f_+ (upper data set) and f_0 for $B \to \pi \ell \nu$ decays (taken from Ref. [224]). The data are taken from Refs. [225] (UKQCD), [226] (Abada et al.), [227] (El-Khadra et al.), [228] (JLQCD) and [229] (FNAL04). The unquenched results by HPQCD have been updated [230]

ansatz for the shape of the form factor. Although a parameterization of the q^2-dependence which goes beyond vector pole dominance and is also consistent with the expected heavy-quark scaling laws has been proposed [223], the extrapolation to small momentum transfers typically introduces some model dependence in the result for $|V_{ub}|$.

Figure 5.21 shows a compilation of lattice data for the form factors as a function of q^2 together with the curves which represent the extrapolations to $q^2 = 0$. The problem of the model dependence introduced by the extrapolation to small momentum transfer can be avoided by combining form factors from lattice simulations with the decay rate measured in restricted intervals of q^2, which overlap with the range of momentum transfers that are directly accessible in simulations. Such a procedure has been performed by the CLEO Collaboration [231]. The result for $|V_{ub}|$ obtained in this way is somewhat smaller compared to the standard method based on form factor extrapolations, but the uncertainties are still quite large. For the actual estimates of $|V_{ub}|$ obtained in this way, the reader may consult the original papers.

Semi-leptonic heavy-to-heavy decays such as $\bar{B} \to (D, D^*)\ell\bar{\nu}_\ell$ offer a way to determine $|V_{cb}|$. In this case it is convenient to use the four-velocities of the two mesons as the kinematical variables instead of the four-momenta. The decay amplitudes are then parameterized in terms of six form factors, i.e.

$$\begin{aligned}
\frac{\langle D(v')\,|(\bar{c}\gamma^\mu b)|\,B(v)\rangle}{\sqrt{m_{\mathrm{B}} m_{\mathrm{D}}}} &= (v+v')^\mu h_+(\omega) + (v-v')^\mu h_-(\omega) \\
\frac{\langle D^*(v',\epsilon)\,|(\bar{c}\gamma^\mu b)|\,B(v)\rangle}{\sqrt{m_{\mathrm{B}} m_{\mathrm{D}}^*}} &= \mathrm{i}\epsilon^{\mu\nu\alpha\beta}\epsilon_\nu^* v'_\alpha v_\beta h_{\mathrm{V}}(\omega) \qquad (5.245) \\
\frac{\langle D^*(v',\epsilon)\,|(\bar{c}\gamma^\mu \gamma_5 b)|\,B(v)\rangle}{\sqrt{m_{\mathrm{B}} m_{\mathrm{D}}^*}} &= (\omega+1)\epsilon^{*\mu} h_{\mathrm{A}_1}(\omega) - \epsilon^* \cdot v\left[v^\mu h_{\mathrm{A}_2}(\omega) + v'^\mu h_{\mathrm{A}_3}(\omega)\right],
\end{aligned}$$

where $\omega = v{\cdot}v'$. In the limit of infinite heavy quark mass, four out of these six form factors can be replaced by a single, universal form factor, $\xi(\omega)$, which is called the Isgur-Wise function [232]

$$m_b,\, m_c \to \infty \quad \Rightarrow \quad h_+(\omega) = h_{A_1}(\omega) = h_{A_3}(\omega) = h_V(\omega) = \xi(\omega), \tag{5.246}$$

while $h_-(\omega)$ and $h_{A_2}(\omega)$ vanish as m_b, m_c become infinitely heavy. Outside the exact heavy-quark limit, the relation between the Isgur-Wise function and the form factors is modified. For instance,

$$h_+(\omega) = (1 + \beta_+(\omega) + \gamma_+(\omega))\, \xi(\omega), \tag{5.247}$$

where β_+, γ_+ parameterize radiative corrections and corrections arising from operators of higher dimension, which are suppressed by additional inverse powers of the heavy quark mass. Similar relations hold for h_{A_1}, h_{A_3} and h_V. Another important result, known as Luke's Theorem [233], states that at zero recoil, $v = v'$, i.e. $\omega = 1$, the leading corrections to the form factors h_+ and h_{A_1} are quadratic in the inverse heavy quark mass.

With this setup one may devise a strategy to determine $|V_{cb}|$ by combining the experimentally determined decay rate with lattice calculations of the form factors. The differential decay rate for $\bar{B} \to D^*\ell\bar{\nu}_\ell$ in the limit of zero recoil reads

$$\lim_{\omega\to 1} \frac{1}{\sqrt{\omega^2-1}} \frac{\mathrm{d}\Gamma(B \to D^*\ell\nu)}{\mathrm{d}\omega} = |V_{cb}|^2 \frac{G_F^2}{4\pi^3}(m_B - m_{D^*})^2 m_{D^*}^3 [h_{A_1}(1)]^2, \tag{5.248}$$

which, owing to Luke's Theorem, receives corrections of order $1/m_c^2$ only. For $\omega > 1$ the single axial form factor h_{A_1} must be replaced by a linear combination of several form factors. Thus, the theoretical uncertainties appear to be controlled best at zero recoil. Since the rate is suppressed near $\omega = 1$, the measured decay rate must be extrapolated to that value to determine $|V_{cb}|$. Most of the published lattice calculations of the form factors and the Isgur-Wise function [234–238] are therefore focused on the determination of the slope of $\xi(\omega)$ at $\omega = 1$. The measured decay rate can then be extrapolated to zero recoil using a particular parameterization of $\xi(\omega)$, with its slope constrained via the lattice calculation. After taking radiative and power corrections into account, a value for $|V_{cb}|$ can be extracted.

A different but related strategy is to compute the form factors $h_+(1)$ and $h_{A_1}(1)$ directly via suitably chosen double ratios of hadronic matrix elements in which many systematic effects can be expected to cancel [239, 240]. Using the "Fermilab approach" for the heavy quarks in the quenched approximation, the authors of Ref. [240] find

$$h_{A_1}(1) = 0.913^{+0.024\,+0.017}_{-0.017\,-0.030}, \tag{5.249}$$

where the first error is statistical, while the second represents an estimate of various systematic uncertainties added in quadrature. Again, this result can be combined with the experimental decay rate to determine $|V_{cb}|$. More details can be found in [240].

Most lattice studies of heavy-to-heavy semi-leptonic B-decays have been restricted to the quenched approximation. However, results for the form factors from dynamical simulations can be expected in the near future. Clearly, in order to have maximum impact on the determination of $|V_{cb}|$, systematic effects arising from lattice artefacts and the formulation used to treat the heavy quark must be controlled to a high degree.

5.8 Concluding Remarks

In this article we have introduced the lattice approach to QCD and discussed a variety of applications, which range from hadron spectroscopy, confinement, quark masses and the running coupling, to spontaneous chiral symmetry breaking and hadronic matrix elements for flavour physics. This illustrates not only the versatility of the lattice method, but also indicates that lattice calculations have become ever more important for making quantitative predictions in the notoriously difficult sector of non-perturbative QCD. Still, a great number of other applications have not even been covered here, including nucleon structure functions and form factors, calculations at finite temperature and/or chemical potential, or detailed investigations of the QCD vacuum structure.

That lattice calculations have reached this standing is owed to the enormous progress which been made in developing more efficient algorithms for dynamical fermions, better discretizations, as well as a number of new theoretical concepts such as non-perturbative renormalization. These developments, in conjunction with the availability of ever more powerful computers, shall allow for precise computations of many phenomenologically relevant quantities, which previously seemed virtually intractable.

5.9 Addendum: QCD on the Lattice

5.9.1 Introduction

Since the first edition of this article [241] the field of lattice QCD has undergone a huge transformation. While the actual methodology was well established at the time of writing (2007), few simulations employing dynamical quarks had produced results with controlled errors, having a direct impact on phenomenology and experiment. During the past ten years or so this has changed dramatically.

Simulations with light dynamical quarks, whose masses correspond to the physical value of the pion mass, have become the state of the art, and the effects of dynamical strange and charm quarks are now routinely included as well. In fact, lattice calculations of certain observables have reached (or are aiming for) a level of precision where the effects of the breaking of isospin symmetry can no longer be ignored. This necessitates that lattice QCD must account not only for the effects of unequal u and d quark masses but also for corrections due to electromagnetism, owing to the different electric charges of up- and down-type quarks.

In this context it is interesting to quote a remark by Ken Wilson, made at the 1989 International Conference on Lattice Field Theory [242]: *"I still believe that an extraordinary increase in computing power* (10^8 *is I think not enough) and equally powerful algorithmic advances will be necessary before a full interaction with experiment takes place."* Given that, in 1989, the most powerful supercomputers could sustain 10 GFlops (i.e. 10^{10} floating point operations per second), Wilson's estimate was tantamount to requiring ExaFlops capabilities (10^{18} Flops) for lattice QCD to make an impact, a performance figure that has only been reached very recently by less than a handful of machines. The enormous progress that the field of lattice QCD has already seen over the past decade proves that Wilson's view was far too pessimistic.[21] For instance, results from lattice calculations for the decay constants and form factors of mesons and baryons containing heavy quarks are vital input for global analyses of observables in flavour physics, designed to constrain the elements of the Cabibbo–Kobayashi–Maskawa matrix. Furthermore, lattice QCD yields precise values for the masses of the light (u, d, s) quarks [244].

An impressive testimony to the importance of lattice QCD for the entire field of particle physics is the regular report provided by the Flavour Lattice Averaging Group (FLAG). Since its inception in 2007, FLAG has been charting the progress in lattice QCD, by collecting results for a range of phenomenologically relevant quantities. Taking inspiration from the Particle Data Group, FLAG assesses the quality of individual calculations and produces world averages by combining those results that satisfy a defined set of requirements regarding the overall control over systematic effects. Three editions of the FLAG report, published in 2010 [245], 2013 [246] and 2016 [247], have appeared until now, and a fourth one has been published in 2019 [248]. In fact, the current status of lattice calculations of many observables that have been reviewed in the first edition of this article can be found in these comprehensive reports.

This short review is organized as follows. In Sects. 5.9.2 and 5.9.3 we give an update of lattice calculations applied to hadron spectroscopy, weak hadronic matrix elements and the determination of Standard Model parameters such as quark masses and the strong coupling constant. These quantities were covered extensively in the original edition of [241]. Then, in Sect. 5.9.4 we extend the discussion to the determination of quantities that describe structural and other properties of the nucleon, such as form factors and the axial charge. Finally, in Sect. 5.9.5 we discuss lattice calculations of the hadronic contributions to the muon anomalous magnetic

[21] Even Wilson himself acknowledged, at least partially, that this was the case [243].

moment, which is a key quantity to study possible deviations from the Standard Model. The review concludes with a few remarks on the progress achieved over the past decade and an outlook for future calculations.

5.9.2 *Hadron Spectroscopy*

The calculation of the light hadron spectrum, i.e. the masses of the lowest-lying mesons and baryons has long been regarded a benchmark for lattice QCD. In the quenched approximation, i.e. in the absence of dynamical quarks, a significant deviation between the calculated spectrum and experiment at the level of 10–15% was observed. When the light hadron spectrum could eventually be accurately reproduced within the overall uncertainty after the inclusion of light dynamical quarks [249–252] (see Fig. 5.22), this was hailed as a major success of lattice QCD. Thanks to these milestone results, the credibility of lattice calculations was firmly established throughout the particle and hadron physics communities.

Calculations of the light hadron spectrum have since been further refined, by taking the effects of isospin breaking into account. Strong isospin breaking arises from the mass splitting between the u and d quarks, $m_u \neq m_d$. Since the electric charges of u and d quarks differ as well, electromagntic interactions are another source of isospin breaking. The formulation of QED on a lattice of finite volume poses considerable technical challenges since the photon is massless. There are several strategies to address the problem of the associated zero mode, and we refer the reader to recent reviews of the subject [253–255], which also serve as a guide to the literature.

After the inclusion of strong and electromagnetic isospin breaking effects, it became possible to perform another benchmark calculation, namely the accurate determination of the neutron-proton mass difference, as well as the mass splittings of other baryonic iso-multiplets [256–259]. The ability to determine isospin breaking effects arising from QED was also instrumental for calculations of the electromagnetic mass splittings of pions and kaons [260–265], which can be used to study violations of Dashen's theorem [266]. The latter states that the electromagnetic self-energies of the charged pions and kaons are identical, while those of their neutral

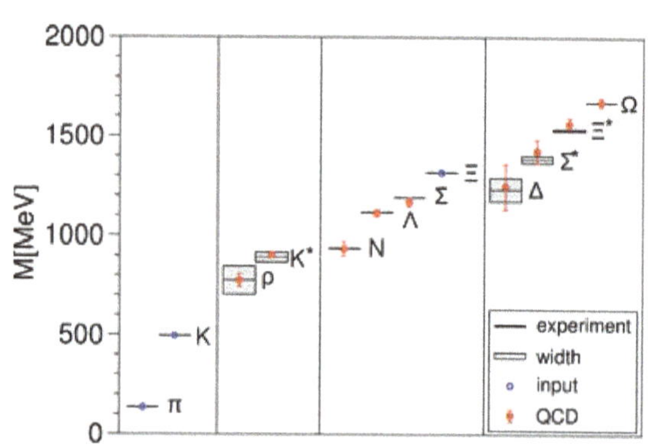

Fig. 5.22 The spectrum of the lowest-lying hadrons as computed in Ref. [250], to be compared to Fig. 5.5 of the original review [241]

counterparts vanish. More details are found in section 3.1.1 of the FLAG report [247].

Another recent focus of lattice spectroscopy has been the determination of the excitation spectrum and the properties of hadronic resonances. This is a major refinement of previous calculations in which the masses of resonances (the simplest being the ρ-meson) were extracted naively from the exponential decay of the vector correlation function, thereby ignoring the fact that resonances are characterized both by a mass and a width. The general framework for the study of resonance properties in lattice QCD was developed by Lüscher already in the 1980s and 1990s [267–270], and it is only now that the potential of this elegant and powerful formalism can be fully exploited. The key idea that underlies the Lüscher method is the realization that computing the energy levels of multi-particle states in a finite volume gives access to the scattering phase shifts in infinite volume, provided that the spectrum (including excited states) can be determined sufficiently well for a range of kinematical situations. The latter are typically determined by the lattice volume and/or the total momentum of the multi-particle system in question.

To be more specific, let us consider the simplest resonance, the ρ-meson, whose properties can be accessed in p-wave $\pi\pi$ scattering. For energies below the inelastic threshold, the Lüscher condition reads

$$\phi(q) + \delta_1(k) = 0 \mod \pi, \qquad q = \frac{kL}{2\pi}, \tag{5.250}$$

where $\phi(q)$ is a known kinematic function of the scaled scattering momentum in units of the box size, $q = kL/2\pi$ and δ_1 is the scattering phase shift. The scattering momentum k is determined from the nth energy level ω_n in a finite volume, according to

$$\omega_n = \sqrt{m_\pi^2 + k^2}, \tag{5.251}$$

where m_π is the pion mass. Figure 5.23 shows an example of a calculation of the p-wave scattering phase shift as a function of the centre-of-mass energy [271].

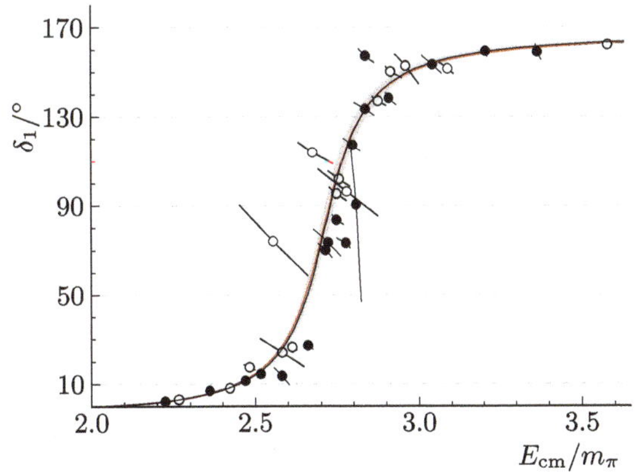

Fig. 5.23 The p-wave scattering phase shift of the ρ-meson, computed for $m_\pi = 280\,\mathrm{MeV}$ as a function of the centre-of-mass energy [271]. Data obtained for two different values of the lattice spacing (open and filled grey symbols) are shown. The solid line is obtained from a fit to a Breit-Wigner *ansatz* for the resonance

A crucial ingredient for the reliable determination of not just the energy level of the ground state but also the excitation spectrum is the use of correlator matrices computed using a suitable basis of interpolating operators (see Section 5.3 in Ref. [241]). The diagonalization of the correlator matrix can be achieved by solving a generalized eigenvalue problem from which the energy levels in a given channel can be determined [272–274]. The sometimes arduous task of constructing efficient interpolators for multi-particle states has been helped enormously by practical methods to compute "all-to-all" quark propagators [275] and, in particular, the so-called "distillation" technique [276, 277]. With these new developments it has been possible to perform lattice investigations of $\pi\pi$ scattering and the ρ resonance [278–291], as well as determinations of $K\pi$ [292, 293] and KK scattering lengths [294, 295]. The formalism has also been used to study meson-baryon [296–300] and baryon-baryon [301, 302] interactions.

While the original Lüscher formalism was derived for the case of elastic two-particle scattering, it has now been generalized to coupled-channel systems [303–307], including the treatment of three-particle thresholds [308–315]. It also opens the possibility to study weak non-leptonic kaon decays [316] and compute form factors for timelike momentum transfers [317–320].

Moreover, the experimental discovery of new charmonium-like resonances, commonly referred to as the X, Y and Z states, has kindled a new interest in hadron spectroscopy. A distinctive feature of the new resonances is their closeness to particle thresholds, and efforts are underway to gain a detailed understanding of the resonance structure in the charm sector. Using the formalism described above, there have been many calculations of a variety of charmonium-like resonances in lattice QCD. In view of the vast literature, we refer the reader to several recent reviews of the subject [321–323].

5.9.3 Parameters of the Standard Model

The Standard Model (SM) contains 19 parameters (excluding the neutrino sector) whose values are not predicted by the theory itself but must instead be fixed using experimental input. In many cases the relations between experimentally accessible observables and SM parameters involve quantities that encode the effects of the strong interactions. A well-known example is the kaon B-parameter B_K that enters the relation between the quantity ϵ_K, which is a measure of indirect CP violation, and a particular combination of Cabibbo–Kobayashi–Maskawa (CKM) matrix elements V_{td}, V_{ts}, i.e.

$$\epsilon_K \propto \hat{B}_K \,\mathrm{Im}\,(V_{td}\, V_{ts}^*). \tag{5.252}$$

While ϵ_K can be determined experimentally from a ratio of decay amplitudes of long- and short-lived K-mesons, $K_{L,S} \to (\pi\pi)_{I=0}$, the parameter $\hat{B}_K$ must be extracted from the hadronic matrix element of a four-quark operator between

K^0 and $\bar{K}^0$ states. Obviously, such a calculation must be performed in the non-perturbative regime of QCD since it involves typical hadronic scales.

Other CKM matrix elements, such as V_{us}, V_{ub} and V_{cb} are related to weak processes involving kaons, D- and B-mesons, which are described by a variety of leptonic decay constants (and their ratios), form factors of semi-leptonic meson and baryon decays, as well as the B-parameters that encode strong interaction contributions to $B^0-\bar{B}^0$ and $B_s^0-\bar{B}_s^0$ mixing. All these quantities have been studied in lattice QCD for many years, and increasingly precise estimates with controlled systematic errors have appeared over the past decade. They have been instrumental for recent analyses of the unitarity of the CKM matrix [324–327].

Similar considerations apply to SM parameters such as the strong coupling constant α_s and the masses of the quarks. While the asymptotic scaling behaviour of α_s gives rise to the dimensionful Λ-parameter that encodes the intrinsic scale of QCD, the quark masses are external parameters. Providing the link between experimentally accessible quantities and quark masses, as well as expressing the Λ-parameter in units of some measurable low-energy quantity has been a primary task for lattice QCD. Lattice calculations have also be instrumental for determining the coupling constants of effective descriptions of QCD, such as the low-energy constants of Chiral Perturbation Theory.

The importance of accurate, model-independent determinations of SM parameters and input quantities for flavour physics has led to the foundation of the Flavour Lattive Averaging Group (FLAG). Updates of the FLAG report have appeared at regular intervals since the publication of its first edition in 2010 [245]. As part of its mission, FLAG issues global estimates and averages of lattice results, provided that they satisfy a set of defined quality criteria. FLAG estimates are quoted separately according to the sea quark content of the calculations that enter the global analyses, i.e. whether they have been obtained with a degenerate doublet of u, d quarks ($N_f = 2$) or with an additional dynamical strange ($N_f = 2+1$) and charm quark ($N_f = 2+1+1$). The current status of lattice QCD calculations of quark masses, the strong coupling, decay constants, form factors, mixing parameters and low-energy constants is summarized in Tables 1 and 2 of the 2016 FLAG report [247]. The FLAG webpage[22] contains additional updates. Below we comment on the current status of a few selected quantities.

Quark Masses According to FLAG, the strange quark mass is known to 1% precision, while the accuracy in the determination of the average u and d quark mass, $\hat{m} \equiv \frac{1}{2}(m_u + m_d)$, varies between 1–5 %, depending on the sea quark content [328–332, 332–340]. Thanks to the recent progress in including the effects of isospin breaking in lattice QCD calculations, estimates for the masses of the individual u and d quarks could also be obtained, typically with 2 – 5 % precision [261, 262, 264, 330]. Furthermore, the masses of the heavy quarks have been determined with excellent precision [328, 330–332, 335, 337, 341–348].

---

[22] http://flag.unibe.ch/.

Running Coupling A milestone was achieved by the ALPHA collaboration, who published [349] an estimate for $\alpha_s(M_Z^2)$ obtained by tracing the scale evolution of the strong coupling non-perturbatively over several orders of magnitude into an energy range where the application of perturbation theory can be considered safe (at least as far as the quoted precision is concerned). Their main result is the determination of the Λ-parameter in three-flavour QCD, i.e. $\Lambda_{\overline{\rm MS}}^{(3)} = 341(12)$ MeV, which can be matched to the Λ-parameter in the five-flavour theory using perturbation theory, giving $\Lambda_{\overline{\rm MS}}^{(5)} = 215(10)(3)$ MeV. Finally, this is translated into the result for the strong coupling [349]:

$$\alpha_s^{\overline{\rm MS}}(M_Z^2) = 0.11852(84). \tag{5.253}$$

The quoted error is 30% smaller than that of the 2016 PDG estimate of $\alpha_s = 0.1181(11)$ [244]. The latter includes lattice results from Refs. [331, 335, 350–354].

Kaon Weak Matrix Elements The kaon B-parameter B_K is now known with an overall accuracy of 1.3% [336, 355–359]. Moreover, the calculations of matrix elements relevant for $K^0 - \bar{K}^0$ mixing have been extended to include operators that arise in extensions of the Standard Model [355, 358–363].

Lattice QCD results for kaon leptonic decay constants (more precisely: the ratio f_{K^+}/f_{π^+}) and the form factor $f_+(0)$ describing semi-leptonic $K \to \pi\ell\nu_\ell$ decays have now reached a level of precision that enables a competitive and model-independent determination of V_{us} (see Sect. 5.7.1 of the original review article). Moreover, it is possible to test the unitarity of the first row in the CKM matrix, i.e. the relation

$$|V_{ud}|^2 + |V_{us}|^2 + |V_{ub}|^2 = 1, \tag{5.254}$$

by combining experimental information with lattice results for $f_+(0)$ and f_{K^+}/f_{π^+}. Neglecting the contribution from $|V_{ub}|^2 \approx 1.7 \cdot 10^{-5}$, one finds that $|V_{ud}|^2 + |V_{us}|^2$ can be determined with a total precision at the percent level, by combining the FLAG estimates[23] for $f_+(0)$ and f_{K^+}/f_{π^+} with the experimentally accessible combinations $|V_{us}|f_+(0) = 0.2165(4)$ and $|V_{us}/V_{ud}|f_{K^+}/f_{\pi^+} = 0.2760(4)$ [244, 364]. In QCD with dynamical light, strange and charm quarks ($N_f = 2+1+1$) the result is $|V_{ud}|^2 + |V_{us}|^2 = 0.9797(74)$, which signals a slight tension of 2.7 standard deviations with the Standard Model. The precision of the unitarity test can be sharpened considerably by replacing $|V_{ud}|$ with the value extracted from neutron β-decay, i.e. $|V_{ud}| = 0.97417(21)$ [365]. It is then sufficient to provide one additional constraint from lattice QCD, either in the form of $f_+(0)$ or the ratio f_{K^+}/f_{π^+}. Inserting the lattice result for $f_+(0)$ yields $|V_{ud}|^2 + |V_{us}|^2 = 0.99884(53)$, which again differs from unitarity by about 2σ. Using instead the

[23] See the web update at http://flag.unibe.ch/.

lattice result for f_{K^+}/f_{π^+} implies $|V_{ud}|^2 + |V_{us}|^2 = 0.99986(46)$. Thus, first-row unitarity can be probed with permil-level precision [247].

Heavy-Light Decay Constants and Form Factors The treatment of heavy quarks on the lattice presents additional significant challenges: since the mass of the charm quark is close to typical values of the inverse lattice spacing, which acts as the ultraviolet cutoff, lattice results are prone to suffering from large discretisation errors. Moreover, the mass of the bottom quark exceeds currently accessible values of a^{-1}, and specially designed methods are required for a consistent treatment. This has been discussed extensively in Sect. 5.7.2 of the original review.

The overall precision of lattice estimates for weak hadronic matrix elements involving charm and bottom quarks has vastly improved over the past decade. As shown in Table 2 of FLAG 2016 [247], the leptonic decay constants of the B and B_s mesons are now known at the level of 2%, while ratios such as f_{B_s}/f_B have been determined with even better accuracy [347, 366–373]. Since the 2016 edition of the FLAG report, new results obtained with $N_f = 2{+}1{+}1$ flavours of dynamical quarks [343, 374, 375] have pushed the overall precision to the sub-percent level, which is an impressive achievement. Also the estimates of the individual B-parameters $\hat{B}_B$ and $\hat{B}_{B_s}$, their ratios and combinations with the leptonic decay constants are now known with overall errors at the percent level [347, 370, 376, 377].

Results for form factors describing semi-leptonic decays of hadrons containing b-quarks, such as $B \rightarrow (D, D^*)\ell\nu$, or even $\Lambda_b \rightarrow p\ell\nu$ have reached a level of precision that is sufficient for competitive determinations of the CKM matrix elements V_{cb} and V_{ub} from exclusive processes. An extensive discussion is presented in the web update of the FLAG report.

5.9.4 Nucleon Matrix Elements

The understanding of the internal structure of the nucleon in terms of the fundamental interactions between its constituents, the quarks and gluons, has become a major activity within the field of lattice QCD. Structural information is encoded in quantities such as form factors, structure functions and (generalized) parton distribution functions (PDFs). An open problem in this context is the decomposition of the proton's spin in terms of the spins of quarks and gluons, as well as their angular momentum [378, 379]. Another important issue is the so-called "proton radius puzzle" [380], which arises due to the observed discrepancy between the proton radius extracted from the Lamb shift in muonic hydrogen [381, 382] compared to the more traditional determinations from electron-proton scattering [383] or the Lamb shift in electronic hydrogen [384]. Accurate knowledge of the electromagnetic form factors of the proton are indispensable in order to resolve—or corroborate—this puzzle.

The determination of quantities such as nucleon form factors in lattice QCD proceeds by calculating the corresponding hadronic matrix elements between nucleon initial and final states. A strong motivation for computing such quantities is provided by the fact that fundamental interactions are often probed in scattering experiments involving nuclear targets. For instance, probing the neutrino sector requires accurate knowledge of the scattering cross sections of neutrinos with nuclear targets. Similar considerations apply to the search for dark matter candidates. Therefore, precise determinations of the corresponding nucleon matrix elements are indispensable for exploring the limits of the SM.

The past decade has seen a huge rise in the number of publications describing lattice calculations of nucleon matrix elements. Quantities that have been studied include

- the electromagnetic form factors of the nucleon, $G_E(Q^2)$ and $G_M(Q^2)$, which give access to the electric and magnetic charge radii of the nucleon and its magnetic moment [385–397];
- the iso-vector axial charge of the nucleon, g_A, which is a measure of the strength of weak interaction in neutron β-decay [386, 387, 389, 392, 397–414], as well as the scalar and tensor charges, g_S and g_T [386, 393, 404–406, 411, 412, 414–419];
- axial and induced pseudoscalar form factors of the nucleon [397, 407, 409, 420, 421], as well as the strange electromagnetic and axial form factors [421–426, 529] which probe the quark sea inside the nucleon;
- the pion-nucleon σ-term $\sigma_{\pi N}$ [412, 427–438] and the strange content of the nucleon σ_s [412, 429–431, 435–444]. These σ-terms are proportional to nucleon matrix element of the flavour-diagonal scalar density, $\overline{q}q$, which parameterizes the rate of change in the nucleon mass due to a non-zero value of the corresponding quark mass.

Recent reviews, presented at the annual conference on lattice field theory, can be found in Refs. [445–447]. Some results on nucleon form factors and other matrix elements are reviewed in section 3.2.5 of [448], and a dedicated chapter has been prepared for the 2019 edition of the FLAG report. In addition, there has been a community effort in the form of a white paper [449] in which lattice results are used to reduce the overall uncertainties in polarized and unpolarized proton PDFs and their moments.

The relevant nucleon hadronic matrix elements are extracted from suitable three-point correlation functions of quark bilinears between interpolating operators representing the initial and final-state nucleons. Examples of the corresponding diagrams, with the initial-state nucleon placed at Euclidean time $t = 0$ (the source), the final-state nucleon at time t_s (the sink) and the operator insertion at time t, are shown in Fig. 5.24. In addition to the quark-connected diagram, in which the operator is inserted on a valence quark line, there are also quark-disconnected diagrams in which the operator probes the quark sea. The latter class of diagrams must be computed to determine, for instance, iso-scalar quantities, the strangeness form factors and the σ-terms.

Precise determinations of nucleon matrix elements with controlled statistical and systematic errors are particularly challenging. This is a consequence of the fact that the noise-to-signal ratio in three-point correlation functions corresponding to the

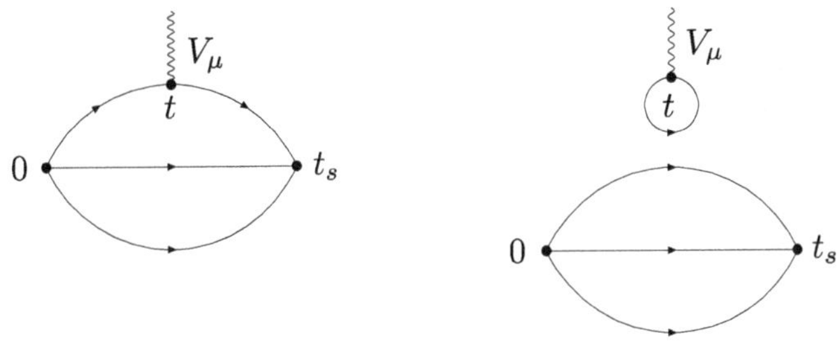

Fig. 5.24 Quark-connected (left) and disconnected (right) diagrams representing the interaction of the vector current with the nucleon

diagrams in Fig. 5.24 grows exponentially with a rate proportional to $\exp\{(m_N - \frac{3}{2}m_\pi)t_s\}$, where m_N and m_π denote the nucleon and pion masses, respectively, and t_s is the source-sink separation. Techniques designed to enhance the statistical signal at affordable numerical cost have been developed and applied, including the truncated solver method [450] and "all-mode-averaging" [451]. Furthermore, a technique to achieve an exponential error reduction via domain decomposition and multi-level integration has been proposed and tested in [452, 453]. So far, it has not been employed in actual calculations of nucleon matrix elements with dynamical quarks.

Quark-disconnected diagrams of the type shown on the right of Fig. 5.24 are intrinsically even noisier than their quark-connected counterparts and require special techniques that balance statistical accuracy against numerical cost. Commonly applied variance reduction techniques for quark-disconnected diagrams include hierarchical probing [454, 455], the coherent source sequential propagator method [389, 456] low-mode averaging [457, 458], the hopping parameter expansion [450, 459–461] and partitioning/dilution [275, 462]).

Despite these improvements, typical values of the source-sink separation t_s for which the signal has not yet disappeared into the noise are limited to $t_s \gtrsim 1.5\,\text{fm}$. Since the correlation function is dominated by the ground state for $t, (t_s - t) \to \infty$, it is then not guaranteed that the matrix element of interest can be extracted without incurring a bias from unsuppressed excited state contributions, as long as one cannot probe the region $t_s > 1.5\,\text{fm}$. Hence, in addition to "standard" systematic effects such as lattice artefacts or finite-volume effects, one must also ensure that the asymptotic regime of nucleon correlation functions has been correctly isolated. Indeed, controlling excited state effects has become perhaps the most important issue in current lattice calculations of nucleon matrix elements. The commonly used strategies include

- fits to three-point correlation functions or suitably defined ratios of correlators including sub-leading contributions from excited states [393, 394];
- calculations of three-point correlators summed over the operator insertion time t [463–467]. Contributions from excited states can be shown to be parametrically more strongly suppressed than in the standard case [468];

- increasing the projection of nucleon interpolators onto the ground state [404, 469], as well as the construction of an operator basis for the variational method, which allows for the projection onto the approximate ground state [456, 469, 470].

The first two approaches proceed by fitting data obtained in a finite interval of source-sink separations t_s to a function that describes the approach to the asymptotic behaviour. To be able to resolve the sub-leading contributions from excited states in such a fit obviously requires sufficiently precise input data.

Another challenge for lattice calculations of nucleon matrix elements is the accurate description of the pion mass dependence. Although simulations at or near the physical pion mass are now routinely performed, the result at the physical point is often obtained via an extrapolation in the pion mass. The fit *ansatz* for the pion mass dependence is usually derived from chiral effective theory. However, the convergence properties of baryonic chiral perturbation theory are not as well understood as in the mesonic sector, and it is still unclear whether the predicted functional form provides a good description in the pion mass range over which it is applied. It is thus mandatory to gather sufficiently precise results at small enough pion mass, in order to control the systematic uncertainty associated with the chiral extrapolation.

Instead of performing a detailed survey of a variety of nucleon observables, we single out one particular quantity—the iso-vector axial charge of the nucleon, g_A, which is perhaps the most widely studied of nucleon matrix elements in lattice QCD and serves to illustrate the current state of the art. The axial charge describes the coupling of the W boson to the nucleon. In Minkowski space notation it is defined by

$$\langle \mathrm{p}(k,s')|\,\overline{u}\gamma^{\mu}\gamma_5 d\,|\mathrm{n}(k,s)\rangle = g_A\,\overline{u}_p(k,s')\,\gamma^{\mu}\gamma_5\,u_n(k,s), \tag{5.255}$$

where $u_n(k,s)$ and $u_p(k,s')$ denote the Dirac spinors of the neutron and proton with four-momentum k and spins s and s', respectively. The axial charge has been measured experimentally in neutron β-decay, and the current world average quoted in the PDG is $g_A = 1.2724 \pm 0.0023$ [471]. Provided that the experimental sensitivity is sufficient, it may be possible to probe for scalar and tensor interactions that are generated by loop effects or arise due to new forces in extensions of the SM. The definitions of the associated scalar and tensor charges, g_S and g_T are derived from Eq. (5.255) by replacing the axial current $\overline{u}\gamma^{\mu}\gamma_5 d$ by the scalar density $\overline{u}d$ and the tensor current $\overline{u}\sigma^{\mu\nu}d$, respectively.

The calculation of g_A is facilitated by the fact that it is derived from a forward matrix element without any momentum transfer and, secondly, since the contributions from quark-disconnected diagrams cancel in the iso-vector combination, for mass-degenerate up and down quarks. Coupled with the fact that a precise experimental value is known, the iso-vector axial charge is a benchmark quantity for lattice calculations of nucleon matrix elements. Obviously, the ability of state-of-the-art lattice calculations to reproduce the experimental result will enhance the

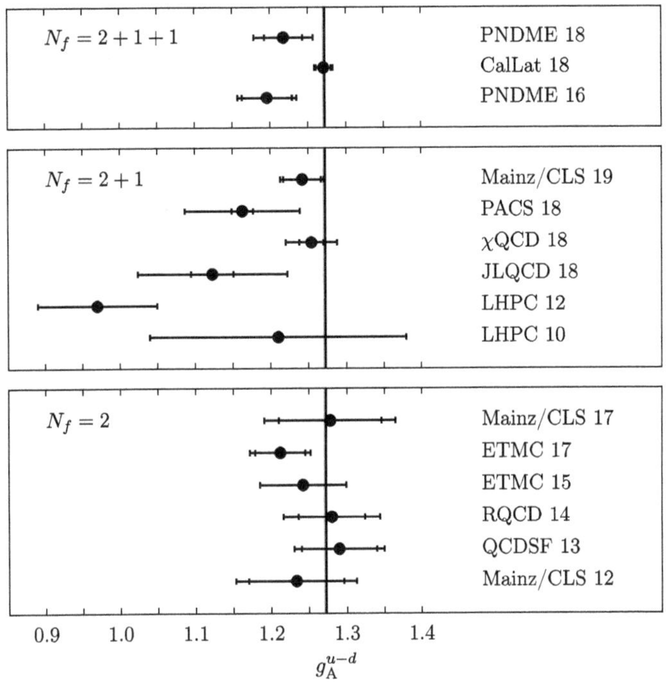

Fig. 5.25 Compilation of recent results for the isovector axial charge. The vertical red band indicates the PDG average [471]. Lattice results are labelled by PNDME 18 [411], CalLat 18 [410], PNDME 16 [406], Mainz/CLS 19 [414], PACS 18 [397], χQCD 18 [413], JLQCD 18 [412], LHPC 12 [392], LHPC 10 [389], Mainz/CLS 17 [409], ETMC 17 [407], ETMC 15 [405], RQCD 14 [404], QCDSF 13 [403] and Mainz/CLS 12 [402]

credibility of lattice predictions for the unmeasured charges g_S and g_T. Figure 5.25 shows a compilation of recent results for g_A, obtained in lattice QCD with $N_f = 2, 2+1$ and $2+1+1$ flavours of dynamical quarks. While most estimates agree with the experimental result within errors, it is clear that the overall precision of current lattice calculations does not match that of the experiments. To state this observation more precisely, we note that the typical total error of current lattice results is at the level of 1–3% while experiment is an order of magnitude more precise. It should also be mentioned that, more often than not, lattice results tend to be slightly lower that the PDG average. Whether this is due to a remnant bias from excited state contributions or indeed to any other systematic effect, must be investigated in future calculations able to realize larger source-sink separations.

The tendency to underestimate g_A in early lattice calculations of g_A has been attributed to unsuppressed excited state effects. In this context it is interesting to note that recent analyses of the contributions from $N\pi$ states to nucleon matrix elements based on chiral effective theory [472, 473] suggest that the asymptotic (physical) value of g_A is approached from above. The different conclusions drawn from numerical and analytic studies can only be reconciled if one succeeds in simulating significantly larger source-sink separations at affordable cost.

Given that lattice QCD calculations reproduce the experimental value of benchmark quantities such as the axial charge at the level of a few percent, it is interesting to look at quantities that have not been measured so far. Results for the (iso-vector) scalar and tensor charges have been reported in [386, 393, 404–406, 411, 412, 414–419]. For both quantities one obtains $g_S, g_T \approx 1$, and while the typical overall uncertainty in g_S is at the level of 10%, the tensor charge is determined with 3% precision, similar to that of g_A. The 2019 edition of the FLAG report contains a detailed compilation and comparison of results for the axial, scalar and tensor charges, as well as flavour-singlet charges and σ-terms. Calculations of these quantities have matured to a level which allows for global averages to be determined.

Lattice calculations of nucleon matrix elements is a rich subject, and while a comprehensive discussion of other quantities such as form factors and moments of PDFs is beyond the scope of this short review, we refer the reader to recent reviews [445–447], specific sections of [448] and the white paper on PDFs [449].

5.9.5 Hadronic Contributions to the Muon Anomalous Magnetic Moment

The SM describes with great accuracy and precision the properties of the constituents of the visible matter in the universe but leaves several profound questions unanswered. For instance, it cannot account for the matter-antimatter asymmetry and does not explain the vast hierarchy between the electroweak scale and the Planck mass. Most prominently, the SM cannot account for the presence of dark matter in the universe for which there is overwhelming observational evidence. Against this backdrop, the exploration of the limits of the SM and the search for "new physics" has become a major activity in particle physics. Traditionally, high-energy particle colliders have had the highest discovery potential. However, despite the fact that the LHC is the most powerful accelerator in the world, new particles that can, for instance, explain the dark matter puzzle have not been observed in the expected region. Therefore, additional search strategies must be pursued to detect evidence for physics beyond the SM.

Observables that can be measured with very high precision and for which similarly accurate theoretical predictions exist at the same time, play an increasingly important rôle for exploring the limits of the SM. One such quantity is the anomalous magnetic moment of the muon, $a_\mu \equiv \frac{1}{2}(g_\mu - 2)$, where g_μ denotes the muon's gyromagnetic ratio. There has been a persistent tension of about 3.5 standard deviations between the measured value and the SM prediction [244]:

$$a_\mu^{\text{exp}} - a_\mu^{\text{SM}} = (266 \pm 76) \cdot 10^{-11}. \tag{5.256}$$

As described in detail in the extensive reviews in Refs. [474] and [475], the SM estimate of the anomalous magnetic moment receives contributions from QED, the weak and the strong interactions, i.e.

$$a_\mu^{\rm SM} = a_\mu^{\rm QED} + a_\mu^{\rm weak} + a_\mu^{\rm strong}. \quad (5.257)$$

While QED effects account for about 99.994% of the absolute value of $a_\mu^{\rm SM}$, its total uncertainty is completely dominated by the contribution from $a_\mu^{\rm strong}$. Since the latter is mostly due to hadronic effects that are intrinsically non-perturbative, it is clear that special attention must be paid to their reliable evaluation.

The most important quantum corrections to $a_\mu^{\rm SM}$ arising from strong interaction physics are the leading hadronic vacuum polarization (HVP) and hadronic light-by-light scattering (HLbL) contributions. The HVP contribution, $a_\mu^{\rm hvp}$, which arises at order α^2 (where α is the fine structure constant), can be expressed in terms of a dispersion integral of the cross section ratio $R(s) = \sigma(e^+e^- \to \text{hadrons})/\sigma(e^+e^- \to \mu^+\mu^-)$, multiplied by a known kernel function. At small values of the centre-of-mass energy s, the dispersion integral is evaluated using experimental data for the R-ratio $R(s)$ as input [476–480]. For instance, the recent analysis of Ref. [479], which is based on the available data for $e^+e^- \to$ hadrons, produced an estimate of $a_\mu^{\rm hvp} = (693.1 \pm 3.4) \cdot 10^{-10}$. While the total error is at the level of 0.5%, it is clear that experimental uncertainties enter the SM prediction for a_μ in this approach.

The HLbL contribution has been quantified mostly using hadronic models, although efforts are under way to formulate and apply a dispersive or data-driven framework to treat some of the dominant sub-processes [481–491]. The current SM estimate $a_\mu^{\rm SM}$ is based on model calculations such as the "Glasgow consensus", i.e. $a_\mu^{\rm hlbl} = (105 \pm 26) \cdot 10^{-11}$ [492]. Other studies, which have produced consistent results, can be found in Refs. [474, 478, 493].

Given the importance of a_μ for testing the limits of the SM, it is crucial to verify the current estimates of $a_\mu^{\rm hvp}$ and $a_\mu^{\rm hlbl}$ and possibly reduce their overall errors using an ab initio approach such as lattice QCD. Given that two new experiments (E989 at Fermilab and E34 at J-PARC) are set to improve the precision of the measurement of a_μ by a factor four, the importance of reliably estimating the hadronic contributions has become even higher. In order to make an impact, lattice QCD must be able to constrain $a_\mu^{\rm hvp}$ with sub-percent accuracy, while an estimate of $a_\mu^{\rm hlbl}$ at the level of 10% would already be a major step forward. Both tasks, however, present a considerable challenge to lattice QCD. The current status of lattice calculations of $a_\mu^{\rm hvp}$ and $a_\mu^{\rm hlbl}$ was reviewed extensively in Ref. [494], which can be consulted for details. Here we present merely an overview of the main issues and a guide to the literature.

The hadronic vacuum polarization contribution, $a_\mu^{\rm hvp}$, is accessible in lattice QCD via different integral representations involving the correlator of the electromagnetic current. The first possibility is to consider a convolution integral over Euclidean momenta Q^2 of the subtracted vacuum polarization function [500, 501]. The second possibility is the so-called time-momentum representation defined in Ref. [502], in which the product of the spatially summed vector correlator $G(x_0)$ and a kernel function is integrated over the Euclidean time x_0. A variant of the time-momentum representation uses the time moments of $G(x_0)$ [503]. Finally, there also exists a Lorentz-covariant formulation in coordinate space [504] involving the point-to-point vector correlator $G(x, y)$.

In order to meet the precision goal of sub-percent uncertainty, it is mandatory to have good control over the infrared regime which makes a sizeable contribution to a_μ^{hvp}. In the formulation of Refs. [500, 501] this implies that momenta corresponding to $Q^2 \lesssim m_\mu^2$ must be included, since this is where the convolution integral receives its dominant contribution. Instead, in the time-momentum representation or the Lorentz-covariant formulation one must constrain the long-distance regime of the correlator sufficiently well. The statistical accuracy that one can attain for a_μ^{hvp} is affected by the well-known noise problem encountered for the vector correlator, i.e. the fact that the signal-to-noise ratio increases exponentially at large distances.[24] Another limiting factor for the overall precision of a_μ^{hvp} in lattice QCD is the knowledge of the lattice scale [499, 505]. At first sight this may seem surprising, given that a_μ^{hvp} is a dimensionless quantity. However, employing the time-momentum representation, one easily sees that the lattice scale enters through the combination $(x_0 m_\mu)^2$ in the kernel function. Similar arguments exist for the other representations of a_μ^{hvp}. Furthermore, at the level of sub-percent precision, it is necessary to include the contributions from quark-disconnected diagrams and the effects from isospin breaking (see Sect. 5.9.2). All of this is explained in great detail in Ref. [494].

First exploratory calculations of a_μ^{hvp} in full QCD were published in 2008 [506], and in the following years several studies appeared [497, 507–509], employing a range of different discretisations of the quark action, which were mostly aimed at investigating systematic effects. The most recent calculations are focussed on reducing the overall uncertainties [495, 496, 498, 499, 510–515, 530]. A comparison of recent estimates for a_μ^{hvp} from lattice QCD to results obtained via the dispersive approach is shown in Fig. 5.26. As of now, current calculations cannot match the accuracy of the dispersive approach, but efforts are under way to reduce the uncertainties to a level that makes the lattice approach competitive with data-driven methods [494, 516].

In order to determine the hadronic light-by-light scattering contribution, it is necessary to formulate the problem in such a way that a_μ^{hlbl} is expressed in terms of quantities that can be computed on the lattice with affordable effort. Several different strategies have been proposed and are currently being pursued:

In a first method, the matrix element of the electromagnetic current between explicit muon initial and final states is computed is QCD+QED [517]. In order to isolate the desired light-by-light scattering contribution, one has to perform a non-perturbative subtraction. While the method has produced estimates in the expected

[24] This is similar to, but less severe, than the noise problem encountered in nucleon correlation functions discussed in Sect. 5.9.4 of this review.

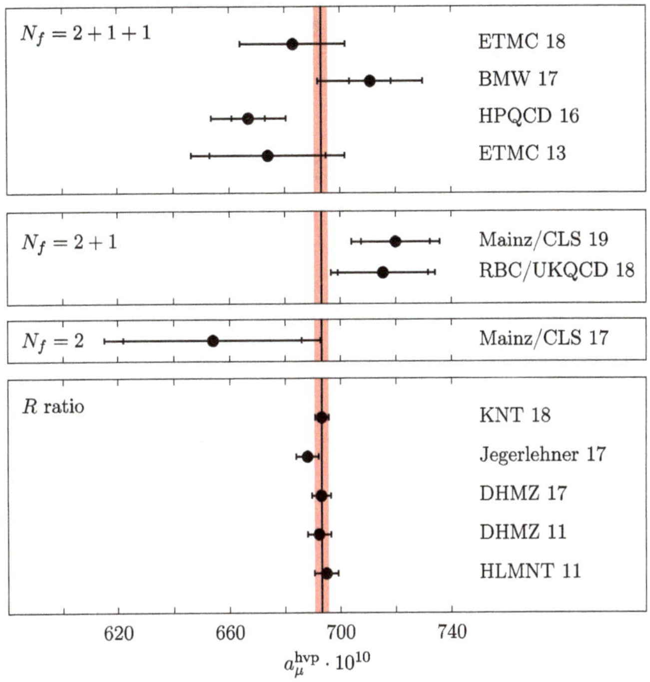

Fig. 5.26 Compilation of recent results for the hadronic vacuum polarisation contribution in units of 10^{-10}. The three panels represent calculations with different numbers of sea quarks. Lattice results are labelled by ETMC 18 [515], BMW 17 [495], HPQCD 16 [496], ETMC 13 [497], Mainz/CLS 19 [530], RBC/UKQCD 18 [498], and Mainz/CLS 17 [499]. The phenomenological determinations based on the R-ratio are labelled as HLMNT 11 [477], DHMZ 11 [476], Jegerlehner 17 [478] and KNT 18 [480]. The red vertical band denotes the estimate from dispersion theory quoted in KNT 18 [480]

range, statistical errors are large, as a result of the cancellation between two large numbers [518].

In another method proposed by the RBC/UKQCD Collaboration [519, 520], the light-by-light scattering diagram is evaluated by inserting three explicit photon propagators. The positions of the insertion of these propagators are then sampled stochastically. In this way, results for the quark-connected and the leading quark-disconnected contributions have been obtained, i.e.

$$(a_\mu^{\rm hlbl})^{\rm conn} = (116.0 \pm 9.6) \cdot 10^{-11}, \quad (a_\mu^{\rm hlbl})^{\rm disc} = (-62.5 \pm 8.0) \cdot 10^{-11}. \tag{5.258}$$

The sum of the two contributions gives $a_\mu^{\rm hlbl} = (53.5 \pm 13.5) \cdot 10^{-11}$ which differs from the Glasgow consensus by a factor two. However, before jumping to conclusions one must take into account that systematic effects have not yet been fully quantified in these calculations.

The Mainz group has proposed a method in which the QED kernel function is computed semi-analytically in infinite volume [521–524]. This has the advantage that large finite-volume effects arising from the massless photon mode are absent.

The method has yet to produce explicit estimates for $a_\mu^{\rm hlbl}$. A variant was proposed by RBC/UKQCD in Ref. [525]. Another project of the Mainz group has focussed on the forward light-by-light scattering amplitude, which can be linked via the optical theorem and dispersive sum rules to models of the cross section for the process $\gamma^*\gamma^* \to$ hadrons [526, 527]. The results provide an important test for model estimates of $a_\mu^{\rm hlbl}$.

Finally, lattice QCD calculations can also be used to directly test model estimates of the expected dominant contribution to $a_\mu^{\rm hlbl}$ from the pion pole, which requires knowledge of the transition form factor for $\pi^0 \to \gamma^*\gamma^*$. The calculation of Ref. [528], which was performed in two-flavour QCD, gives

$$(a_\mu^{\rm hlbl})^{\pi^0} = (65.0 \pm 8.3) \cdot 10^{-11} \qquad (5.259)$$

which is in very good agreement with model estimates [491]. It will be interesting to extend this calculation by including the corresponding contributions of the η and η' mesons.

This brief survey demonstrates that lattice QCD contributes in many different and complementary ways to constrain the hadronic contributions to the muon $g-2$ more precisely.

5.9.6 Concluding Remarks

In this short review we have charted the progress of lattice QCD calculations over more than a decade, i.e. since the publication of the original review article. Back in 2007, lattice QCD was on the verge of providing estimates for hadronic observables from first principles, which were of immediate phenomenological relevance. In the meantime, lattice QCD has become an indispensable tool in particle and hadron physics: In addition to to providing accurate estimates of SM parameters and input quantities for analyses in flavour physics, lattice QCD is now also making inroads into field such as nucleon structure and precision observables. This underlines the important role of lattice calculations for exploring the limits of the SM and searches for new physics.

Furthermore, studying hadronic interactions, i.e. the physics of resonances and multi-hadron systems, has become a major activity in lattice QCD and also serves as a basis for the understanding of light nuclei from first principles. Other important applications of the lattice formulation that have not been covered in this article are studies of matter under extreme conditions. Indeed, many features of the QCD phase diagram and properties of the quark-gluon plasma that are otherwise inaccessible can nowadays be obtained reliably from lattice calculations. Perhaps the most significant development since Ken Wilson's 1989 remark, quoted in the introduction, is the fact that there is now a vigorous interaction between lattice QCD and experiment.

Acknowledgments I am grateful to Guido Altarelli, Andreas Jüttner, Stefan Scherer and Rainer Sommer for a careful reading of the original manuscript and many helpful suggestions.

References

1. M. Creutz, *Quarks, Gluons and Lattices*, Cambridge University Press (1983), Cambridge, UK.
2. I. Montvay and G. Münster, *Quantum fields on a lattice*, Cambridge University Press (1994), Cambridge, UK.
3. J. Smit, *Introduction to quantum fields on a lattice: A robust mate*, Cambridge Lect. Notes Phys. 15 (2002) 1.
4. H.J. Rothe, *Lattice gauge theories: An Introduction*, World Sci. Lect. Notes Phys. 74 (2005) 1.
5. K. Osterwalder and R. Schrader, Commun. Math. Phys. 31 (1973) 83; Commun. Math. Phys. 42 (1975) 281.
6. K.G. Wilson, Phys. Rev. D10 (1974) 2445.
7. P. Weisz, Nucl. Phys. B212 (1983) 1; P. Weisz and R. Wohlert, Nucl. Phys. B236 (1984) 397.
8. Y. Iwasaki, Nucl. Phys. B258 (1985) 141.
9. L. Susskind, Phys. Rev. D16 (1977) 3031.
10. D.B. Kaplan, Phys. Lett. B288 (1992) 342, hep-lat/9206013.
11. V. Furman and Y. Shamir, Nucl. Phys. B439 (1995) 54, hep-lat/9405004.
12. H. Neuberger, Phys. Lett. B417 (1998) 141, hep-lat/9707022; Phys. Lett. B427 (1998) 353, hep-lat/9801031.
13. P. Hasenfratz, Nucl. Phys. B (Proc. Suppl.) 63 (1998) 53, hep-lat/9709110.
14. K. Symanzik, Nucl. Phys. B226 (1983) 187; Nucl. Phys. B226 (1983) 205.
15. B. Sheikholeslami and R. Wohlert, Nucl. Phys. B259 (1985) 572.
16. M. Lüscher, S. Sint, R. Sommer and P. Weisz, Nucl. Phys. B478 (1996) 365, hep-lat/9605038.
17. M. Lüscher, S. Sint, R. Sommer, P. Weisz and U. Wolff, Nucl. Phys. B491 (1997) 323, hep-lat/9609035.
18. H. Kluberg-Stern, A. Morel, O. Napoly and B. Petersson, Nucl. Phys. B220 (1983) 447.
19. G.P. Lepage, Phys. Rev. D59 (1999) 074502, hep-lat/9809157.
20. MILC Collaboration, K. Orginos, D. Toussaint and R.L. Sugar, Phys. Rev. D60 (1999) 054503, hep-lat/9903032.
21. H.B. Nielsen and M. Ninomiya, Nucl. Phys. B185 (1981) 20; Nucl. Phys. B193 (1981) 173.
22. P.H. Ginsparg and K.G. Wilson, Phys. Rev. D25 (1982) 2649.
23. P. Hasenfratz, V. Laliena and F. Niedermayer, Phys. Lett. B427 (1998) 125, hep-lat/9801021.
24. M. Lüscher, Phys. Lett. B428 (1998) 342, hep-lat/9802011.
25. P. Hernández, K. Jansen and M. Lüscher, Nucl. Phys. B552 (1999) 363, hep-lat/9808010.
26. ALPHA Collaboration, R. Frezzotti, P.A. Grassi, S. Sint and P. Weisz, JHEP 08 (2001) 058, hep-lat/0101001.
27. R. Frezzotti and G.C. Rossi, JHEP 08 (2004) 007, hep-lat/0306014.
28. S. Sint, *Lattice QCD with a chiral twist*, (2007), hep-lat/0702008.
29. A. Shindler, *Twisted mass lattice QCD*, (2007), arXiv:0707.4093 [hep-lat].
30. N. Metropolis, A.W. Rosenbluth, M.N. Rosenbluth, A.H. Teller and E. Teller, J. Chem. Phys. 21 (1953) 1087.
31. S. Duane, A.D. Kennedy, B.J. Pendleton and D. Roweth, Phys. Lett. B195 (1987) 216.
32. M. Lüscher, Commun. Math. Phys. 54 (1977) 283.
33. R. Sommer, Nucl. Phys. B411 (1994) 839, hep-lat/9310022.
34. ALPHA Collaboration, M. Guagnelli, R. Sommer and H. Wittig, Nucl. Phys. B535 (1998) 389, hep-lat/9806005.
35. ALPHA & UKQCD Collaborations, J. Garden, J. Heitger, R. Sommer and H. Wittig, Nucl. Phys. B571 (2000) 237, hep-lat/9906013.

36. M. Lüscher, Commun. Math. Phys. 104 (1986) 177.
37. H. Wittig, Nucl. Phys. B (Proc. Suppl.) 119 (2003) 59, hep-lat/0210025.
38. H. Leutwyler, Phys. Lett. B378 (1996) 313, hep-ph/9602366.
39. S.R. Sharpe, PoS LAT2006 (2006) 022, hep-lat/0610094.
40. HPQCD Collaboration, C.T.H. Davies et al., Phys. Rev. Lett. 92 (2004) 022001, hep-lat/0304004.
41. ETM Collaboration, P. Boucaud et al., Phys. Lett. B650 (2007) 304, hep-lat/0701012.
42. M. Hasenbusch, Phys. Lett. B519 (2001) 177, hep-lat/0107019.
43. C. Urbach, K. Jansen, A. Shindler and U. Wenger, Comput. Phys. Commun. 174 (2006) 87, hep-lat/0506011.
44. M. Lüscher, Comput. Phys. Commun. 165 (2005) 199, hep-lat/0409106.
45. M.A. Clark and A.D. Kennedy, Phys. Rev. Lett. 98 (2007) 051601, hep-lat/0608015.
46. F. Butler, H. Chen, J. Sexton, A. Vaccarino and D. Weingarten, Nucl. Phys. B430 (1994) 179, hep-lat/9405003.
47. CP-PACS Collaboration, S. Aoki et al., Phys. Rev. Lett. 84 (2000) 238, hep-lat/9904012; Phys. Rev. D67 (2003) 034503, hep-lat/0206009.
48. UKQCD Collaboration, K.C. Bowler et al., Phys. Rev. D62 (2000) 054506, hep-lat/9910022.
49. MILC Collaboration, C.W. Bernard et al., Phys. Rev. Lett. 81 (1998) 3087, hep-lat/9805004.
50. C.W. Bernard et al., Phys. Rev. D64 (2001) 054506, hep-lat/0104002.
51. BGR Collaboration, C. Gattringer et al., Nucl. Phys. B677 (2004) 3, hep-lat/0307013.
52. CP-PACS Collaboration, A. Ali Khan et al., Phys. Rev. D65 (2002) 054505, hep-lat/0105015.
53. C.J. Morningstar and M.J. Peardon, Phys. Rev. D60 (1999) 034509, hep-lat/9901004.
54. APE Collaboration, M. Albanese et al., Phys. Lett. 192B (1987) 163.
55. M. Teper, Phys. Lett. B183 (1987) 345.
56. M. Lüscher and U. Wolff, Nucl. Phys. B339 (1990) 222.
57. Y. Chen et al., Phys. Rev. D73 (2006) 014516, hep-lat/0510074.
58. UKQCD Collaboration, G.S. Bali et al., Phys. Lett. B309 (1993) 378, hep-lat/9304012.
59. J. Sexton, A. Vaccarino and D. Weingarten, Phys. Rev. Lett. 75 (1995) 4563, hep-lat/9510022.
60. M.J. Teper, *Glueball masses and other physical properties of SU(N) gauge theories in D = (3+1): A Review of lattice results for theorists*, (1998), hep-th/9812187.
61. Particle Data Group, W.M. Yao et al., J. Phys. G33 (2006) 1.
62. E. Klempt and A. Zaitsev, Phys. Rept. 454 (2007) 1, arXiv:0708.4016 [hep-ph].
63. M. Lüscher, K. Symanzik and P. Weisz, Nucl. Phys. B173 (1980) 365; M. Lüscher, Nucl. Phys. B180 (1981) 317.
64. G. Parisi, R. Petronzio and F. Rapuano, Phys. Lett. 128B (1983) 418.
65. S. Necco and R. Sommer, Nucl. Phys. B622 (2002) 328, hep-lat/0108008.
66. G.S. Bali, Phys. Rept. 343 (2001) 1, hep-ph/0001312.
67. M. Lüscher and P. Weisz, JHEP 07 (2002) 049, hep-lat/0207003.
68. O. Philipsen and H. Wittig, Phys. Rev. Lett. 81 (1998) 4056, hep-lat/9807020.
69. ALPHA Collaboration, F. Knechtli and R. Sommer, Phys. Lett. B440 (1998) 345, hep-lat/9807022; Nucl. Phys. B590 (2000) 309, hep-lat/0005021.
70. SESAM Collaboration, G.S. Bali, H. Neff, T. Duessel, T. Lippert and K. Schilling, Phys. Rev. D71 (2005) 114513, hep-lat/0505012.
71. T. van Ritbergen, J.A.M. Vermaseren and S.A. Larin, Phys. Lett. B400 (1997) 379, hep-ph/9701390; K.G. Chetyrkin, Phys. Lett. B404 (1997) 161, hep-ph/9703278; J.A.M. Vermaseren, S.A. Larin and T. van Ritbergen, Phys. Lett. B405 (1997) 327, hep-ph/9703284.
72. M. Lüscher, R. Narayanan, P. Weisz and U. Wolff, Nucl. Phys. B384 (1992) 168, hep-lat/9207009.
73. S. Sint, Nucl. Phys. B421 (1994) 135, hep-lat/9312079.
74. S. Sint, Nucl. Phys. B451 (1995) 416, hep-lat/9504005.
75. ALPHA Collaboration, S. Capitani, M. Lüscher, R. Sommer and H. Wittig, Nucl. Phys. B544 (1999) 669, hep-lat/9810063.
76. G. Martinelli, C. Pittori, C.T. Sachrajda, M. Testa and A. Vladikas, Nucl. Phys. B445 (1995) 81, hep-lat/9411010.

77. ALPHA Collaboration, M. Della Morte et al., Nucl. Phys. B713 (2005) 378, hep-lat/0411025.
78. ALPHA Collaboration, M. Della Morte et al., Nucl. Phys. B729 (2005) 117, hep-lat/0507035.
79. Y. Taniguchi, JHEP 10 (2006) 027, hep-lat/0604002.
80. S. Sint, PoS LAT2005 (2006) 235, hep-lat/0511034.
81. M. Lüscher, JHEP 05 (2006) 042, hep-lat/0603029.
82. G. Parisi, Presented at 20th Int. Conf. on High Energy Physics, Madison, Wis., Jul 17–23, 1980.
83. G.P. Lepage and P.B. Mackenzie, Phys. Rev. D48 (1993) 2250, hep-lat/9209022.
84. M. Lüscher, R. Sommer, P. Weisz and U. Wolff, Nucl. Phys. B413 (1994) 481, hep-lat/9309005.
85. S. Sint and R. Sommer, Nucl. Phys. B465 (1996) 71, hep-lat/9508012.
86. HPQCD Collaboration, Q. Mason et al., Phys. Rev. Lett. 95 (2005) 052002, hep-lat/0503005.
87. Y. Schröder, Phys. Lett. B447 (1999) 321, hep-ph/9812205.
88. S. Bethke, Nucl. Phys. Proc. Suppl. 135 (2004) 345, hep-ex/0407021.
89. J. Gasser and H. Leutwyler, Ann. Phys. 158 (1984) 142.
90. J. Gasser and H. Leutwyler, Nucl. Phys. B250 (1985) 465.
91. M. Lüscher, S. Sint, R. Sommer and H. Wittig, Nucl. Phys. B491 (1997) 344, hep-lat/9611015.
92. M. Della Morte, R. Hoffmann, F. Knechtli, R. Sommer and U. Wolff, JHEP 07 (2005) 007, hep-lat/0505026.
93. $SPQ_{cd}R$ Collaboration, D. Bećirević, V. Lubicz and C. Tarantino, Phys. Lett. B558 (2003) 69, hep-lat/0208003.
94. JLQCD Collaboration, S. Aoki et al., Phys. Rev. Lett. 82 (1999) 4392, hep-lat/9901019.
95. JLQCD Collaboration, T. Ishikawa et al., Phys. Rev. D78 (2008) 011502.
96. HPQCD Collaboration, Q. Mason, H.D. Trottier, R. Horgan, C.T.H. Davies and G.P. Lepage, Phys. Rev. D73 (2006) 114501, hep-ph/0511160.
97. M. Göckeler et al., Phys. Rev. D73 (2006) 054508, hep-lat/0601004.
98. D. Bećirević et al., Nucl. Phys. B734 (2006) 138, hep-lat/0510014.
99. CP-PACS Collaboration, A. Ali Khan et al., Phys. Rev. Lett. 85 (2000) 4674, hep-lat/0004010.
100. ETM Collaboration, B. Blossier et al., JHEP 04 (2008) 020.
101. J. Goldstone, Nuovo Cim. 19 (1961) 154.
102. S. Scherer, Adv. Nucl. Phys. 27 (2003) 277, hep-ph/0210398.
103. V. Bernard and U.G. Meißner, *Chiral perturbation theory*, (2006), hep-ph/0611231.
104. D.B. Kaplan and A.V. Manohar, Phys. Rev. Lett. 56 (1986) 2004.
105. M. Gell-Mann, R.J. Oakes and B. Renner, Phys. Rev. 175 (1968) 2195.
106. T. Banks and A. Casher, Nucl. Phys. B169 (1980) 103.
107. J.J.M. Verbaarschot and I. Zahed, Phys. Rev. Lett. 70 (1993) 3852, hep-th/9303012.
108. H. Leutwyler and A. Smilga, Phys. Rev. D46 (1992) 5607.
109. E.V. Shuryak and J.J.M. Verbaarschot, Nucl. Phys. A560 (1993) 306, hep-th/9212088.
110. L. Giusti, C. Hoelbling, M. Lüscher and H. Wittig, Comput. Phys. Commun. 153 (2003) 31, hep-lat/0212012.
111. L. Giusti, M. Lüscher, P. Weisz and H. Wittig, JHEP 11 (2003) 023, hep-lat/0309189.
112. W. Bietenholz, K. Jansen and S. Shcheredin, JHEP 07 (2003) 033, hep-lat/0306022.
113. H. Fukaya et al., Phys. Rev. D76 (2007) 054503, arXiv:0705.3322 [hep-lat].
114. J. Wennekers and H. Wittig, JHEP 09 (2005) 059, hep-lat/0507026.
115. P. Hernández, K. Jansen, L. Lellouch and H. Wittig, JHEP 07 (2001) 018, hep-lat/0106011.
116. L. Giusti, F. Rapuano, M. Talevi and A. Vladikas, Nucl. Phys. B538 (1999) 249, hep-lat/9807014.
117. P. Hernández, K. Jansen and L. Lellouch, Phys. Lett. B469 (1999) 198, hep-lat/9907022.
118. T. Blum et al., Phys. Rev. D69 (2004) 074502, hep-lat/0007038.
119. MILC Collaboration, T.A. DeGrand, Phys. Rev. D64 (2001) 117501, hep-lat/0107014.
120. L. Giusti, C. Hoelbling and C. Rebbi, Phys. Rev. D64 (2001) 114508, hep-lat/0108007.
121. P. Hernández, K. Jansen, L. Lellouch and H. Wittig, Nucl. Phys. B (Proc. Suppl.) 106 (2002) 766, hep-lat/0110199.

122. P. Hasenfratz, S. Hauswirth, T. Jörg, F. Niedermayer and K. Holland, Nucl. Phys. B643 (2002) 280, hep-lat/0205010.
123. D. Bećirević and V. Lubicz, Phys. Lett. B600 (2004) 83, hep-ph/0403044.
124. V. Gimenez, V. Lubicz, F. Mescia, V. Porretti and J. Reyes, Eur. Phys. J. C41 (2005) 535, hep-lat/0503001.
125. C. McNeile, Phys. Lett. B619 (2005) 124, hep-lat/0504006.
126. H.G. Dosch and S. Narison, Phys. Lett. B417 (1998) 173, hep-ph/9709215.
127. S. Narison, (2002), hep-ph/0202200.
128. M.R. Pennington, (2002), hep-ph/0207220.
129. L. Wolfenstein, Phys. Rev. Lett. 51 (1983) 1945.
130. UTfit Collaboration, M. Bona et al., JHEP 10 (2006) 081, hep-ph/0606167.
131. C. Dawson, PoS LAT2005 (2006) 007.
132. M. Okamoto, PoS LAT2005 (2006) 013, hep-lat/0510113.
133. T. Onogi, PoS LAT2006 (2006) 017, hep-lat/0610115.
134. A. Jüttner, PoS LAT2007 (2007) 014, arXiv:0711.1239 [hep-lat].
135. M. Della Morte, PoS LAT2007 (2007) 008, arXiv:0711.3160 [hep-lat].
136. L. Conti et al., Phys. Lett. B421 (1998) 273, hep-lat/9711053.
137. JLQCD Collaboration, S. Aoki et al., Phys. Rev. D60 (1999) 034511, hep-lat/9901018.
138. D. Bećirević et al., Phys. Lett. B487 (2000) 74, hep-lat/0005013.
139. R.S. Van de Water and S.R. Sharpe, Phys. Rev. D73 (2006) 014003, hep-lat/0507012.
140. M. Guagnelli, J. Heitger, C. Pena, S. Sint and A. Vladikas, Nucl. Phys. B (Proc. Suppl.) 106 (2002) 320, hep-lat/0110097.
141. R. Frezzotti and G.C. Rossi, JHEP 10 (2004) 070, hep-lat/0407002.
142. A. Donini, V. Gimenez, G. Martinelli, M. Talevi and A. Vladikas, Eur. Phys. J. C10 (1999) 121, hep-lat/9902030.
143. ALPHA Collaboration, M. Guagnelli, J. Heitger, C. Pena, S. Sint and A. Vladikas, JHEP 03 (2006) 088, hep-lat/0505002.
144. RBC and UKQCD Collaborations, D.J. Antonio et al., Phys. Rev. Lett. 100 (2008) 032001, hep-ph/0702042.
145. HPQCD and UKQCD Collaborations, E. Gamiz et al., Phys. Rev. D73 (2006) 114502, hep-lat/0603023.
146. Y. Aoki et al., Phys. Rev. D72 (2005) 114505, hep-lat/0411006.
147. UKQCD Collaboration, J.M. Flynn, F. Mescia and A.S.B. Tariq, JHEP 11 (2004) 049, hep-lat/0406013.
148. RBC Collaboration, T. Blum et al., Phys. Rev. D68 (2003) 114506, hep-lat/0110075.
149. CP-PACS Collaboration, A. Ali Khan et al., Phys. Rev. D64 (2001) 114506, hep-lat/0105020.
150. N. Garron, L. Giusti, C. Hoelbling, L. Lellouch and C. Rebbi, Phys. Rev. Lett. 92 (2004) 042001, hep-ph/0306295.
151. MILC Collaboration, T.A. DeGrand, Phys. Rev. D69 (2004) 014504, hep-lat/0309026.
152. ALPHA Collaboration, P. Dimopoulos et al., Nucl. Phys. B749 (2006) 69, hep-ph/0601002; Nucl. Phys. B776 (2007) 258, hep-lat/0702017.
153. D. Bećirević, P. Boucaud, V. Gimenez, V. Lubicz and M. Papinutto, Eur. Phys. J. C37 (2004) 315, hep-lat/0407004.
154. JLQCD Collaboration, S. Aoki et al., Phys. Rev. Lett. 80 (1998) 5271, hep-lat/9710073.
155. D. Bećirević, D. Meloni and A. Retico, JHEP 01 (2001) 012, hep-lat/0012009.
156. J. Gasser and H. Leutwyler, Nucl. Phys. B250 (1985) 517.
157. D. Bećirević et al., Nucl. Phys. B705 (2005) 339, hep-ph/0403217.
158. H. Leutwyler and M. Roos, Z. Phys. C25 (1984) 91.
159. UKQCD Collaboration, P.A. Boyle, J.M. Flynn, A. Jüttner, C.T. Sachrajda and J.M. Zanotti, JHEP 05 (2007) 016, hep-lat/0703005.
160. P.F. Bedaque, Phys. Lett. B593 (2004) 82, nucl-th/0402051.
161. G.M. de Divitiis, R. Petronzio and N. Tantalo, Phys. Lett. B595 (2004) 408, hep-lat/0405002.
162. C.T. Sachrajda and G. Villadoro, Phys. Lett. B609 (2005) 73, hep-lat/0411033.

163. UKQCD Collaboration, J.M. Flynn, A. Jüttner and C.T. Sachrajda, Phys. Lett. B632 (2006) 313, hep-lat/0506016.
164. F.J. Jiang and B.C. Tiburzi, Phys. Lett. B645 (2007) 314, hep-lat/0610103.
165. C. Dawson, T. Izubuchi, T. Kaneko, S. Sasaki and A. Soni, Phys. Rev. D74 (2006) 114502, hep-ph/0607162.
166. UKQCD and RBC Collaborations, D.J. Antonio et al., (2007), hep-lat/0702026; UKQCD and RBC Collaborations, P.A. Boyle et al., Phys. Rev. Lett. 100 (2008) 141601, arXiv:0710.5136 [hep-lat].
167. J. Bijnens and P. Talavera, Nucl. Phys. B669 (2003) 341, hep-ph/0303103.
168. M. Jamin, J.A. Oller and A. Pich, JHEP 02 (2004) 047, hep-ph/0401080.
169. V. Cirigliano et al., JHEP 04 (2005) 006, hep-ph/0503108.
170. W.J. Marciano, Phys. Rev. Lett. 93 (2004) 231803, hep-ph/0402299.
171. ALPHA Collaboration, J. Heitger, R. Sommer and H. Wittig, Nucl. Phys. B588 (2000) 377, hep-lat/0006026.
172. JLQCD Collaboration, S. Aoki et al., Phys. Rev. D68 (2003) 054502, hep-lat/0212039.
173. MILC Collaboration, C. Aubin et al., Phys. Rev. D70 (2004) 114501, hep-lat/0407028.
174. S.R. Beane, P.F. Bedaque, K. Orginos and M.J. Savage, Phys. Rev. D75 (2007) 094501, hep-lat/0606023.
175. RBC and UKQCD Collaborations, C. Allton et al., Phys. Rev. D76 (2007) 014504, hep-lat/0701013.
176. HPQCD Collaboration, E. Follana, C.T.H. Davies, G.P. Lepage and J. Shigemitsu, Phys. Rev. Lett. 100 (2008) 062002, arXiv:0706.1726 [hep-lat].
177. D. Bećirević and G. Villadoro, Phys. Rev. D69 (2004) 054010, hep-lat/0311028.
178. G. Colangelo, S. Durr and C. Haefeli, Nucl. Phys. B721 (2005) 136, hep-lat/0503014.
179. E. Eichten, Nucl. Phys. B (Proc. Suppl.) 4 (1988) 170.
180. B.A. Thacker and G.P. Lepage, Phys. Rev. D43 (1991) 196; G.P. Lepage, L. Magnea, C. Nakhleh, U. Magnea and K. Hornbostel, Phys. Rev. D46 (1992) 4052, hep-lat/9205007.
181. A.X. El-Khadra, A.S. Kronfeld and P.B. Mackenzie, Phys. Rev. D55 (1997) 3933, hep-lat/9604004.
182. ALPHA Collaboration, J. Heitger and R. Sommer, JHEP 02 (2004) 022, hep-lat/0310035.
183. M. Guagnelli, F. Palombi, R. Petronzio and N. Tantalo, Phys. Lett. B546 (2002) 237, hep-lat/0206023.
184. E. Eichten and B.R. Hill, Phys. Lett. B234 (1990) 511.
185. M. Della Morte, A. Shindler and R. Sommer, JHEP 08 (2005) 051, hep-lat/0506008.
186. R. Sommer, *Non-perturbative QCD: Renormalization, O(a)-improvement and matching to heavy quark effective theory*, (2006), hep-lat/0611020.
187. S. Aoki, Y. Kuramashi and S.i. Tominaga, Prog. Theor. Phys. 109 (2003) 383, hep-lat/0107009.
188. N.H. Christ, M. Li and H.W. Lin, Phys. Rev. D76 (2007) 074505, hep-lat/0608006.
189. ALPHA Collaboration, J. Heitger, M. Kurth and R. Sommer, Nucl. Phys. B669 (2003) 173, hep-lat/0302019.
190. B. Grinstein, E.E. Jenkins, A.V. Manohar, M.J. Savage and M.B. Wise, Nucl. Phys. B380 (1992) 369, hep-ph/9204207; J.L. Goity, Phys. Rev. D46 (1992) 3929, hep-ph/9206230; M.J. Booth, Phys. Rev. D51 (1995) 2338, hep-ph/9411433; S.R. Sharpe and Y. Zhang, Phys. Rev. D53 (1996) 5125, hep-lat/9510037.
191. A.S. Kronfeld and S.M. Ryan, Phys. Lett. B543 (2002) 59, hep-ph/0206058.
192. HPQCD Collaboration, A. Gray et al., Phys. Rev. Lett. 95 (2005) 212001, hep-lat/0507015.
193. N. Tantalo, *Lattice calculations for B and K mixing*, (2007), hep-ph/0703241.
194. D. Bećirević et al., Nucl. Phys. B618 (2001) 241, hep-lat/0002025.
195. UKQCD Collaboration, L. Lellouch and C.J.D. Lin, Phys. Rev. D64 (2001) 094501, hep-ph/0011086.
196. UKQCD Collaboration, K.C. Bowler et al., Nucl. Phys. B619 (2001) 507, hep-lat/0007020.
197. G.M. de Divitiis, M. Guagnelli, F. Palombi, R. Petronzio and N. Tantalo, Nucl. Phys. B672 (2003) 372, hep-lat/0307005.

198. M. Della Morte et al., JHEP 0802 (2008) 078, arXiv:0710.2201 [hep-lat].
199. D. Guazzini, R. Sommer and N. Tantalo, JHEP 0801 (2008) 076, arXiv:0710.2229 [hep-lat].
200. C.W. Bernard et al., Phys. Rev. Lett. 81 (1998) 4812, hep-ph/9806412.
201. A.X. El-Khadra, A.S. Kronfeld, P.B. Mackenzie, S.M. Ryan and J.N. Simone, Phys. Rev. D58 (1998) 014506, hep-ph/9711426.
202. JLQCD Collaboration, S. Aoki et al., Phys. Rev. Lett. 80 (1998) 5711.
203. CP-PACS Collaboration, A. Ali Khan et al., Phys. Rev. D64 (2001) 034505, hep-lat/0010009.
204. MILC Collaboration, C. Bernard et al., Phys. Rev. D66 (2002) 094501, hep-lat/0206016.
205. A. Ali Khan et al., Phys. Lett. B427 (1998) 132, hep-lat/9801038.
206. JLQCD Collaboration, K.I. Ishikawa et al., Phys. Rev. D61 (2000) 074501, hep-lat/9905036.
207. CP-PACS Collaboration, A. Ali Khan et al., Phys. Rev. D64 (2001) 054504, hep-lat/0103020.
208. S. Collins et al., Phys. Rev. D60 (1999) 074504, hep-lat/9901001.
209. JLQCD Collaboration, S. Aoki et al., Phys. Rev. Lett. 91 (2003) 212001, hep-ph/0307039.
210. M. Wingate, C.T.H. Davies, A. Gray, G.P. Lepage and J. Shigemitsu, Phys. Rev. Lett. 92 (2004) 162001, hep-ph/0311130.
211. D0 Collaboration, V.M. Abazov et al., Phys. Rev. Lett. 97 (2006) 021802, hep-ex/0603029.
212. CDF Collaboration, A. Abulencia et al., Phys. Rev. Lett. 97 (2006) 062003, hep-ex/0606027.
213. V. Gadiyak and O. Loktik, Phys. Rev. D72 (2005) 114504, hep-lat/0509075.
214. V. Gimenez and G. Martinelli, Phys. Lett. B398 (1997) 135, hep-lat/9610024.
215. V. Gimenez and J. Reyes, Nucl. Phys. B545 (1999) 576, hep-lat/9806023.
216. UKQCD Collaboration, A.K. Ewing et al., Phys. Rev. D54 (1996) 3526, hep-lat/9508030.
217. J.C. Christensen, T. Draper and C. McNeile, Phys. Rev. D56 (1997) 6993, hep-lat/9610026.
218. D. Bećirević et al., Nucl. Phys. B618 (2001) 241, hep-lat/0002025.
219. D. Bećirević, V. Gimenez, G. Martinelli, M. Papinutto and J. Reyes, JHEP 04 (2002) 025, hep-lat/0110091.
220. F. Palombi, M. Papinutto, C. Pena and H. Wittig, JHEP 08 (2006) 017, hep-lat/0604014.
221. F. Palombi, M. Papinutto, C. Pena and H. Wittig, JHEP 09 (2007) 062, arXiv:0706.4153 [hep-lat].
222. P. Dimopoulos et al., PoS LAT2007 (2007) 368, arXiv:0710.2862 [hep-lat]; ALPHA Collaboration, P. Dimopoulos et al., JHEP 0805 (2008) 065, arXiv:0712.2429 [hep-lat].
223. D. Bećirević and A.B. Kaidalov, Phys. Lett. B478 (2000) 417, hep-ph/9904490.
224. S. Hashimoto, Int. J. Mod. Phys. A20 (2005) 5133, hep-ph/0411126.
225. UKQCD Collaboration, K.C. Bowler et al., Phys. Lett. B486 (2000) 111, hep-lat/9911011.
226. A. Abada et al., Nucl. Phys. B619 (2001) 565, hep-lat/0011065.
227. A.X. El-Khadra, A.S. Kronfeld, P.B. Mackenzie, S.M. Ryan and J.N. Simone, Phys. Rev. D64 (2001) 014502, hep-ph/0101023.
228. JLQCD Collaboration, S. Aoki et al., Phys. Rev. D64 (2001) 114505, hep-lat/0106024.
229. M. Okamoto et al., Nucl. Phys. B (Proc. Suppl.) 140 (2005) 461, hep-lat/0409116.
230. HPQCD Collaboration, E. Dalgic et al., Phys. Rev. D73 (2006) 074502, hep-lat/0601021.
231. CLEO Collaboration, S.B. Athar et al., Phys. Rev. D68 (2003) 072003, hep-ex/0304019.
232. N. Isgur and M.B. Wise, Phys. Lett. B232 (1989) 113; Phys. Lett. B237 (1990) 527.
233. M.E. Luke, Phys. Lett. B252 (1990) 447.
234. C.W. Bernard, Y. Shen and A. Soni, Phys. Lett. B317 (1993) 164, hep-lat/9307005.
235. UKQCD Collaboration, S.P. Booth et al., Phys. Rev. Lett. 72 (1994) 462, hep-lat/9308019.
236. UKQCD Collaboration, K.C. Bowler et al., Phys. Rev. D52 (1995) 5067, hep-ph/9504231.
237. UKQCD Collaboration, K.C. Bowler, G. Douglas, R.D. Kenway, G.N. Lacagnina and C.M. Maynard, Nucl. Phys. B637 (2002) 293, hep-lat/0202029.
238. G.M. de Divitiis, E. Molinaro, R. Petronzio and N. Tantalo, Phys. Lett. B655 (2007) 45, arXiv:0707.0582 [hep-lat]; G.M. de Divitiis, R. Petronzio and N. Tantalo, JHEP 0710 (2007) 062, arXiv:0707.0587 [hep-lat].
239. S. Hashimoto et al., Phys. Rev. D61 (2000) 014502, hep-ph/9906376.
240. S. Hashimoto, A.S. Kronfeld, P.B. Mackenzie, S.M. Ryan and J.N. Simone, Phys. Rev. D66 (2002) 014503, hep-ph/0110253.
241. H. Wittig, (2008).

242. K.G. Wilson, Nucl. Phys. Proc. Suppl. 17 (1990) 82.
243. K.G. Wilson, Nucl. Phys. Proc. Suppl. 140 (2005) 3, hep-lat/0412043.
244. Particle Data Group, C. Patrignani et al., Chin. Phys. C40 (2016) 100001.
245. G. Colangelo et al., Eur. Phys. J. C71 (2011) 1695, 1011.4408.
246. S. Aoki et al., Eur. Phys. J. C74 (2014) 2890, 1310.8555.
247. S. Aoki et al., Eur. Phys. J. C77 (2017) 112, 1607.00299.
248. S. Aoki et al., Eur. Phys. J. C80 (2020) 113. https://doi.org/10.1140/epjc/s10052-019-7354-7
249. PACS-CS, S. Aoki et al., Phys. Rev. D79 (2009) 034503, 0807.1661.
250. S. Dürr et al., Science 322 (2008) 1224, 0906.3599.
251. ETM, C. Alexandrou et al., Phys. Rev. D78 (2008) 014509, 0803.3190.
252. MILC, A. Bazavov et al., Rev. Mod. Phys. 82 (2010) 1349, 0903.3598.
253. N. Tantalo, PoS LATTICE2013 (2014) 007, 1311.2797.
254. A. Portelli, PoS LATTICE2014 (2015) 013, 1505.07057.
255. A. Patella, PoS LATTICE2016 (2017) 020, 1702.03857.
256. T. Blum, R. Zhou, T. Doi, M. Hayakawa, T. Izubuchi, S. Uno and N. Yamada, Phys. Rev. D82 (2010) 094508, 1006.1311.
257. Budapest-Marseille-Wuppertal, S. Borsanyi et al., Phys. Rev. Lett. 111 (2013) 252001, 1306.2287.
258. S. Borsanyi et al., Science 347 (2015) 1452, 1406.4088.
259. R. Horsley et al., J. Phys. G43 (2016) 10LT02, 1508.06401.
260. T. Blum, T. Doi, M. Hayakawa, T. Izubuchi and N. Yamada, Phys. Rev. D76 (2007) 114508, 0708.0484.
261. RM123, G.M. de Divitiis et al., Phys. Rev. D87 (2013) 114505, 1303.4896.
262. D. Giusti, V. Lubicz, C. Tarantino, G. Martinelli, S. Sanfilippo, S. Simula and N. Tantalo, Phys. Rev. D95 (2017) 114504, 1704.06561.
263. R. Horsley et al., JHEP 04 (2016) 093, 1509.00799.
264. Z. Fodor et al., Phys. Rev. Lett. 117 (2016) 082001, 1604.07112.
265. MILC, S. Basak et al., Phys. Rev. D99 (2019) 034503, 1807.05556.
266. R.F. Dashen, Phys. Rev. 183 (1969) 1245.
267. M. Lüscher, Commun.Math.Phys. 104 (1986) 177.
268. M. Lüscher, Commun.Math.Phys. 105 (1986) 153.
269. M. Lüscher, Nucl. Phys. B354 (1991) 531.
270. M. Lüscher, Nucl. Phys. B364 (1991) 237.
271. C. Andersen, J. Bulava, B. Hörz and C. Morningstar, Nucl. Phys. B939 (2019) 145, 1808.05007.
272. C. Michael, Nucl. Phys. B259 (1985) 58.
273. M. Lüscher and U. Wolff, Nucl. Phys. B339 (1990) 222.
274. B. Blossier, M. Della Morte, G. von Hippel, T. Mendes and R. Sommer, JHEP 04 (2009) 094, 0902.1265.
275. J. Foley, K. Jimmy Juge, A. O'Cais, M. Peardon, S.M. Ryan and J.I. Skullerud, Comput. Phys. Commun. 172 (2005) 145, hep-lat/0505023.
276. Hadron Spectrum, M. Peardon et al., Phys. Rev. D80 (2009) 054506, 0905.2160.
277. C. Morningstar, J. Bulava, J. Foley, K.J. Juge, D. Lenkner, M. Peardon and C.H. Wong, Phys. Rev. D83 (2011) 114505, 1104.3870.
278. CP-PACS, T. Yamazaki et al., Phys. Rev. D70 (2004) 074513, hep-lat/0402025.
279. NPLQCD, S.R. Beane, P.F. Bedaque, K. Orginos and M.J. Savage, Phys. Rev. D73 (2006) 054503, hep-lat/0506013.
280. S.R. Beane, T.C. Luu, K. Orginos, A. Parreno, M.J. Savage, A. Torok and A. Walker-Loud, Phys. Rev. D77 (2008) 014505, 0706.3026.
281. X. Feng, K. Jansen and D.B. Renner, Phys. Lett. B684 (2010) 268, 0909.3255.
282. T. Yagi, S. Hashimoto, O. Morimatsu and M. Ohtani, (2011), 1108.2970.
283. Z. Fu, Phys. Rev. D87 (2013) 074501, 1303.0517.
284. PACS-CS, K. Sasaki, N. Ishizuka, M. Oka and T. Yamazaki, Phys. Rev. D89 (2014) 054502, 1311.7226.

285. D.J. Wilson, R.A. Briceño, J.J. Dudek, R.G. Edwards and C.E. Thomas, Phys. Rev. D92 (2015) 094502, 1507.02599.
286. ETM, C. Helmes et al., JHEP 09 (2015) 109, 1506.00408.
287. RQCD, G.S. Bali, S. Collins, A. Cox, G. Donald, M. Göckeler, C.B. Lang and A. Schäfer, Phys. Rev. D93 (2016) 054509, 1512.08678.
288. J. Bulava, B. Fahy, B. Hörz, K.J. Juge, C. Morningstar and C.H. Wong, Nucl. Phys. B910 (2016) 842, 1604.05593.
289. L. Liu et al., Phys. Rev. D96 (2017) 054516, 1612.02061.
290. D. Guo, A. Alexandru, R. Molina and M. Döring, Phys. Rev. D94 (2016) 034501, 1605.03993.
291. C. Morningstar, J. Bulava, B. Singha, R. Brett, J. Fallica, A. Hanlon and B. Hörz, Nucl. Phys. B924 (2017) 477, 1707.05817.
292. S. Prelovsek, L. Leskovec, C.B. Lang and D. Mohler, Phys. Rev. D88 (2013) 054508, 1307.0736.
293. R. Brett, J. Bulava, J. Fallica, A. Hanlon, B. Hörz and C. Morningstar, Nucl. Phys. B932 (2018) 29, 1802.03100.
294. C. Helmes, C. Jost, B. Knippschild, B. Kostrzewa, L. Liu, C. Urbach and M. Werner, PoS LATTICE2016 (2016) 135, 1611.09584.
295. C. Helmes, C. Jost, B. Knippschild, B. Kostrzewa, L. Liu, C. Urbach and M. Werner, Phys. Rev. D96 (2017) 034510, 1703.04737.
296. A. Torok, S.R. Beane, W. Detmold, T.C. Luu, K. Orginos, A. Parreno, M.J. Savage and A. Walker-Loud, Phys. Rev. D81 (2010) 074506, 0907.1913.
297. C.B. Lang and V. Verduci, Phys. Rev. D87 (2013) 054502, 1212.5055.
298. W. Detmold and A. Nicholson, Phys. Rev. D93 (2016) 114511, 1511.02275.
299. C.B. Lang, L. Leskovec, M. Padmanath and S. Prelovsek, Phys. Rev. D95 (2017) 014510, 1610.01422.
300. C.W. Andersen, J. Bulava, B. Hörz and C. Morningstar, Phys. Rev. D97 (2018) 014506, 1710.01557.
301. K. Orginos, A. Parreno, M.J. Savage, S.R. Beane, E. Chang and W. Detmold, Phys. Rev. D92 (2015) 114512, 1508.07583.
302. A. Francis, J.R. Green, P.M. Junnarkar, C. Miao, T.D. Rae and H. Wittig, (2018), 1805.03966.
303. S. He, X. Feng and C. Liu, JHEP 07 (2005) 011, hep-lat/0504019.
304. V. Bernard, M. Lage, U.G. Meissner and A. Rusetsky, JHEP 01 (2011) 019, 1010.6018.
305. M.T. Hansen and S.R. Sharpe, Phys. Rev. D86 (2012) 016007, 1204.0826.
306. R.A. Briceño and Z. Davoudi, Phys. Rev. D88 (2013) 094507, 1204.1110.
307. P. Guo, J. Dudek, R. Edwards and A.P. Szczepaniak, Phys. Rev. D88 (2013) 014501, 1211.0929.
308. L. Roca and E. Oset, Phys. Rev. D85 (2012) 054507, 1201.0438.
309. K. Polejaeva and A. Rusetsky, Eur. Phys. J. A48 (2012) 67, 1203.1241.
310. R.A. Briceño and Z. Davoudi, Phys. Rev. D87 (2013) 094507, 1212.3398.
311. M.T. Hansen and S.R. Sharpe, Phys. Rev. D90 (2014) 116003, 1408.5933.
312. M.T. Hansen and S.R. Sharpe, Phys. Rev. D93 (2016) 096006, 1602.00324, [Erratum: Phys. Rev. D96, no. 3, 039901(2017)].
313. M.T. Hansen and S.R. Sharpe, Phys. Rev. D95 (2017) 034501, 1609.04317.
314. R.A. Briceño, M.T. Hansen and S.R. Sharpe, Phys. Rev. D95 (2017) 074510, 1701.07465.
315. R.A. Briceño, M.T. Hansen and S.R. Sharpe, Phys. Rev. D98 (2018) 014506, 1803.04169.
316. L. Lellouch and M. Lüscher, Commun. Math. Phys. 219 (2001) 31, hep-lat/0003023.
317. H.B. Meyer, Phys. Rev. Lett. 107 (2011) 072002, 1105.1892.
318. X. Feng, S. Aoki, S. Hashimoto and T. Kaneko, Phys. Rev. D91 (2015) 054504, 1412.6319.
319. J. Bulava, B. Hörz, B. Fahy, K.J. Juge, C. Morningstar and C.H. Wong, PoS LATTICE2015 (2016) 069, 1511.02351.
320. F. Erben, J.R. Green, D. Mohler and H. Wittig, Phys. Rev. D101 (2020) 054504, 1910.01083.
321. S. Prelovsek, PoS LATTICE2014 (2014) 015, 1411.0405.
322. C. Liu, PoS LATTICE2016 (2017) 006, 1612.00103.
323. D. Mohler, EPJ Web Conf. 181 (2018) 01027.

324. CKMfitter Group, J. Charles et al., Eur. Phys. J. C41 (2005) 1, hep-ph/0406184.
325. J. Charles et al., Phys. Rev. D91 (2015) 073007, 1501.05013.
326. UTfit, M. Bona et al., JHEP 10 (2006) 081, hep-ph/0606167.
327. UTfit, M. Bona et al., JHEP 03 (2008) 049, 0707.0636.
328. Fermilab Lattice, MILC, TUMQCD, A. Bazavov et al., Phys. Rev. D98 (2018) 054517, 1802.04248.
329. Y. Maezawa and P. Petreczky, Phys. Rev. D94 (2016) 034507, 1606.08798.
330. ETM, N. Carrasco et al., Nucl. Phys. B887 (2014) 19, 1403.4504.
331. B. Chakraborty et al., Phys. Rev. D91 (2015) 054508, 1408.4169.
332. Fermilab Lattice, MILC, A. Bazavov et al., Phys. Rev. D90 (2014) 074509, 1407.3772.
333. S. Dürr et al., Phys. Lett. B701 (2011) 265, 1011.2403.
334. S. Dürr et al., JHEP 08 (2011) 148, 1011.2711.
335. C. McNeile, C.T.H. Davies, E. Follana, K. Hornbostel and G.P. Lepage, Phys. Rev. D82 (2010) 034512, 1004.4285.
336. RBC/UKQCD, T. Blum et al., Phys. Rev. D93 (2016) 074505, 1411.7017.
337. ETM, B. Blossier, P. Dimopoulos, R. Frezzotti, V. Lubicz, M. Petschlies, F. Sanfilippo, S. Simula and C. Tarantino, Phys. Rev. D82 (2010) 114513, 1010.3659.
338. P. Fritzsch, F. Knechtli, B. Leder, M. Marinkovic, S. Schaefer, R. Sommer and F. Virotta, Nucl. Phys. B865 (2012) 397, 1205.5380.
339. A. Bazavov et al., PoS LATTICE2010 (2010) 083, 1011.1792.
340. T. Burch et al., Phys. Rev. D81 (2010) 034508, 0912.2701.
341. P. Gambino, A. Melis and S. Simula, Phys. Rev. D96 (2017) 014511, 1704.06105.
342. K. Nakayama, B. Fahy and S. Hashimoto, Phys. Rev. D94 (2016) 054507, 1606.01002.
343. ETM, A. Bussone et al., Phys. Rev. D93 (2016) 114505, 1603.04306.
344. Y.B. Yang et al., Phys. Rev. D92 (2015) 034517, 1410.3343.
345. C.T.H. Davies et al., Phys. Rev. Lett. 104 (2010) 132003, 0910.3102.
346. B. Colquhoun, R.J. Dowdall, C.T.H. Davies, K. Hornbostel and G.P. Lepage, Phys. Rev. D91 (2015) 074514, 1408.5768.
347. ETM, N. Carrasco et al., JHEP 03 (2014) 016, 1308.1851.
348. F. Bernardoni et al., Phys. Lett. B730 (2014) 171, 1311.5498.
349. ALPHA, M. Bruno et al., Phys. Rev. Lett. 119 (2017) 102001, 1706.03821.
350. K. Maltman, D. Leinweber, P. Moran and A. Sternbeck, Phys. Rev. D78 (2008) 114504, 0807.2020.
351. PACS-CS, S. Aoki et al., JHEP 10 (2009) 053, 0906.3906.
352. B. Blossier et al., Phys. Rev. Lett. 108 (2012) 262002, 1201.5770.
353. ETM, B. Blossier, P. Boucaud, M. Brinet, F. De Soto, V. Morenas, O. Pene, K. Petrov and J. Rodriguez-Quintero, Phys. Rev. D89 (2014) 014507, 1310.3763.
354. A. Bazavov, N. Brambilla, X. Garcia i Tormo, P. Petreczky, J. Soto and A. Vairo, Phys. Rev. D90 (2014) 074038, 1407.8437.
355. ETM, N. Carrasco, P. Dimopoulos, R. Frezzotti, V. Lubicz, G.C. Rossi, S. Simula and C. Tarantino, Phys. Rev. D92 (2015) 034516, 1505.06639.
356. S. Dürr et al., Phys. Lett. B705 (2011) 477, 1106.3230.
357. J. Laiho and R.S. Van de Water, PoS LATTICE2011 (2011) 293, 1112.4861.
358. SWME, B.J. Choi et al., Phys. Rev. D93 (2016) 014511, 1509.00592.
359. ETM, V. Bertone et al., JHEP 03 (2013) 089, 1207.1287, [Erratum: JHEP07,143(2013)].
360. RBC/UKQCD, P.A. Boyle, N. Garron and R.J. Hudspith, Phys. Rev. D86 (2012) 054028, 1206.5737.
361. RBC/UKQCD, N. Garron, R.J. Hudspith and A.T. Lytle, JHEP 11 (2016) 001, 1609.03334.
362. SWME, T. Bae et al., Phys. Rev. D88 (2013) 071503, 1309.2040.
363. SWME, J. Leem et al., PoS LATTICE2014 (2014) 370, 1411.1501.
364. M. Moulson, PoS CKM2016 (2017) 033, 1704.04104.
365. J.C. Hardy and I.S. Towner, Phys. Rev. C91 (2015) 025501, 1411.5987.
366. ALPHA, F. Bernardoni et al., Phys. Lett. B735 (2014) 349, 1404.3590.
367. ETM, N. Carrasco et al., PoS LATTICE2013 (2014) 382, 1310.1851.

368. H. Na, C.J. Monahan, C.T.H. Davies, R. Horgan, G.P. Lepage and J. Shigemitsu, Phys. Rev. D86 (2012) 034506, 1202.4914.
369. N.H. Christ, J.M. Flynn, T. Izubuchi, T. Kawanai, C. Lehner, A. Soni, R.S. Van de Water and O. Witzel, Phys. Rev. D91 (2015) 054502, 1404.4670.
370. Y. Aoki, T. Ishikawa, T. Izubuchi, C. Lehner and A. Soni, Phys. Rev. D91 (2015) 114505, 1406.6192.
371. C. McNeile, C.T.H. Davies, E. Follana, K. Hornbostel and G.P. Lepage, Phys. Rev. D85 (2012) 031503, 1110.4510.
372. Fermilab Lattice, MILC, A. Bazavov et al., Phys. Rev. D85 (2012) 114506, 1112.3051.
373. HPQCD, R.J. Dowdall, C.T.H. Davies, R.R. Horgan, C.J. Monahan and J. Shigemitsu, Phys. Rev. Lett. 110 (2013) 222003, 1302.2644.
374. A. Bazavov et al., Phys. Rev. D98 (2018) 074512, 1712.09262.
375. C. Hughes, C.T.H. Davies and C.J. Monahan, Phys. Rev. D97 (2018) 054509, 1711.09981.
376. HPQCD, E. Gamiz, C.T.H. Davies, G.P. Lepage, J. Shigemitsu and M. Wingate, Phys. Rev. D80 (2009) 014503, 0902.1815.
377. Fermilab Lattice, MILC, A. Bazavov et al., Phys. Rev. D93 (2016) 113016, 1602.03560.
378. R.L. Jaffe and A. Manohar, Nucl. Phys. B337 (1990) 509.
379. X.D. Ji, Phys. Rev. Lett. 78 (1997) 610, hep-ph/9603249.
380. R. Pohl, R. Gilman, G.A. Miller and K. Pachucki, Ann. Rev. Nucl. Part. Sci. 63 (2013) 175, 1301.0905.
381. R. Pohl et al., Nature 466 (2010) 213.
382. R. Pohl et al., Science 353 (2016) 669.
383. A1, J.C. Bernauer et al., Phys. Rev. C90 (2014) 015206, 1307.6227.
384. P.J. Mohr, D.B. Newell and B.N. Taylor, Rev. Mod. Phys. 88 (2016) 035009, 1507.07956.
385. C. Alexandrou, G. Koutsou, J.W. Negele and A. Tsapalis, Phys. Rev. D74 (2006) 034508, hep-lat/0605017.
386. H.W. Lin, T. Blum, S. Ohta, S. Sasaki and T. Yamazaki, Phys. Rev. D78 (2008) 014505, 0802.0863.
387. T. Yamazaki, Y. Aoki, T. Blum, H.W. Lin, S. Ohta, S. Sasaki, R. Tweedie and J. Zanotti, Phys. Rev. D79 (2009) 114505, 0904.2039.
388. S. Syritsyn et al., Phys. Rev. D81 (2010) 034507, 0907.4194.
389. LHP, J. Bratt et al., Phys. Rev. D82 (2010) 094502, 1001.3620.
390. S. Collins et al., Phys. Rev. D84 (2011) 074507, 1106.3580.
391. C. Alexandrou et al., Phys. Rev. D83 (2011) 094502, 1102.2208.
392. J.R. Green, M. Engelhardt, S. Krieg, J.W. Negele, A.V. Pochinsky and S.N. Syritsyn, Phys. Lett. B734 (2014) 290, 1209.1687.
393. T. Bhattacharya, S.D. Cohen, R. Gupta, A. Joseph, H.W. Lin and B. Yoon, Phys. Rev. D89 (2014) 094502, 1306.5435.
394. S. Capitani et al., Phys. Rev. D92 (2015) 054511, 1504.04628.
395. QCDSF, UKQCD, CSSM, A.J. Chambers et al., Phys. Rev. D96 (2017) 114509, 1702.01513.
396. C. Alexandrou, M. Constantinou, K. Hadjiyiannakou, K. Jansen, C. Kallidonis, G. Koutsou and A. Vaquero Aviles-Casco, Phys. Rev. D96 (2017) 034503, 1706.00469.
397. PACS, K.I. Ishikawa, Y. Kuramashi, S. Sasaki, N. Tsukamoto, A. Ukawa and T. Yamazaki, Phys. Rev. D98 (2018) 074510, 1807.03974.
398. R.G. Edwards et al., Phys. Rev. Lett. 96 (2006) 052001, hep-lat/0510062.
399. A. Ali Khan et al., Phys. Rev. D74 (2006) 094508, hep-lat/0603028.
400. T. Yamazaki et al., Phys. Rev. Lett. 100 (2008) 171602, 0801.4016.
401. ETM, C. Alexandrou et al., Phys. Rev. D83 (2011) 045010, 1012.0857.
402. S. Capitani et al., Phys. Rev. D86 (2012) 074502, 1205.0180.
403. R. Horsley et al., Phys. Lett. B732 (2014) 41, 1302.2233.
404. G.S. Bali et al., Phys. Rev. D91 (2015) 054501, 1412.7336.
405. A. Abdel-Rehim et al., Phys. Rev. D92 (2015) 114513, 1507.04936, [Erratum: Phys. Rev.D93,no.3,039904(2016)].

406. T. Bhattacharya, V. Cirigliano, S. Cohen, R. Gupta, H.W. Lin and B. Yoon, Phys. Rev. D94 (2016) 054508, 1606.07049.
407. C. Alexandrou, M. Constantinou, K. Hadjiyiannakou, K. Jansen, C. Kallidonis, G. Koutsou and A. Vaquero Aviles-Casco, Phys. Rev. D96 (2017) 054507, 1705.03399.
408. E. Berkowitz et al., (2017), 1704.01114.
409. S. Capitani et al., Int. J. Mod. Phys. A34 (2019) 1950009, 1705.06186.
410. C.C. Chang et al., Nature 558 (2018) 91, 1805.12130.
411. R. Gupta, Y.C. Jang, B. Yoon, H.W. Lin, V. Cirigliano and T. Bhattacharya, (2018), 1806.09006.
412. JLQCD, N. Yamanaka, S. Hashimoto, T. Kaneko and H. Ohki, Phys. Rev. D98 (2018) 054516, 1805.10507.
413. J. Liang, Y.B. Yang, T. Draper, M. Gong and K.F. Liu, Phys. Rev. D98 (2018) 074505, 1806.08366.
414. T. Harris, G. von Hippel, P. Junnarkar, H.B. Meyer, K. Ottnad, J. Wilhelm, H. Wittig and L. Wrang, Phys. Rev. D100 (2019) 034513, 1905.01291.
415. Y. Aoki, T. Blum, H.W. Lin, S. Ohta, S. Sasaki, R. Tweedie, J. Zanotti and T. Yamazaki, Phys. Rev. D82 (2010) 014501, 1003.3387.
416. J.R. Green, J.W. Negele, A.V. Pochinsky, S.N. Syritsyn, M. Engelhardt and S. Krieg, Phys. Rev. D86 (2012) 114509, 1206.4527.
417. PNDME, T. Bhattacharya, V. Cirigliano, S. Cohen, R. Gupta, A. Joseph, H.W. Lin and B. Yoon, Phys. Rev. D92 (2015) 094511, 1506.06411.
418. T. Bhattacharya, V. Cirigliano, R. Gupta, H.W. Lin and B. Yoon, Phys. Rev. Lett. 115 (2015) 212002, 1506.04196.
419. C. Alexandrou et al., Phys. Rev. D95 (2017) 114514, 1703.08788, [erratum: Phys. Rev.D96,no.9,099906(2017)].
420. R. Gupta, Y.C. Jang, H.W. Lin, B. Yoon and T. Bhattacharya, Phys. Rev. D96 (2017) 114503, 1705.06834.
421. J. Green et al., Phys. Rev. D95 (2017) 114502, 1703.06703.
422. T. Doi, M. Deka, S.J. Dong, T. Draper, K.F. Liu, D. Mankame, N. Mathur and T. Streuer, Phys. Rev. D80 (2009) 094503, 0903.3232.
423. R. Babich, R.C. Brower, M.A. Clark, G.T. Fleming, J.C. Osborn, C. Rebbi and D. Schaich, Phys. Rev. D85 (2012) 054510, 1012.0562.
424. P.E. Shanahan et al., Phys. Rev. Lett. 114 (2015) 091802, 1403.6537.
425. J. Green et al., Phys. Rev. D92 (2015) 031501, 1505.01803.
426. R.S. Sufian, Y.B. Yang, A. Alexandru, T. Draper, J. Liang and K.F. Liu, Phys. Rev. Lett. 118 (2017) 042001, 1606.07075.
427. H. Ohki et al., Phys. Rev. D78 (2008) 054502, 0806.4744.
428. PACS-CS, K.I. Ishikawa et al., Phys. Rev. D80 (2009) 054502, 0905.0962.
429. J. Martin Camalich, L.S. Geng and M.J. Vicente Vacas, Phys. Rev. D82 (2010) 074504, 1003.1929.
430. S. Dürr et al., Phys. Rev. D85 (2012) 014509, 1109.4265, [Erratum: Phys. Rev.D93,no.3,039905(2016)].
431. QCDSF-UKQCD, R. Horsley et al., Phys. Rev. D85 (2012) 034506, 1110.4971.
432. G.S. Bali et al., Nucl. Phys. B866 (2013) 1, 1206.7034.
433. P.E. Shanahan, A.W. Thomas and R.D. Young, Phys. Rev. D87 (2013) 074503, 1205.5365.
434. C. Alexandrou, V. Drach, K. Jansen, C. Kallidonis and G. Koutsou, Phys. Rev. D90 (2014) 074501, 1406.4310.
435. S. Dürr et al., Phys. Rev. Lett. 116 (2016) 172001, 1510.08013.
436. χQCD, Y.B. Yang, A. Alexandru, T. Draper, J. Liang and K.F. Liu, Phys. Rev. D94 (2016) 054503, 1511.09089.
437. RQCD, G.S. Bali, S. Collins, D. Richtmann, A. Schäfer, W. Söldner and A. Sternbeck, Phys. Rev. D93 (2016) 094504, 1603.00827.

438. ETM, A. Abdel-Rehim, C. Alexandrou, M. Constantinou, K. Hadjiyiannakou, K. Jansen, C. Kallidonis, G. Koutsou and A. Vaquero Aviles-Casco, Phys. Rev. Lett. 116 (2016) 252001, 1601.01624.
439. MILC, D. Toussaint and W. Freeman, Phys. Rev. Lett. 103 (2009) 122002, 0905.2432.
440. JLQCD, H. Ohki, K. Takeda, S. Aoki, S. Hashimoto, T. Kaneko, H. Matsufuru, J. Noaki and T. Onogi, Phys. Rev. D87 (2013) 034509, 1208.4185.
441. M. Engelhardt, Phys. Rev. D86 (2012) 114510, 1210.0025.
442. MILC, W. Freeman and D. Toussaint, Phys. Rev. D88 (2013) 054503, 1204.3866.
443. χQCD, M. Gong et al., Phys. Rev. D88 (2013) 014503, 1304.1194.
444. P. Junnarkar and A. Walker-Loud, Phys. Rev. D87 (2013) 114510, 1301.1114.
445. S. Syritsyn, PoS LATTICE2013 (2014) 009, 1403.4686.
446. M. Constantinou, PoS LATTICE2014 (2015) 001, 1411.0078.
447. J. Green, 36th International Symposium on Lattice Field Theory (Lattice 2018) East Lansing, MI, United States, July 22–28, 2018, 2018, 1812.10574.
448. N. Brambilla et al., Eur. Phys. J. C74 (2014) 2981, 1404.3723.
449. H.W. Lin et al., Prog. Part. Nucl. Phys. 100 (2018) 107, 1711.07916.
450. G.S. Bali, S. Collins and A. Schafer, Comput. Phys. Commun. 181 (2010) 1570, 0910.3970.
451. T. Blum, T. Izubuchi and E. Shintani, Phys. Rev. D88 (2013) 094503, 1208.4349.
452. M. Cè, L. Giusti and S. Schaefer, Phys. Rev. D93 (2016) 094507, 1601.04587.
453. M. Cè, L. Giusti and S. Schaefer, Phys. Rev. D95 (2017) 034503, 1609.02419.
454. A. Stathopoulos, J. Laeuchli and K. Orginos, (2013), 1302.4018.
455. A.S. Gambhir, A. Stathopoulos, K. Orginos, B. Yoon, R. Gupta and S. Syritsyn, PoS LATTICE2016 (2016) 265, 1611.01193.
456. B. Yoon et al., Phys. Rev. D93 (2016) 114506, 1602.07737.
457. T.A. DeGrand and S. Schaefer, Comput. Phys. Commun. 159 (2004) 185, hep-lat/0401011.
458. L. Giusti, P. Hernandez, M. Laine, P. Weisz and H. Wittig, JHEP 04 (2004) 013, hep-lat/0402002.
459. R. Gupta, A. Patel, C.F. Baillie, G. Guralnik, G.W. Kilcup and S.R. Sharpe, Phys. Rev. D40 (1989) 2072.
460. C. Thron, S.J. Dong, K.F. Liu and H.P. Ying, Phys. Rev. D57 (1998) 1642, hep-lat/9707001.
461. V. Gülpers, G. von Hippel and H. Wittig, Phys. Rev. D89 (2014) 094503, 1309.2104.
462. S. Bernardson, P. McCarty and C. Thron, Comput. Phys. Commun. 78 (1993) 256.
463. L. Maiani, G. Martinelli, M.L. Paciello and B. Taglienti, Nucl. Phys. B293 (1987) 420.
464. S. Güsken, K. Schilling, R. Sommer, K.H. Mütter and A. Patel, Phys. Lett. B212 (1988) 216.
465. S. Güsken, U. Löw, K.H. Mütter, R. Sommer, A. Patel and K. Schilling, Phys. Lett. B227 (1989) 266.
466. S.J. Dong, K.F. Liu and A.G. Williams, Phys. Rev. D58 (1998) 074504, hep-ph/9712483.
467. S. Capitani, B. Knippschild, M. Della Morte and H. Wittig, PoS LATTICE2010 (2010) 147, 1011.1358.
468. J. Bulava, M. Donnellan and R. Sommer, JHEP 01 (2012) 140, 1108.3774.
469. B.J. Owen, J. Dragos, W. Kamleh, D.B. Leinweber, M.S. Mahbub, B.J. Menadue and J.M. Zanotti, Phys. Lett. B723 (2013) 217, 1212.4668.
470. J. Dragos et al., Phys. Rev. D94 (2016) 074505, 1606.03195.
471. Particle Data Group, M. Tanabashi et al., Phys. Rev. D98 (2018) 030001.
472. O. Bär, Int. J. Mod. Phys. A32 (2017) 1730011, 1705.02806.
473. O. Bär, EPJ Web Conf. 175 (2018) 01007, 1708.00380.
474. F. Jegerlehner and A. Nyffeler, Phys. Rept. 477 (2009) 1, 0902.3360.
475. F. Jegerlehner, Springer Tracts Mod. Phys. 274 (2017) pp.1.
476. M. Davier, A. Höcker, B. Malaescu and Z. Zhang, Eur. Phys. J. C71 (2011) 1515, 1010.4180.
477. K. Hagiwara, R. Liao, A.D. Martin, D. Nomura and T. Teubner, J. Phys. G38 (2011) 085003, 1105.3149.
478. F. Jegerlehner, EPJ Web Conf. 166 (2018) 00022, 1705.00263.
479. M. Davier, A. Hoecker, B. Malaescu and Z. Zhang, Eur. Phys. J. C77 (2017) 827, 1706.09436.
480. A. Keshavarzi, D. Nomura and T. Teubner, Phys. Rev. D97 (2018) 114025, 1802.02995.

481. G. Colangelo, M. Hoferichter, M. Procura and P. Stoffer, JHEP 09 (2014) 091, 1402.7081.
482. G. Colangelo, M. Hoferichter, B. Kubis, M. Procura and P. Stoffer, Phys. Lett. B738 (2014) 6, 1408.2517.
483. G. Colangelo, M. Hoferichter, M. Procura and P. Stoffer, JHEP 09 (2015) 074, 1506.01386.
484. G. Colangelo, M. Hoferichter, M. Procura and P. Stoffer, Phys. Rev. Lett. 118 (2017) 232001, 1701.06554.
485. G. Colangelo, M. Hoferichter, M. Procura and P. Stoffer, JHEP 04 (2017) 161, 1702.07347.
486. V. Pascalutsa and M. Vanderhaeghen, Phys. Rev. Lett. 105 (2010) 201603, 1008.1088.
487. V. Pascalutsa, V. Pauk and M. Vanderhaeghen, Phys. Rev. D85 (2012) 116001, 1204.0740.
488. V. Pauk and M. Vanderhaeghen, (2014), 1403.7503.
489. V. Pauk and M. Vanderhaeghen, Phys. Rev. D90 (2014) 113012, 1409.0819.
490. I. Danilkin and M. Vanderhaeghen, Phys. Rev. D95 (2017) 014019, 1611.04646.
491. A. Nyffeler, Phys. Rev. D94 (2016) 053006, 1602.03398.
492. J. Prades, E. de Rafael and A. Vainshtein, Adv. Ser. Direct. High Energy Phys. 20 (2009) 303, 0901.0306.
493. A. Nyffeler, Phys. Rev. D79 (2009) 073012, 0901.1172.
494. H.B. Meyer and H. Wittig, Prog. Part. Nucl. Phys. 104 (2019) 46, 1807.09370.
495. Budapest-Marseille-Wuppertal, S. Borsanyi et al., Phys. Rev. Lett. 121 (2018) 022002, 1711.04980.
496. B. Chakraborty, C.T.H. Davies, P.G. de Oliviera, J. Koponen, G.P. Lepage and R.S. Van de Water, Phys. Rev. D96 (2017) 034516, 1601.03071.
497. ETM, F. Burger, X. Feng, G. Hotzel, K. Jansen, M. Petschlies and D.B. Renner, JHEP 02 (2014) 099, 1308.4327.
498. RBC/UKQCD, T. Blum et al., Phys. Rev. Lett. 121 (2018) 022003, 1801.07224.
499. M. Della Morte et al., JHEP 10 (2017) 020, 1705.01775.
500. B.e. Lautrup, A. Peterman and E. de Rafael, Phys. Rept. 3 (1972) 193.
501. T. Blum, Phys. Rev. Lett. 91 (2003) 052001, hep-lat/0212018.
502. D. Bernecker and H.B. Meyer, Eur. Phys. J. A47 (2011) 148, 1107.4388.
503. HPQCD, B. Chakraborty, C.T.H. Davies, G.C. Donald, R.J. Dowdall, J. Koponen, G.P. Lepage and T. Teubner, Phys. Rev. D89 (2014) 114501, 1403.1778.
504. H.B. Meyer, Eur. Phys. J. C77 (2017) 616, 1706.01139.
505. M. Della Morte et al., EPJ Web Conf. 175 (2018) 06031, 1710.10072.
506. C. Aubin and T. Blum, Phys. Rev. D75 (2007) 114502, hep-lat/0608011.
507. X. Feng, K. Jansen, M. Petschlies and D.B. Renner, Phys. Rev. Lett. 107 (2011) 081802, 1103.4818.
508. P. Boyle, L. Del Debbio, E. Kerrane and J. Zanotti, Phys. Rev. D85 (2012) 074504, 1107.1497.
509. M. Della Morte, B. Jäger, A. Jüttner and H. Wittig, JHEP 1203 (2012) 055, 1112.2894.
510. T. Blum et al., Phys. Rev. Lett. 116 (2016) 232002, 1512.09054.
511. RBC/UKQCD, T. Blum et al., JHEP 04 (2016) 063, 1602.01767.
512. S. Borsanyi et al., Phys. Rev. D96 (2017) 074507, 1612.02364.
513. D. Giusti, V. Lubicz, G. Martinelli, F. Sanfilippo and S. Simula, JHEP 10 (2017) 157, 1707.03019.
514. Fermilab Lattice, LATTICE-HPQCD, MILC, B. Chakraborty et al., Phys. Rev. Lett. 120 (2018) 152001, 1710.11212.
515. D. Giusti, F. Sanfilippo and S. Simula, Phys. Rev. D98 (2018) 114504, 1808.00887.
516. K. Miura, PoS LATTICE2018 (2019) 010, 1901.09052.
517. M. Hayakawa, T. Blum, T. Izubuchi and N. Yamada, PoS LAT2005 (2006) 353, hep-lat/0509016.
518. T. Blum, S. Chowdhury, M. Hayakawa and T. Izubuchi, Phys. Rev. Lett. 114 (2015) 012001, 1407.2923.
519. T. Blum, N. Christ, M. Hayakawa, T. Izubuchi, L. Jin and C. Lehner, Phys. Rev. D93 (2016) 014503, 1510.07100.
520. T. Blum, N. Christ, M. Hayakawa, T. Izubuchi, L. Jin, C. Jung and C. Lehner, Phys. Rev. Lett. 118 (2017) 022005, 1610.04603.

521. J. Green, N. Asmussen, O. Gryniuk, G. von Hippel, H.B. Meyer, A. Nyffeler and V. Pascalutsa, PoS LATTICE2015 (2016) 109, 1510.08384.
522. N. Asmussen, J. Green, H.B. Meyer and A. Nyffeler, PoS LATTICE2016 (2016) 164, 1609.08454.
523. N. Asmussen, A. Gérardin, H.B. Meyer and A. Nyffeler, EPJ Web Conf. 175 (2018) 06023, 1711.02466.
524. N. Asmussen et al., EPJ Web Conf. 179 (2018) 01017, 1801.04238.
525. T. Blum, N. Christ, M. Hayakawa, T. Izubuchi, L. Jin, C. Jung and C. Lehner, Phys. Rev. D96 (2017) 034515, 1705.01067.
526. J. Green, O. Gryniuk, G. von Hippel, H.B. Meyer and V. Pascalutsa, Phys. Rev. Lett. 115 (2015) 222003, 1507.01577.
527. A. Gérardin, J. Green, O. Gryniuk, G. von Hippel, H.B. Meyer, V. Pascalutsa and H. Wittig, Phys. Rev. D98 (2018) 074501, 1712.00421.
528. A. Gérardin, H.B. Meyer and A. Nyffeler, Phys. Rev. D94 (2016) 074507, 1607.08174.
529. D. Djukanovic, K. Ottnad, J. Wilhelm and H. Wittig, Phys. Rev. Lett. 123, 21 (2019), 212001.
530. A. Gérardin, M. Cè, G. von Hippel, B. Hörz, H.B. Meyer, D. Mohler, K. Ottnad, J. Wilhelm and H. Wittig, Phys. Rev. D100 (2019) 014510, 1904.03120.

6

A Study of the Large Hadronic Collider: Higgs Boson and its Discovery

Peter Jenni and Tejinder S. Virdee

6.1 Introduction and the Standard Model

The standard model of particle physics (SM) is a theory that is based upon principles of great beauty and simplicity. The theory comprises the building blocks of visible matter, the fundamental fermions: quarks and leptons, and the fundamental bosons that mediate three of the four fundamental interactions; photons for electromagnetism, the W and Z bosons for the weak interaction and gluons for the strong interaction (Fig. 6.1).

The SM provides a very successful description of the visible universe and has been verified in many experiments to a very high precision. It has an enormous range of applicability and validity. So far no significant deviations have been observed experimentally.

The possibility of installing a proton-proton accelerator in the LEP tunnel, after the e^+e^- programme, was being discussed in the 1980's. At the time there were many profound open questions in particle physics, and several are still present. In simple terms these are: what is the origin of mass i.e. how do fundamental particles acquire mass, and why do they have the masses that they have? Why is there more matter than anti-matter? What is dark matter? What is the path towards unification of all forces? Do we live in a world with more space-time dimensions than the familiar four? The LHC [1, 2] was conceived to address or shed light on these questions.

P. Jenni
CERN, Geneva, Switzerland

Albert-Ludwigs University Freiburg, Freiburg im Breisgau, Germany

T. S. Virdee (✉)
Imperial College London, London, UK
e-mail: Tejinder.Virdee@cern.ch

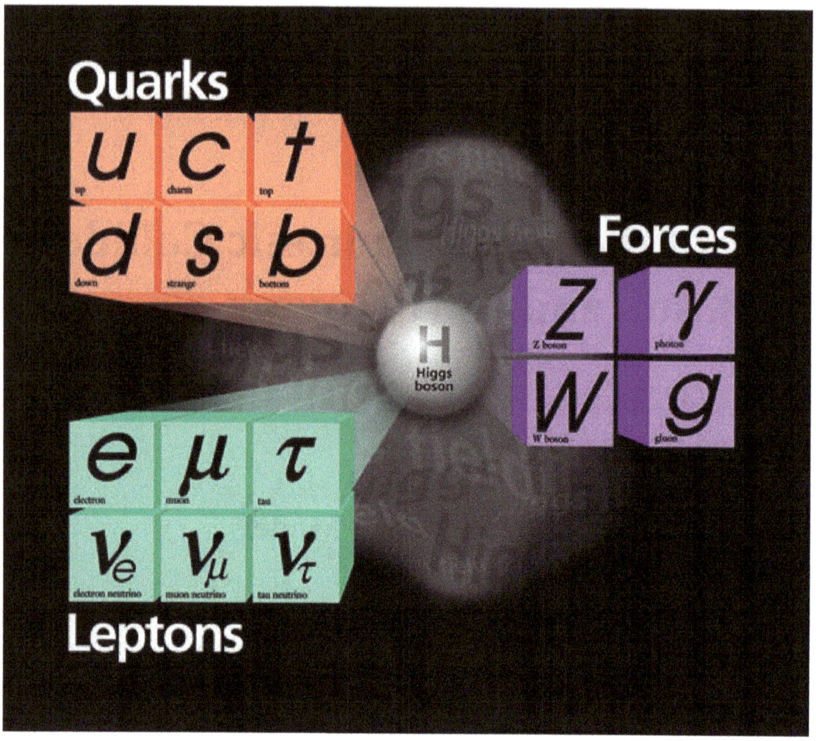

Fig. 6.1 Particle content of the SM, including the Higgs boson considered to be the keystone of the SM

The question of how fundamental particles acquire mass was first posed in the following form: how does the photon remain massless, giving the electromagnetic force an infinite range, whilst the W and Z bosons acquire a seemingly large mass, explaining the short-range of the weak nuclear force.

In 1964 three groups of physicists, Englert and Brout; Higgs; and Guralnik, Hagen, and Kibble [3–7], proposed that there exists an omnipresent field, pervading the universe, and fundamental particles can acquire mass by interacting with this field. At the heart of the mechanism endowing mass was spontaneous symmetry breaking of a local gauge symmetry, through the field's non-zero vacuum expectation value. The new field being a quantum field had an associated quantum, which became known as the Higgs boson.

Today, it seems remarkable that not much attention was paid to the papers [3–7], and even less to the associated Higgs boson. This was partly due to the fact that in the early 1960's most particle physicists were trying to make sense of a plethora of new particles being discovered.

In 1967 Kibble [8] generalized his earlier work with Guralnik and Hagen and brought the mechanism of spontaneous symmetry breaking closer to its application to the description of the real world, one in which the photon remains massless and the W and Z particles become massive [9]. This vein of work reached fruition in the seminal papers of Weinberg [10] and Salam [11], which raised the prospect of the unification of electromagnetism and weak interactions, now labeled the electro-weak theory. Earlier work on a similar model had been carried out by S. Glashow [12]. Weinberg and Salam assumed that W and Z bosons acquired mass by interacting with the field introduced in the earlier papers [3–7].

Both Weinberg and Salam conjectured that such a model would be renormalizable i.e. calculations would give finite answers. The key prediction of their theory was the existence of the Z^0 boson, in addition to the long-known charged W bosons. Again not much attention was paid to these papers.

The situation changed dramatically in 1971. t'Hooft in a *tour de force*, using methods developed by Veltman, outlined the proof that, indeed, the electro-weak theory would be renormalizable [13]. The electro-weak theory started being taken very seriously, so much so that Weinberg's paper [10] has now become the most cited paper in physics.

Experimentally, the 1973 discovery of weak neutral currents [14], mediated by the Z^0 boson, provided strong evidence for the verity of the electro-weak theory.

In parallel much progress had been made in understanding the particles that were being discovered in the 1950s and 1960s. Eventually, these were understood through an underlying gauge field theory, where the "charge" of strong interactions was labeled "colour", and the interactions of coloured quarks are mediated by gluons. The theory [15, 16] displayed two main properties: colour confinement, resulting in the hadrons being colourless, and asymptotic freedom, leading to a steady decrease in the strength of the interaction between quarks and gluons as the interaction energy scale increases. The latter enabled the use of perturbation theory for calculating strong interaction processes at high energies, which has been key to understanding the physics at the LHC.

Further major discoveries included those of new quarks and the gluon meant that the discovery, in 1983, of the W and Z bosons [17, 18] at CERN set the stage for the search for the Higgs boson. The Higgs boson, that earlier had been considered to be a minor and uninteresting feature of the spontaneous breaking mechanism, became to assume a role of central importance as the still missing key particle of the SM. The SM worked so well that the Higgs boson, or something else doing the same job, more or less had to be present.

In 1984, one year after the discovery of the W and Z bosons, a workshop was held in Lausanne where first ideas were discussed about a possible proton-proton collider and associated experiments to make a search for such a particle. The aim was to reuse the LEP tunnel after the end of the electron-positron programme. Amongst the leading protagonists were the scientists from UA1 and UA2 experiments. An exploratory machine was required to cover the wide range of mass values possible for the SM Higgs boson, its diverse decay signatures and production mechanisms and to discover any new high-mass particles at a centre-of-mass energy ten times higher than previously probed. A hadron (proton-proton) collider is such a machine as long as the proton energy is high enough and the instantaneous luminosity, $\mathscr{L}$, measured in $cm^{-2}\,s^{-1}$, is sufficiently large. The rate of production of a given particle is determined by $\mathscr{L} \times \sigma$ where σ is the cross section of the production reaction, measured in units of cm^2. The most interesting and easily detectable final states at a hadron collider involve charged leptons and photons and have a low $\sigma \times$ BR, where BR is the branching ratio into the decay mode of interest.

A major goal of the LHC thus became the elucidation of the mechanism for electroweak symmetry breaking. It also was clear that a search had to be made for new physics at the TeV energy scale as the SM is logically incomplete; it does not incorporate gravity. A promising avenue is the superstring theory, an attempt towards a unified theory with dramatic predictions of extra space dimensions and supersymmetry.

The LHC and its experiments were designed to find new particles, new forces and new symmetries amongst which could be the Higgs boson(s), supersymmetric particles, Z' bosons, or evidence of extra space dimensions. An experiment that could cover the detection of all these hypothesized but yet undiscovered particles would provide the best opportunity to discover whatever else might be produced at LHC energies.

In July 2012 the ATLAS and CMS collaborations discovered a Higgs boson [19, 20].

This paper is based on the previous articles [1, 21–23] written by the authors, some with M. Della Negra, using the recently published results from the ATLAS and CMS Collaborations on the measurements of the properties of the Higgs boson.

6.2 The SM Higgs Boson

In the early 1990's the search for the SM Higgs boson played a pivotal role in the design of the ATLAS and CMS experiments. The mass of the Higgs boson (m_H) is not predicted by theory, but from general considerations, $m_H < 1$ TeV. At the start of the LHC operation, direct searches for the Higgs boson carried out at the LEP collider led to a lower bound of $m_H > 114.4$ GeV at 95% CL [24], whilst precision electroweak constraints, including LEP data, implied that $m_H < 152$ GeV at 95% confidence level (CL) [25]. At time of the discovery at CERN, CDF and D0 experiments operating the Tevatron proton antiproton collider, detected an excess of events in the range 120–135 GeV [26].

It is known that quantum corrections make the mass of any fundamental scalar particle, such as the SM Higgs boson, float up to the next highest mass scale present in the theory, which in the absence of extensions to the SM, can be as high as 10^{15} GeV. Hence finding the scalar Higgs boson would immediately raise a more puzzling question: Why should it have a mass in the range between 100 GeV and 1 TeV? One appealing hypothesis, much discussed at the time, and still being investigated, predicts a new symmetry labeled *supersymmetry*. For every known SM particle there would be a partner with spin differing by half a unit; fermions would have boson superpartners and vice versa, thus doubling the number of fundamental particles. The contributions from the boson and fermion superpartners, and vice a versa, with amplitudes of opposite signs, would lead to their cancellation, and allow a low mass for the Higgs boson. In supersymmetry five Higgs bosons are predicted to exist with one resembling the SM Higgs boson with a mass below about 140 GeV. The lightest of this new species of superparticles could be the candidate for *dark matter* whose presence, by mass, in the universe is around five times more abundant than ordinary matter.

In 1975, physicists had already started to turn their attention to how a putative Higgs boson would manifest itself in experiments [27].

The search for the SM Higgs boson provided a stringent benchmark for evaluating the physics performance of various experiment designs under consideration in the early 1990s and heavily influenced the conceptual design of the general-purpose experiments, ATLAS and CMS.

6.2.1 *Higgs Boson: Production and Decay*

Although the mass of the Higgs boson is not predicted by theory, at a given mass all of its other properties are precisely predicted within the SM. The SM Higgs boson is short-lived (10^{-23} s) and hence the experiments only detect the decay products.

The cross sections for differing production mechanisms and the branching fractions for differing decay modes of the SM Higgs boson, as a function of mass, are illustrated in Fig. 6.2a, b, respectively [28], and the principal ones for $m_H = 125$ GeV and at $\sqrt{s} = 14$ TeV are tabulated in Table 6.1. The uncertainties on these numbers can be found in the twiki in Reference [28].

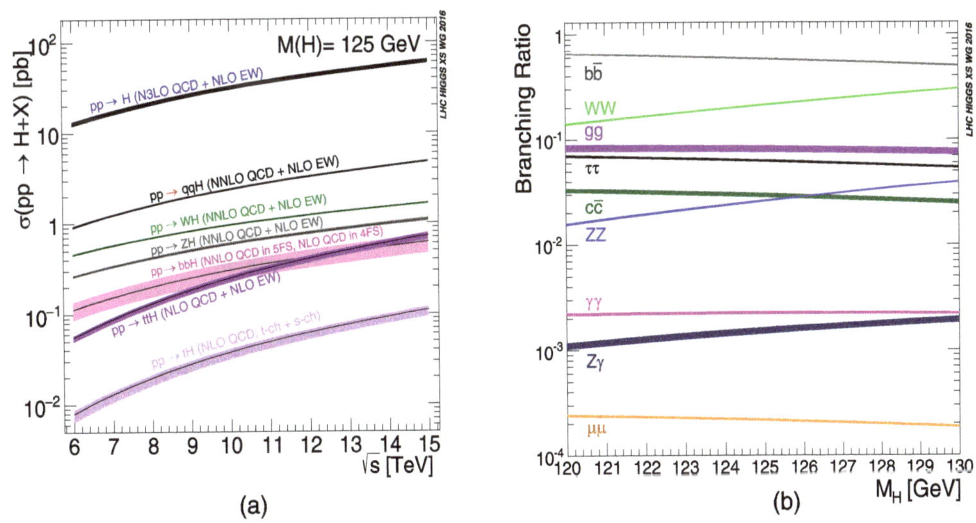

Fig. 6.2 (**a**) SM Higgs boson production cross sections as a function of the centre-of-mass energy, $\sqrt{s}$, for pp. collisions. The VBF process is indicated here as qqH [28]. The theoretical uncertainties are indicated as bands. (**b**) Branching ratios for the main decays of the SM Higgs boson near $m_H = 125$GeV [28]. The theoretical uncertainties are indicated as bands

Table 6.1 The Production cross sections and decay branching fractions at $\sqrt{s} = 14$ TeV

Production process	Cross section (pb)	Decay Mode H→	Branching fraction (%)
gg fusion (pp → H)	54.67	bb	58.24
qqH (VBF)	4.28	$WW^{(*)}$	21.37
WH (associated)	1.51	ττ	6.27
ZH (associated)	0.99	$ZZ^{(*)}$	2.62
ttH	0.61	γγ	2.27×10^{-1}
		μμ	2.18×10^{-2}

The dominant Higgs-boson production mechanism, labeled pp. → H in Fig. 6.2a (for masses up to ≈ 700 GeV) is gluon–gluon fusion.

The vector boson fusion (VBF) mechanism ($WW^{(*)}$ or $ZZ^{(*)}$), labeled pp. → qqH in Fig. 6.2a, becomes important for the production of higher-mass Higgs bosons. Here, the quarks that emit the W or Z bosons have transverse momenta of the order of W and Z masses. The detection of the resulting high-energy jets in the forward pseudorapidity[1] regions, $2.0 < |\eta| < 5.0$, can be used to tag the reaction, improving the signal-to-noise ratio. Tagging of forward jets from the VBF process has turned out to be very important in the measurements of many of the properties of the Higgs boson.

The production of the Higgs boson in association with W and Z boson, labeled pp. → W or Z H in Fig. 6.2a, or the production via the t-tbar fusion, has a much lower cross section, but nevertheless has been important for the final states with large backgrounds such as b-bbar, $\tau^+\tau^-$ or $\mu^+\mu^-$.

The Higgs boson decays in one of several ways (decay modes) into known SM particles, the types depending on its mass. Hence a search had to be envisaged not only over a large range of masses but also many possible decay modes: into pairs of photons, Z bosons, W bosons, τ leptons, and b quarks.

In the mass interval $110 < m_H < 150$ GeV, early detailed studies indicated that the two-photon decay would be the main channel likely to give a significant signal [29]. Detailed studies of another mode, $H \rightarrow ZZ^{(*)} \rightarrow \ell\ell\ell\ell$, where ℓ stands for a charged electron or a muon, dubbed the "golden" mode, suggested that it could be used to cleanly detect the Higgs boson over a wide range of masses starting around $m_H = 130$ GeV [30]. One or both of the Z bosons would be virtual for $m_H < 180$ GeV, and the upper end of the detection range was indicated to be about $m_H < 600$ GeV.

In the region $700 < m_H < 1000$ GeV the cross-section decreases so Higgs boson decays via W and Z decays, where the W and Z decays are to channels with higher branching fractions, have to be employed.

[1]The pseudorapidity $\eta = -\ln[\tan(\theta/2)]$ where and θ is the polar angle measured from the positive z axis (along the beam direction).

6.3 The Large Hadron Collider

6.3.1 *The Road to the LHC*

With the prospect of ground-breaking physics at the LHC, several workshops and conferences followed, where the formidable experimental challenges started to appear manageable, provided that enough R&D work on detectors could be carried out. In 1987 the workshop in La Thuile of the so-called "Rubbia Long-Range Planning Committee" resulted in the recommendation of a proton-proton collider, labeled the Large Hadron Collider (LHC), as the next accelerator for CERN. Meetings of note were the ECFA LHC Workshop in Aachen in 1990 [31], and "Towards the LHC Experimental Programme" [32] which took place in Evian-les-Bains, France in March 1992. At Evian several proto-collaborations presented their designs in "Expressions of Interest". In addition, from the early 1990s, CERN's LHC Detector R&D Committee (DRDC), which reviewed and steered R&D groupings, greatly stimulated innovative developments in detector technology.

Table 6.2 lists the major steps on the long road to the discovery of the Higgs boson.

Table 6.2 The LHC Timeline

1984	Workshop on a Large Hadron Collider in the LEP tunnel, Lausanne, Switzerland.
1987	Workshop on the Physics at Future Accelerators, La Thuile, Italy. The Rubbia "Long-Range Planning Committee" recommends the Large Hadron Collider as the right choice for CERN's future.
1990	LHC Workshop, Aachen, Germany (discussion of physics, technologies and detector design concepts).
1992	General Meeting on LHC Physics and Detectors, Evian-les-Bains, France (with four general-purpose experiment designs presented).
1993	Three Letters of Intent evaluated by the CERN peer review committee LHCC. ATLAS and CMS selected to proceed to a detailed technical proposal.
1994	The LHC accelerator approved for construction, initially in two stages.
1996	ATLAS and CMS Technical Proposals approved.
1997	Formal approval for ATLAS and CMS to move to construction (materials cost ceiling of 475 MCHF).
1997	Construction commences (after approval of detailed Technical Design Reports of detector subsystems).
2000	Assembly of experiments commences, LEP accelerator is closed down to make way for the LHC.
2008	LHC experiments ready for pp. collisions. LHC starts operation. An incident stops LHC operation.
2009	LHC restarts operation, pp. collisions recorded by LHC detectors.
2010	LHC collides protons at high energy (centre of mass energy of 7 TeV).
2012	LHC operates at $\sqrt{s} = 8$ TeV: discovery of a Higgs boson.
2015	LHC operates at $\sqrt{s} = 13$ TeV.

6.3.2 The Challenges of the LHC Accelerator

In this section we outline some of the features and the technological challenges of the LHC [1].

Protons are accelerated by high electric fields generated in superconducting r.f. cavities and are guided around the accelerator by powerful superconducting dipole magnets. The dipole magnets are designed to operate at 8.3 Tesla, allowing the proton beams to be accelerated to 7 TeV, with the current carrying conductor cooled down to 1.9 K in a bath of superfluid helium. The beam pipe in which the protons circulate is under a better vacuum, and at a lower temperature, than that found in inter-planetary space.

The choices of two-in-one high-field superconducting dipole magnets operating at a temperature of 1.9 K, cooled by super-fluid helium were critical to a competitive and affordable design. The LHC could only be competitive with the Superconducting Super Collider (SSC), whose construction had started in the early 1990s in Texas, U.S.A, if the instantaneous luminosity could be an order of magnitude higher (at 10^{34} cm^{-2} s^{-1}). However, the SSC was later cancelled in October 1993.

The main challenges for the accelerator were to build more than one thousand two hundred 15 m long superconducting dipoles able to reach the required magnetic field, the large distributed cryogenic plant to cool the magnets and other superconducting accelerator structures, and the control of the beams, whose stored energy will reach, in design operation, a value of 350 MJ. This magnitude requires extraordinary precautions for beam handling, since if, for any reason this beam is lost in an uncontrolled way, it can do considerable damage to the machine elements, which would result in months of down time.

The counter-rotating LHC beams are organized in 2808 bunches, each of ~10^{11} protons per bunch separated by 25 ns, leading to a bunch crossing rate of ~40 MHz.

Proton beams were first circulated in the LHC in September 2008, and in the days that followed, rapid progress was made in getting a beam to circulate with very good lifetime. Soon after the start a technical incident occurred in the last of the eight sectors to be tested as it was being ramped up to the pre-agreed start-up energy of 5 TeV. The root cause was a failure of one of the 50,000 soldered joints. Substantial damage was done to a large part of the sector involved. After repairs lasting about a year, the LHC started operating again in November 2009. Collisions took place at the injection energy (450 GeV per beam), followed in 2010 and 2011, by a very successful operation at a centre-of-mass energy of 7 TeV. In 2012 the centre-of-mass energy was increased to 8 TeV. The performance surpassed expectations and an integrated luminosity of ~25 fb^{-1}, corresponding to 2×10^{15} proton-proton interactions, was delivered. This is labeled Run 1.

During the period 2015–2018 the LHC operated at a proton-proton centre-of-mass energy of 13 TeV and delivered a total of over 150 fb^{-1} of integrated luminosity. The collider performed close to, or beyond, its design values in many parameters, operating at 13 TeV and reaching peak luminosities of 2×10^{34}

Table 6.3 Some of the LHC parameters of attained/design performance for ATLAS and CMS

	Achieved	Design
Energy	13 TeV	14 TeV
Max. no. of bunches	2556	2808
Bunch spacing (ns)	25	25
Protons/bunch (10^{11})	1.1	1.15
β_* (cm)	30	55
Peak luminosity (10^{34} cm^{-2} s^{-1})	2.1	1.0
Total integrated luminosity (fb^{-1})		
$\sqrt{s} = 7$ TeV (Run 1)	5	
$\sqrt{s} = 8$ TeV (Run 1)	20	
$\sqrt{s} = 13$ TeV (Run 2)	140	

cm^{-2} s^{-1}, twice the design value. This period of operation is labeled Run 2. The achieved performance at the time of writing (2018) can be found in Table 6.3.

6.4 The ATLAS and CMS Experiments

Not only was the putative SM Higgs boson to be rarely produced in the proton collisions, but also it decays into particles (isolated photons, electrons, and muons) that are the best identifiable signatures of its production at the LHC also was expected to be rare. The rarity is illustrated by the fact that Higgs boson production and decay to one such distinguishable signature ($H \rightarrow ZZ^{(*)} \rightarrow 4\,l$) happens roughly once in 10^{13} proton-proton collisions. So a vast number of proton-proton collisions per second have to be delivered by the accelerator and examined by the experiments. At the end of 2018, the LHC was operating at a collision rate of around 10^9 per second. The ATLAS and CMS detectors operate in the harsh environment created by this huge rate of proton-proton collisions. The challenges posed are discussed in reference [33, 34].

6.4.1 The Challenges for ATLAS and CMS Experiments

At the Aachen workshop the physics case for the LHC was thoroughly examined. The experimental search for the Higgs boson across the entire possible range of mass was fully explored for the first time. There was a prevalent prejudice of the protagonists of supersymmetry that m_H should be smaller than 135 GeV. As the decay width of the SM Higgs boson is about 5.5 MeV at $m_H = 100$ GeV, and 8.3 MeV at 150 GeV, the width of the reconstructed invariant ($\gamma\gamma$ or 4 l) mass distribution, and hence the signal/background ratio, would be limited by the electron/photon energy resolution of the electromagnetic calorimeter, and the charged particle momentum resolution of the inner tracker and the muon spectrometer. This lower end of the remaining open mass range was considered to be especially difficult in hadron colliders. Hence the LHC experiments had to pay particular attention to the performance requirements imposed by the search for the Higgs boson in this low mass range. As a consequence much importance was placed on the tracking (inner and muon), as well as the magnetic field strength, and the electromagnetic calorimeters.

The search for the high-mass Higgs boson, particles predicted by SUSY, and other exotic states mentioned above, required excellent resolution for jets and missing transverse momentum (p_T^{miss}), requiring full solid angle calorimeter coverage.

A saying prevalent in the late 1980's and early 1990's captured the challenge: 'We think we know how to build a high energy, high luminosity hadron collider—but we don't have the technology to build a detector for it'. Making discoveries in the unprecedented high collision rate environment, generated by around one billion proton-proton interactions per second, with several tens of simultaneous collisions per bunch crossing, would require extraordinary detectors. Many technical, financial, industrial and human challenges lay ahead, which were all overcome, to yield experiments of unprecedented complexity and power. A flavour can be attained from articles in reference [35].

At the Evian meeting in 1992 four experiment designs were presented: two deploying toroids (one with a superconducting magnet in the barrel) and two deploying superconducting high-field solenoids. The choice of the magnetic field configuration determined the overall design of the experiments.

The collaborations deploying toroids merged to form the ATLAS Collaboration. The ATLAS design [35] was based on a very large superconducting air-core toroid for the measurement of muons, and supplemented by a superconducting 2 Tesla solenoid to provide the magnetic field for inner tracking and by a liquid-argon/lead electromagnetic calorimeter with a novel "accordion" geometry. The CMS design [36] was based on a single large-bore, long, high-field solenoid for analyzing muons, together with powerful microstrip-based inner tracking and an electromagnetic calorimeter comprising scintillating crystals.

On top of the selected event of interest, an average of up to around 40 other proton-proton events are superimposed. These superposed events are referred to as *minimum-bias* events, because no selection is made. Thus thousands of particles emerge from the interaction region every 25 ns where one nanosecond (ns) = 10^{-9} s. Hence the products of an interaction under study can be confused with those from other interactions in the same bunch crossing. This problem, known as *pileup*, clearly becomes more severe if the response time of a detector element and its electronic signal is longer than 25 ns. The effect of pileup can be reduced by using highly granular detectors with fast, short duration, signals, giving low *occupancy* (i.e., a low probability that a detector element will give a signal) at the expense of having large numbers of detector channels. The resulting millions of electronic channels require very good time synchronization.

The large flux of particles emanating from the interaction region creates a high-radiation environment requiring radiation-hard detectors and front-end electronics.

Access for maintenance is very difficult, time consuming, and highly restricted. Hence, a high degree of long-term operational reliability had to be attained, comparable to that which is usually associated with instruments flying on space missions.

The event selection process (called the *trigger*) must select among the billion interactions that occur each second since no more than a thousand events per second can be stored for subsequent analysis. The short time between bunch crossings, 25 ns, has major implications for the design of the readout and trigger systems. It takes a long time to make a trigger decision, yet new events occur in every crossing and a trigger decision must be made for every crossing; the selection process is split in several levels. The first of these is the Level-1 trigger decision, which takes about 3 μs and selects, on average, one crossing out of 400. During this time the data must be stored in *pipelines* integrated into the front-end electronics. In CMS, the data from these selected events are then moved into a commercial farm of CPUs to select and store about one thousand/s of the most interesting events for subsequent analysis.

It cannot be stressed enough how important were the many years of R&D and prototyping that preceded the start of detector construction. Technologies had to be developed far beyond what was the state-of-the-art in early 1990s, in terms of granularity, speed of readout, radiation tolerance, reliability, and very importantly cost. For many detector subsystems, there were initially several technologies considered, as it was far from certain which technologies would be able to attain the required performance. In many cases several variants were developed, prototyped and tested, before choosing the one best able to fulfill the stringent requirements. This involved building and testing increasingly more realistic and larger prototypes, in a process that involved industry from the outset. This took place over a number of years before construction commenced in the second half of the 1990s.

In the 1990's the two collaborations, ATLAS and CMS, grew rapidly in terms of people and institutes. Today each comprises over 3500 scientists and engineers, from over 150 institutions in more than 40 countries. The talents and resources of all these scientists were needed to build the experiments, which are now performing extraordinarily well at the LHC.

The single most important aspect of the experiment design and layout is the magnetic field configuration for the identification of muons and the measurement of their momentum. Large bending power is needed to measure precisely the momentum of charged particles. This forces a choice of superconducting technology for the magnets. The design configurations chosen by ATLAS and CMS are discussed below.

6.4.2 The ATLAS Detector

The design of the ATLAS detector [35], shown in Fig. 6.3 (top), is based on a novel superconducting air-core toroid magnet system, containing ~80 km of

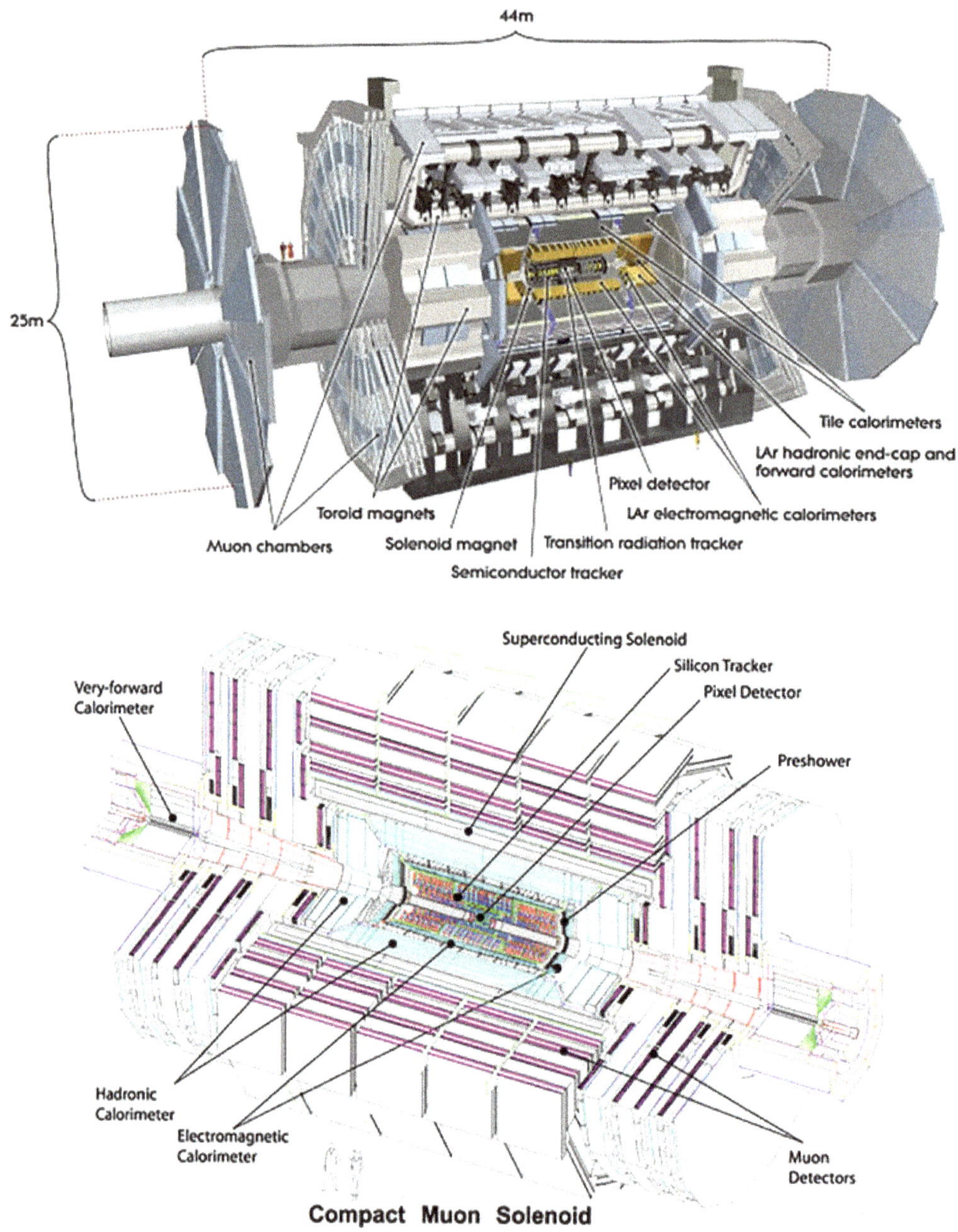

Fig. 6.3 Schematic longitudinal cut-away views of (top) the ATLAS and (bottom) the CMS detectors, showing the different layers around the LHC beam axis, with the collision point in the centre

superconductor cable, in eight separate barrel coils (each $25 \times 5\ m^2$ in a 'racetrack' shape) and two matching endcap toroid systems. A field of ~0.5 Tesla is generated over a large volume. The toroids are complemented with a thin superconducting central solenoid (2.4 m diameter, 5.3 m length) that provides an axial magnetic field of 2 Tesla.

The electromagnetic calorimeter consists of a lead/liquid-argon sampling calorimeter in a novel 'accordion' geometry. A plastic scintillator—iron sampling hadron calorimeter, also with a novel geometry, is used in the barrel part of the experiment. Liquid-argon hadronic calorimeters are employed in the endcap regions near the beam axis. The electromagnetic and hadronic calorimeters have almost 200,000 and 20,000 cells, respectively, and are in an almost field-free region between the toroids and the solenoid.

The momentum of the muons is precisely measured after traversing the calorimeters in the air-core toroid field over a distance of ~5 m. About 1200 large muon chambers of various shapes, with a total area of 5000 m^2, measure the impact position with an accuracy of better than 0.1 mm. Another set of about 4200 fast chambers is used to provide the "trigger".

The reconstruction of all charged particles, and that of displaced vertices, is achieved in the inner detector, which combines highly granular pixel (50 × 400 μm^2 elements, leading to 80 million channels) and microstrip (13 cm × 80 μm elements, leading to six million channels) silicon semiconductor sensors placed close to the beam axis, and a 'straw tube' gaseous detector (350,000 channels) which provides about 30–40 signal hits per track. The latter also helps in the identification of electrons using information from the effects of transition radiation.

The air-core magnet system allows a relatively lightweight overall structure leading to a detector weighing 7000 tons. The muon spectrometer defines the overall diameter of 25 m and length of 44 m of the ATLAS detector.

6.4.3 *The CMS Detector*

The design of the CMS detector [36], shown in Fig. 6.3 (bottom), is based on a state-of-the-art superconducting high-field solenoid, which first reached the design field of 4 Tesla in 2006.

The solenoid generates a uniform magnetic field parallel to the direction of the LHC beams. The field is produced by a current of 20 kA flowing through a reinforced Nb-Ti superconducting coil built in four layers. Economic and transportation constraints limited the outer radius of the coil to 3 m and its length to 13 m. The field is returned through a 1.5 m thick iron yoke, which houses four muon stations to ensure robustness of identification and measurement and full geometric coverage.

The CMS design was first optimized to cleanly identify, trigger and measure muons, e.g. arising from processes such as $H \rightarrow ZZ^{(*)} \rightarrow 4\ \mu$ and few TeV mass $Z' \rightarrow 2\ \mu$, over a wide range of momenta. The muons trace a spiral path in the magnetic field and are identified and reconstructed in ~3000 m^2 of gas chambers interleaved with the iron plates in the return yoke. Another ~500 fast chambers are used to provide a second system of detectors for the Level-1 muon trigger.

The next design priority was driven by the search for the decay of the SM Higgs boson into two photons. A new type of scintillating crystal was selected: lead-tungstate ($PbWO_4$) crystal.

The solution to charged particle tracking was to opt for a small number of precise position measurements of each charged track (~13 each with a position resolution of ~15 μm per measurement) leading to a large number of cells distributed inside a cylindrical volume 5.8 m long and 2.5 m in diameter: 66 million 100 × 150 μm^2 silicon pixels and 9.3 million silicon microstrips ranging from ~10 cm × 80 μm to ~20 cm × 180 μm. The 198 m^2 area of active silicon of the CMS tracker is by far the largest silicon tracker ever built.

Finally the hadron calorimeter, comprising ~3000 projective towers covering almost the full solid angle, is built from alternate plates of ~5 cm brass absorber and ~4 mm thick scintillator plates that sample the energy. The scintillation light is detected by photodetectors (hybrid photodiodes) that can operate in the strong magnetic field.

6.4.4 Installation and Commissioning

The two very different and complementary detector concepts, ATLAS and CMS, resulted in two different strategies for the underground installation of these experiments.

Given its size and its magnet structure, the ATLAS detector had to be assembled directly in the underground cavern. The installation process began in summer 2003 (after the completion of civil engineering work that started in 1998) and ended in summer 2008. Figure 6.4 (top) shows the completion of the barrel toroid magnet system with the insertion of the barrel calorimeters. Figure 6.4 (bottom) shows one end of the cylindrical barrel detector after 3.5 years of installation work, 1.5 years before completion. The ends of four of the barrel toroid coils are visible, illustrating the eightfold symmetry of the structure.

The iron yoke of the CMS detector is divided into five barrel-wheels and three endcap disks at each end, giving a total weight of 12,500 tons. This structure enabled the detector to be assembled and tested in a large surface hall while the underground cavern was being prepared. The sections, weighing between 350 tons and 2000 tons, were then lowered sequentially between October 2006 and January 2008, using a dedicated gantry system equipped with strand jacks: a pioneering use of this technology to simplify the underground assembly of large experiments. Figure 6.5 top shows the lowering of the heaviest and central section, supporting the superconducting coil. Figure 6.5 bottom shows the transverse section of the barrel part of CMS illustrating the successive layers of detection starting from the centre where the collisions occur: the inner tracker, the crystal calorimeter, the hadron calorimeter, the superconducting coil, and the iron yoke instrumented with the four muon stations. The last muon station is at a radius of 7.4 m.

Individual detector components (e.g. chambers) of both experiments were built and assembled in a distributed way all around the globe in the numerous participating institutes and were typically first tested at their production sites, then after delivery to CERN, and finally again after their installation in the underground

Fig. 6.4 (top) Photograph of the barrel toroid magnet system after the completion of the the installation of the eight coils, (bottom) Photograph of one end of the ATLAS detector barrel with the calorimeter end-cap still retracted before its insertion into the barrel toroid magnet structure (February 2007 during the installation phase)

caverns. The collaborations also invested enormous effort in testing representative samples of the detectors in test beams at CERN and other accelerator laboratories around the world. These test-beam campaigns not only verified that performance criteria were met over the several years of production of detector components, but also were used to prepare the calibration and alignment data for LHC operation. The so-called large combined test-beam setups, which represented whole 'slices' of the different detector layers of the final detectors, proved to be very important.

Fig. 6.5 (top) Photograph showing the lowering of the central barrel part and solenoid of the CMS detector during its installation in the cavern in 2007; (bottom) Photograph of transverse section of the barrel part of CMS illustrating the successive layers of detection

During the installation, the experiments made extensive use of the constant flow of cosmic rays impinging on Earth providing a reasonable flux of muons even at a depth of 100 m underground. Typically a few hundred per second traverse the detectors. These muons were used to check the whole chain from sub-detector hardware to analysis programs of the experiments, and to align the detector elements and calibrate their response prior to the proton-proton collisions. In particular, after the LHC incident on 19th September 2008 the experiments used the 15 months LHC down time, before the first collisions on 23rd November 2009, to run the full detectors in very extensive cosmic-ray campaigns, collecting many hundreds of millions of muon events. These runs allowed both ATLAS and CMS to be ready for physics operation, with pre-calibrated and pre-aligned detectors, by the time of the first pp collisions.

6.5 Experiment Software and LHC Worldwide Computing Grid

The experiment collaborations themselves develop the software that enables reconstruction, from raw data, of analyzable objects such as electrons, photons, jets, b jets, muons, and other charged tracks, and their energies or momenta. Algorithms have to be run to calibrate the energy deposits; align the hits from charged particles; and correct for changes in detector response arising from irradiation, variation in environmental parameters such as temperature, or changes in the position of detecting elements. The software packages must also simulate the response of the detectors to the passage of particles generated in simulated events occurring in bunch crossings that contain interesting physics processes, as well as simple backgrounds. These include processes such as the production of W or Z bosons, QCD jets, or Higgs bosons and their decays. Such simulations helped prepare, prior to the first collisions, the experiments' end-to-end processing and analysis chains, which were crucial for the rapid delivery of physics results of outstanding quality and quantity soon after the first collisions.

The LHC computing system, termed the LHC Worldwide Computing Grid (WLCG) [1], was conceived to make effective use of distributed resources, work on a large scale, and enable all the experiments' scientists, wherever they were based, to have access to LHC data, and without regard to the extent of the resources they themselves could afford. The WLCG provided the backbone for the analysis capabilities of the experiments. The global WLCG has continued to grow, now encompassing around 170 computing centers in 42 countries, with an infrastructure that provides access to some 600,000 computing cores, around 500 PB of storage (50% on disk and 50% on tape), and a network that frequently runs at 100 Gb s^{-1} between larger sites and at 10 Gb s^{-1} between smaller sites. The security of access have been instrumental for building a truly federated computing infrastructure for science. Although the individual computing tasks described above were already familiar in particle physics, the scope, scale, and geographical spread of the LHC computing and data analysis are unprecedented.

6.6 Operation of the LHC: The Start of Data Taking

On the tenth of September 2008 first beams circulated in the Large Hadron Collider. Nine days later, during the powering test of the last octant, alarms reached the LHC accelerator's control room and safety systems were activated to protect the accelerator. It turned out that one of the 50,000 soldered joints had malfunctioned. This led to an electrical arc that pierced the vacuum enclosure of a superconducting dipole bending magnet leading a massive escape of helium, the pressure wave of which caused considerable damage. The accelerator went offline for repairs. The ATLAS and CMS experiments continued to run round-the-clock for a few months recording billions of traversals of muons from cosmic rays. These data demonstrated that the experiments were in a good shape to take collision data. After a few tweaks the ATLAS and CMS experiments were even better prepared for first collisions, which came on 23rd November 2009. The first collision data were rapidly distributed, analysed and physics results produced.

Following a preliminary low-energy run in the autumn of 2009, the ATLAS and CMS experiments started recording high-energy proton-proton collisions in March 2010 at $\sqrt{s} = 7$ TeV. Some 45 pb^{-1} of data were recorded, sufficient to demonstrate that the experiments were working well, according to the ambitious design specifications and the results they were producing were consistent with the predictions from known SM physics. Many parameters were examined, including the efficiency of identification and reconstruction of physics objects, the measured energy and momentum resolutions, the resolution of peaks in invariant mass distribution, and more. An example of the performance from the CMS experiment is the comparison of the observed width of Y particle with the design mass resolution. The width is expected to be dominated by instrumental resolution. Figure 6.6 shows that the observed width is measured to be 70 MeV consistent with the design value. Also observed in such di-muon invariant mass distributions is a history of decades of particle physics indicating the excellent performance of the experiments. The next step was to see if known physics could be measured as per the predictions of the SM, extrapolated to the new energies.

6.6.1 Measurement of SM Processes to Verify Experiment Performance

Observation and accurate measurement of the production of known SM particles at the LHC collision energies is a pre-requisite for the exploration of new physics,

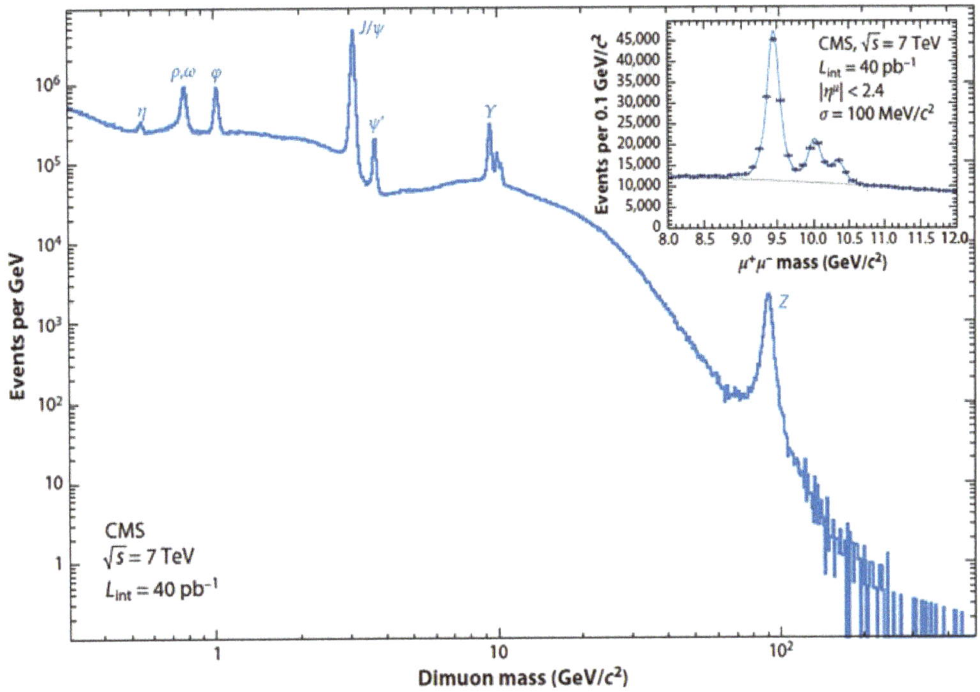

Fig. 6.6 The distribution of the invariant mass for di-muon events, shown here from CMS, displays the various well-known resonant states of the SM. The inset illustrates the excellent mass resolution for the three states of the Υ family. The mass resolutions in the central region are; 28 MeV (0.9%) for J/ψ, 69 MeV (0.7%) for Υ(1S), both dominated by instrumental resolution and $\Gamma = 2.5$ GeV for the Z dominated by its natural width, and are equal to the design values

including the search for the Higgs boson. The SM processes, such as W and Z production, are often considered to be 'standard candles' for the experiments.

In the ATLAS and CMS experiments, SM physics can be studied with unprecedented precision, allowing comparison with the predictions of the SM with small instrumental systematic errors. The data collected so far have enabled many precise measurements of SM processes, including the production of light quarks and gluons, bottom and top quarks, and W and Z bosons, singly and in pairs, and with varying numbers of jets resulting from higher order processes. A summary of such studies is shown in Fig. 6.7, where measurements of cross sections for various selected electroweak and QCD processes are compared with predictions from the SM. These very diverse measurements, probing cross-sections over a range of many orders of magnitude, established that the experiments were "physics commissioned" and ready for discoveries. The detector performance was well understood and known SM processes were correctly observed, crucially important as they often constitute large backgrounds to signatures of new physics, such as those expected for the Higgs boson.

The speed with which these measurements verified the SM predictions for known physics is a tribute to the large amount of work done by many groups, including

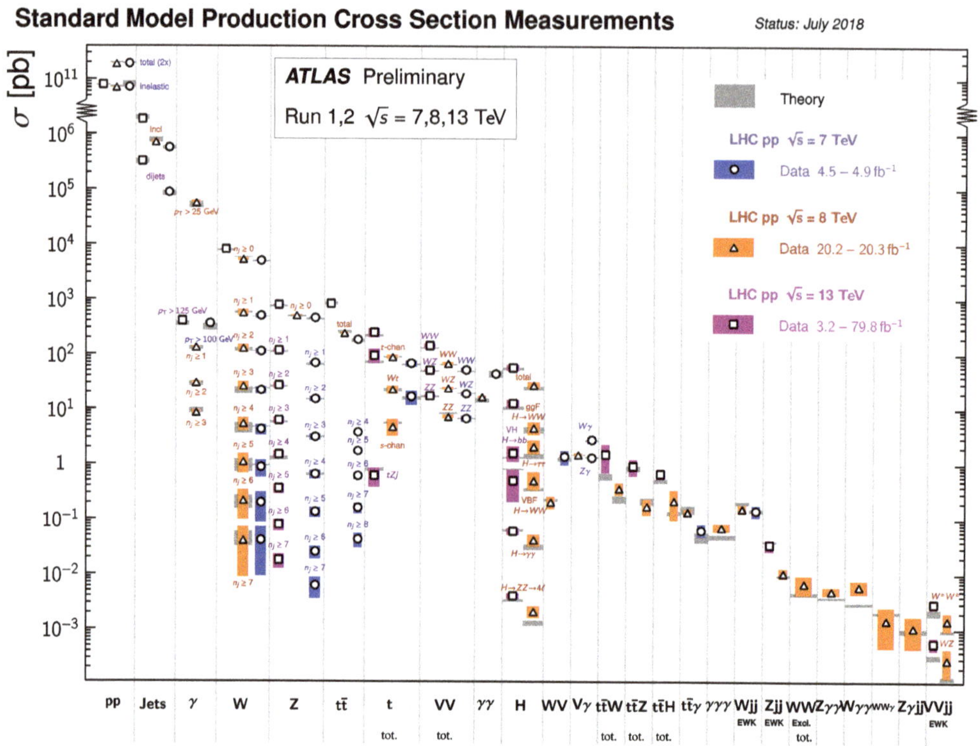

Fig. 6.7 A comparison of cross-section measurements for electroweak and QCD processes with theoretical predictions from the SM, shown here as example from the ATLAS experiment

theorists, other collider experiments at LEP, Tevatron, HERA, b-factories and to the good preparation of the ATLAS and CMS experiments.

In what follows the production of b-bar, $\tau^+\tau^-$, W^+W^-, etc. will be denoted by bb, ττ, $WW^{(*)}$, etc.

Using all the data so far collected extensive searches for new physics, beyond the standard model, have been performed. No new physics beyond the SM has yet been discovered. Limits have been set on e.g. quark substructure, supersymmetric particles (e.g. disfavouring gluino masses below 1.5 TeV in simple models of supersymmetry), potential new bosons (e.g disfavouring new heavy SSM W′ and Z′ bosons with masses below 3 TeV for couplings similar to the ones for the known W and Z bosons) and semi-classical black holes in the context of large extra dimensions (with masses below 10 TeV).

6.7 The Discovery and Properties of a Higgs Boson

Undoubtedly, the most striking result to emerge from the ATLAS and CMS experiments is the discovery of the Higgs boson at a mass of ~125 GeV [19] and [20], respectively.

The SM Higgs boson couples to the different pairs of particles in a set proportion i.e. for fermions (f) proportional to m_f^2, and for bosons (V) proportional m_V^4/v^2 where v is the vacuum expectation value of the scalar field ($v = 246$ GeV). Once produced the Higgs boson disintegrates immediately into known SM particles. Both the production modes and decay modes and rates are precisely predicted in the SM. A search had to be made over a large range of masses and the many possible decay modes with differing branching ratios as shown in Fig. 6.2 and Table 6.1. For example, at $m_H = 125$ GeV the SM boson is predicted to decay into pairs of photons with branching fraction (BR) of 2.2×10^{-3}, into Z bosons and then four electrons or muons or two muons and two electrons with BR $= 1.25 \times 10^{-4}$, into a pair of W bosons and then into $ll\nu\nu$ with BR ~1%, etc.

For a given Higgs boson mass hypothesis, the sensitivity of the search depends on:

- the mass of the Higgs boson
- the Higgs boson production cross section (Fig. 6.2a and Table 6.1),
- the decay branching fraction into the selected final state (Fig. 6.2b and Table 6.1),
- the signal selection efficiency,
- the observed Higgs boson mass resolution, and
- the level of backgrounds with the same or a similar final state.

Comparisons with the expectations from the SM for the various combinations of production and decay modes are usually cast in terms of modifiers such as signal strength, μ, that is the ratio of the measured production $\times$ decay rate of the signal and the SM expectation i.e. $\mu = \sigma.BR/(\sigma.BR)_{SM}$. Signal strength of one would be indicative of the SM Higgs boson.

CMS and ATLAS increasingly use global event reconstruction algorithms, labeled particle-flow reconstruction, that attempt to identify, reconstruct and provide the measurement of the energy of particles by combining information from the inner tracker, the calorimeters and the muon system in an optimized manner. Hadronic jets are clustered from the reconstructed particles using the infrared- and collinear-safe anti-k_T algorithm with a distance parameter usually set at 0.4. The jet momenta are measured by summing vectorially the momenta of all particles in the jet. Jets originating from b-jets are identified by discriminants that include the presence of particles originating from vertices displaced from the primary interaction vertex. A typical b-jet efficiency of around 70% is attained for a 1% misidentification probability for light quarks and gluons. The missing transverse momentum vector is taken as the negative of the vector sum of the momenta of all reconstructed particles in the event; its magnitude is labeled p_T^{miss}.

By the end of 2012 (LHC Run 1) the total amount of data that had been examined corresponded to an integrated luminosities of ~5 fb^{-1} at $\sqrt{s} = 7$ TeV and ~ 20 fb^{-1} at $\sqrt{s} = 8$ TeV, equating to the examination of some 2×10^{15} proton-proton collisions.

By the end of the Run 2 (2018) the total amount of data that had been recorded corresponded to an integrated luminosities ~150 fb^{-1} at $\sqrt{s} = 13$ TeV, equating to some 1.5×10^{16} proton-proton collisions.

6.7.1 Event and Physics Objects Reconstruction and Analysis Techniques

It is convenient to subdivide the analysis of data relating to Higgs bosons according to the decay channel, using datasets where the data have been selected to contain a particular set of final state particles. To improve the sensitivity, the events in this dataset are usually separated into categories that are intended to reflect the expected signal-to-background ratio. Several multivariate methods are used in the analyses

- to improve event reconstruction, estimates of the energies/momenta of physics objects (e.g. photons and electrons, etc.),
- to identify physics objects (such as electrons, photons, b-quarks, tau leptons, etc.)
- to categorize events according to particular production (e.g. ggH, VBF, VH, ttH etc.), decay mode, or expected signal-to-background ratio.

The reader can find the exact description of the multivariate methods used within the individual papers referenced in the sections below.

Charged leptons and photons originating from the fundamental partonic processes tend to be "isolated" i.e. no other particles surround the one of interest. A relative isolation condition is applied on such particles. The sum of transverse momenta of accompanying particles, within an angular radius of approximately 0.3, around the particle of interest, is divided by the transverse momentum of the particle of interest. A cut on this ratio is made, the value of which is separately optimized for electrons, muons or photons. A correction to the accompanying energy is applied when the instantaneous luminosity is high and undesirable energy from pileup interactions is accidentally captured in the region.

6.7.2 The Discovery: Results from the 2011 and Partial 2012 Datasets

In the 2011 data-taking run the ATLAS and CMS experiments recorded data at $\sqrt{s} = 7$ TeV corresponding to an integrated luminosity of ~5 fb^{-1}. In December 2011, the first "tantalizing hints" of a new particle from both the CMS and ATLAS experiments were shown at CERN. The general conclusion was that both experiments were seeing an excess of unusual events at roughly the same place in mass (in the mass range 120–130 GeV) in two different decay channels. That set the stage for data taking in 2012.

In January 2012 it was decided to slightly increase the energy of the protons from 3.5 to 4 TeV, giving a centre of mass energy of 8 TeV. By June 2012 the number of high-energy collisions examined had doubled and both CMS and ATLAS had greatly improved their analyses so it was decided to look at the area which had shown the excess of events but only after all the algorithms and selection procedures had been agreed, in case a bias was inadvertently introduced. These data led to the discovery of a Higgs boson, independently in both the ATLAS and CMS experiments in July 2012.

In this section we shall concentrate on the region of low mass ($114.4 < m_H < 150$ GeV) where the two channels particularly suited for unambiguous discovery are the decays to two photons and to two Z bosons, where one or both of the Z bosons could be virtual, subsequently decaying into four electrons, four muons or two electrons and two muons. These two decay modes are particularly suited for discovery as the observed mass resolution (~1% of m_H) is the best and the backgrounds are manageable or small.

6.7.2.1 The H → γγ Decay Mode

In the $H \rightarrow \gamma\gamma$ analysis a search is made for a narrow peak in the diphoton invariant mass distribution in the mass range 110–150 GeV, on a large irreducible background from QCD production of two photons (via quark-antiquark annihilation and the gluon-fusion or "box" diagrams). There is also a reducible background where one or more of the reconstructed photon candidates originate from misidentification of jet fragments, with the process of QCD Compton scattering dominating. The relative fractions of these backgrounds in the selected events are illustrated in Fig. 6.9a.

The event selection requires two "isolated" photon candidates satisfying p_T and photon identification criteria. As an example, CMS applies a threshold of $p_T = m_{\gamma\gamma}/3$ ($m_{\gamma\gamma}/4$) to the leading (sub-leading) photon in p_T, where $m_{\gamma\gamma}$ is the diphoton invariant mass. Scaling the p_T thresholds in this way avoids distortion of the shape of the $m_{\gamma\gamma}$ distribution. The background is estimated from data, without the use of MC simulation, by fitting the diphoton invariant mass distribution in a range ($100 < m_{\gamma\gamma} < 180$ GeV).

The results from the CMS experiments are shown in Fig. 6.8a [20]. A clear peak at a diphoton mass of around 125 GeV is seen. A similar result was obtained in the ATLAS experiment [19].

6.7.2.2 The H → ZZ$^{(*)}$ → 4 l Decay Mode

In the $H \rightarrow ZZ^{(*)} \rightarrow 4\,l$ decay mode a search is made for a narrow four-charged lepton mass peak in the presence of a small continuum background. The background sources include an irreducible four-lepton contribution from direct $ZZ^{(*)}$ production via quark-antiquark and gluon–gluon processes. Reducible background contributions arise from Z + bb and tt production where the final states contain two isolated leptons and two b-quark jets producing secondary leptons.

The event selection requires two pairs of same-flavour, oppositely charged isolated leptons. Since there are differences in the reducible background rates and mass resolutions between the sub-channels 4e, 4 μ, and 2e2μ, they are analysed separately. Electrons are typically required to have $p_T > 7$ GeV. The corresponding requirements for muons are $p_T > 5$ GeV. Both electrons and muons are required to

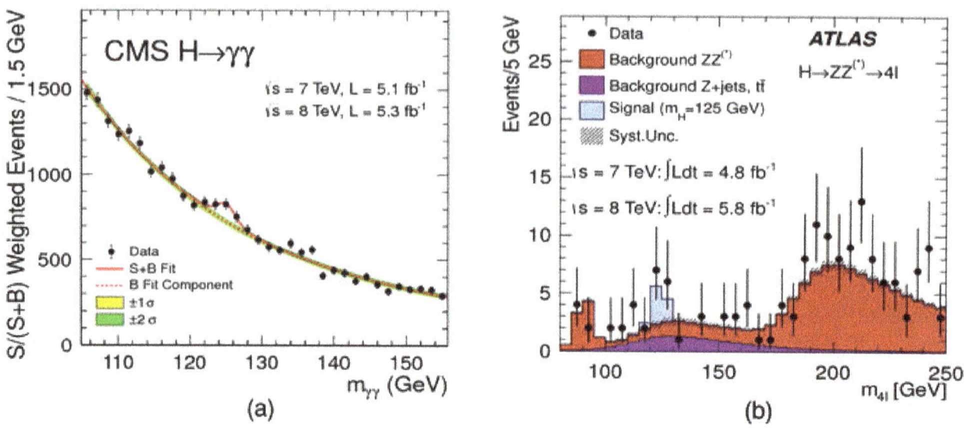

Fig. 6.8 (**a**) The two-photon invariant mass distribution of selected candidates in the CMS experiment, weighted by S/B of the category in which it falls. The lines represent the fitted background and the expected signal contribution ($m_H = 125$ GeV). (**b**) The four-lepton invariant mass distribution in the ATLAS experiment for selected candidates relative to the background expectation. The expected signal contribution ($m_H = 125$ GeV) is also shown

be isolated. The pair with invariant mass closest to the Z boson mass is required to have a mass in the range 40–120 GeV and the other pair is required to have a mass in the range 12–120 GeV. The $ZZ^{(*)}$ background, which is dominant, is evaluated from Monte Carlo simulation studies.

The m_{4l} distribution is shown in Fig. 6.8b for the ATLAS experiment [19]. A clear peak is observed at ~125 GeV in addition to the one at the Z mass. The latter is due to the conversion of an inner bremstrahlung photon emitted simultaneously with the dilepton pair. A similar result was obtained by the CMS experiment [20].

6.7.2.3 Combinations

A search was also made in other decay modes of a possible Higgs boson and combined to yield the final results published in August 2012 by ATLAS [19] and CMS [20] experiments. Both ATLAS and CMS independently discovered a new heavy boson at approximately the same mass, clearly evident in the two different decay modes, $\gamma\gamma$ and $ZZ^{(*)}$. The observed (expected) local significances were 6.0σ (5.0σ) and 5.0σ (5.8σ) in ATLAS and CMS respectively, indicating that a new particle had been discovered.

The decay into two bosons (two photons; two Z bosons or two W bosons) implied that the new particle is a boson with spin different from one, and its decay into two photons that it carries either spin-0 or spin-2.

The results presented by both ATLAS and CMS collaborations were consistent, within uncertainties, with the expectations for a SM Higgs boson. Both noted that collection of more data would enable a more rigorous test of this conclusion and an investigation of whether the properties of the new particle imply physics beyond the SM.

6.7.3 *Results from the Data Recorded Subsequent to the Discovery*

The combined results from the ATLAS and CMS experiment from Run 1 on the Higgs boson production, decay rates and constraints on its couplings were published in 2016 [37]. These results have been superseded by the ones presented below. Results are presented from the most recently published papers (in journals or submitted to the hep arXiv) from the two collaborations. The integrated luminosity differs from one result to another and is indicated in the legends of the plots presented.

The LHC centre of mass energy was increased from $\sqrt{s} = 8$ TeV to $\sqrt{s} = 13$ TeV in 2015. At the higher value of $\sqrt{s}$ the predicted cross-sections for the dominant ggH production mode and the rare ttH production mode increased by factors of ~2.3 and ~ 3.8, respectively. This and the larger datasets from Run 2 allow a more precise comparison of the properties of the Higgs boson with respect to those predicted by the SM. In addition, since the discovery, the theoretical predictions have become more accurate with the inclusion of further (higher) order corrections. Details can be found below in the references included in the individual papers of the two collaborations.

The two collaborations have also improved the reconstruction of physics objects and the methods of analysis. Event categorization and machine learning methods are deployed to study almost all the different production and decay modes. The analyses described below divide events into multiple categories reflecting the different Higgs boson production channels to improve the sensitivity of the measurements. Associated production processes (WH and ZH), or the ttH production process, are tagged by requiring the presence of additional leptons or jets. The VBF process is tagged using distinctive kinematic properties such the presence of two jets with a large separation in pseudorapidity and a large invariant jet-jet mass. In some cases the kinematic characteristics of the whole event, such as large missing p_T, are used to preferentially select events e.g. arising from ZH production where the Z boson decays to neutrinos.

6.7.3.1 The H → $\gamma\gamma$

As the H → $\gamma\gamma$ decay proceeds via W-boson and top-quark loops, it is especially sensitive to the presence of any undiscovered heavy charged fermions and bosons. Any significant deviation from the precise SM prediction for the cross section would be indicative of new physics.

The H $\rightarrow \gamma\gamma$ mode provides good sensitivity to almost all Higgs boson production processes. The interference between W-loop and top-loops provides sensitivity to the relative sign of the fermion and boson couplings.

It is common to use a dedicated boosted decision tree discriminator to select and categorize events; it is constrained using the diphoton kinematic variables, photon isolation and identification variables, and per-event estimated diphoton mass resolution for the pair of photons in the event.

ATLAS has measured the properties of the H $\rightarrow \gamma\gamma$ mode using 79.8 fb^{-1} of collision data recorded at $\sqrt{s} = 13$ TeV [38]. The properties measured include the signal strength, the cross section measurements for the production of a Higgs boson through gluon–gluon fusion, vector boson fusion, and in association with a vector boson or a top-quark pair. They are found to be compatible with the predictions of the SM. The signal strength is measured to be $\mu = 1.06 \pm 0.08$ (stat), $^{+0.08}{}_{-0.07}$ (exp), $^{+0.07}{}_{-0.06}$ (theo), improving on the precision of the previous ATLAS measurement at $\sqrt{s} = 7$ and 8 TeV by over a factor of three. The cross section for the production of the Higgs boson decaying to two isolated photons in the fiducial region of the selection of photons is measured to be 60.4 ± 6.1 (stat) $\pm$ 6.0 (exp) $\pm$ 60.3 (theo) fb, in good agreement with the SM value of 63.5 ± 3.3 fb. The differential cross section, sensitive to higher order QCD corrections and properties of the Higgs boson, such as its spin and CP quantum numbers, is illustrated in Fig. 6.9b and no significant deviation from a wide array of SM predictions is observed.

CMS has reported results from the H $\rightarrow \gamma\gamma$ decay channel based on data collected at $\sqrt{s} = 13$ TeV corresponding to an integrated luminosity of 35.6 fb^{-1} [39]. The diphoton invariant mass distribution, observed in CMS, is shown in

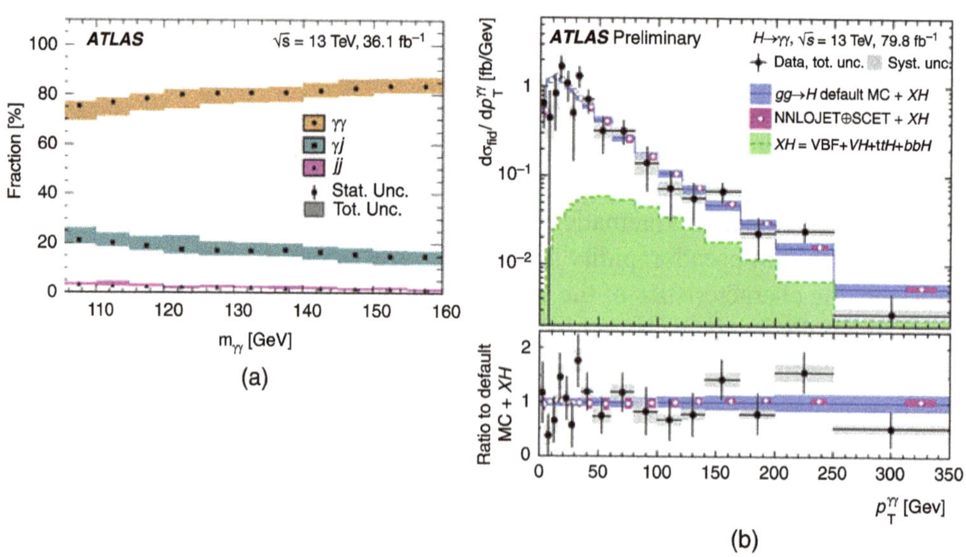

Fig. 6.9 (**a**) The data-driven determination of event fractions for $\gamma\gamma$, γ-jet, and jet-jet events as a function of $m_{\gamma\gamma}$ after the final selection. (**b**) The fiducial differential cross section for pp. $\rightarrow \gamma\gamma$ as a function of $p_T(\gamma\gamma)$ compared to the SM expectations

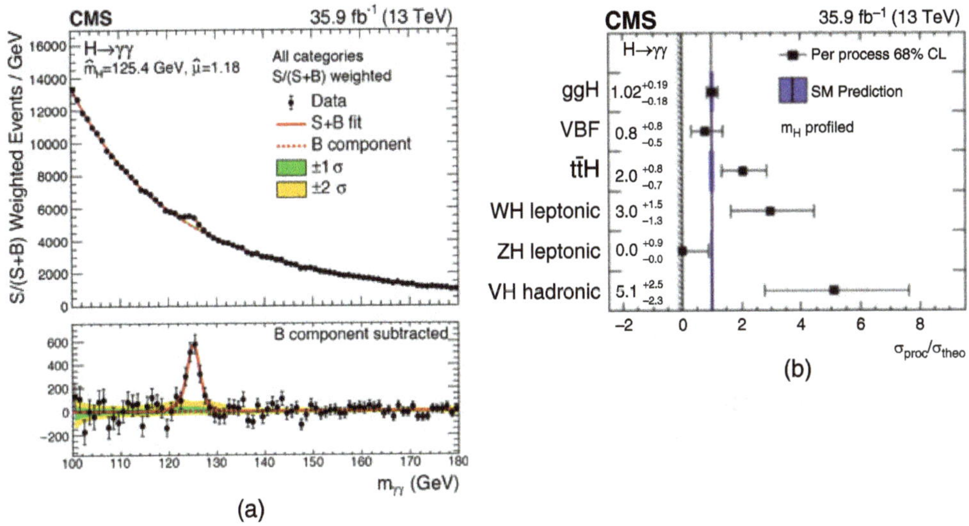

Fig. 6.10 (**a**) Data and signal-plus-background model fits for all categories summed, weighted by their sensitivity. The one (green) and two (yellow) standard deviation bands include the uncertainties in the background component of the fit. The lower panel shows the residuals after the background subtraction. (**b**) Cross section ratios measured for each process (black points) compared to the SM expectations and their uncertainties (blue band). The signal strength modifiers are constrained to be nonnegative, as indicated by the vertical line and hashed pattern at zero

Fig. 6.10a. The measured signal strength is found to be $1.18^{+0.17}_{-0.14}$, largely insensitive to the precise mass value assigned to the Higgs boson. Signal strengths associated with the different Higgs boson production mechanisms are shown in Fig. 6.10b and found to be compatible with the expectations from the SM.

6.7.3.2 $H \rightarrow ZZ^{(*)} \rightarrow 4\,l$ Decay Mode

The Higgs boson decay $H \rightarrow ZZ^{(*)} \rightarrow 4\ l$ is the most significant process in constraining the HZZ coupling. To study the differing production mechanisms involved, the events are categorized on the basis of the presence of jets, b-tagged jets, leptons, p_T^{miss}, and various matrix element discriminants that make use of the information about the additional objects: VBF (1- and 2-jet), VH hadronic, VH leptonic, ttH, VH p_T^{miss}, and untagged categories.

ATLAS has studied the coupling properties of the Higgs boson in the four-lepton (e,μ) decay channel using 36.1 fb^{-1} of pp. collision data recorded at $\sqrt{s} = 13$TeV [40]. The four-lepton invariant mass distribution is illustrated in Fig. 6.11a. Cross sections are measured for the main production modes and the ratio of sigma.BR/(sigma.BR)$_{SM}$ are plotted in Fig. 6.11b. The inclusive cross section times branching fraction for $H \rightarrow ZZ^{(*)}$ decay and for a Higgs boson absolute rapidity below 2.5 is measured to be $1.73^{+0.24}_{-0.23}$ pb, the statistical error dominating, compared to the SM prediction of 1.34 ± 0.09 pb.

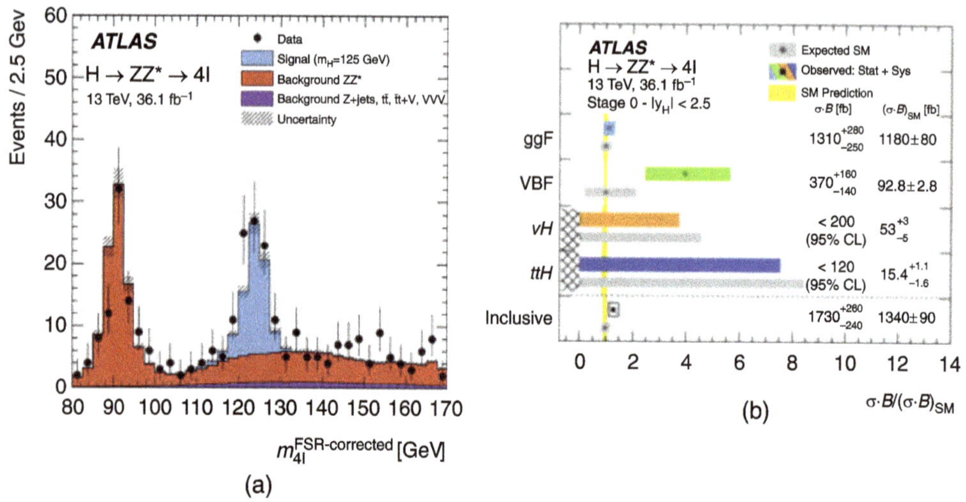

Fig. 6.11 (**a**) Observed and expected four-lepton invariant mass distribution for the selected Higgs boson candidates with a constrained Z boson mass. (**b**) Observed and expected SM values of the cross-section ratios σ.B normalized by the SM expectation $(\sigma.B)_{SM}$ for the inclusive production

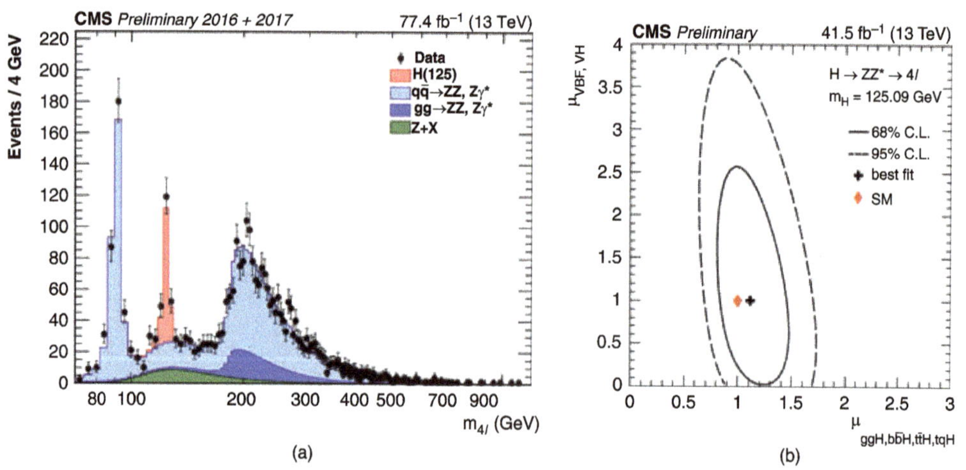

Fig. 6.12 (**a**) Distribution of the four-lepton reconstructed invariant mass m_{4l} in the full mass range, combining data from 2016 an 2017. Points with error bars represent the data and stacked histograms represent expected distributions of the signal and background processes. The $ZZ^{(*)}$ backgrounds are normalized to the SM expectation, the Z + X background to the estimation from data. (**b**) Result of the 2D likelihood scan for the μ(ggH, ttH, bbH, tqH) and μ(VBF,VH) signal-strength modifiers

CMS has studied the coupling properties using 77.4 fb^{-1} of pp. collision data recorded at $\sqrt{s} = 13TeV$ [41]. The four-lepton invariant mass distribution is illustrated in Fig. 6.12a. The signal strength is measured to be $\mu = 1.06^{+0.15}{}_{-0.13}$ at $m_H = 125.09$ GeV, the combined ATLAS and CMS measurement of the Higgs boson mass [42]. The result of a 2D likelihood scan of the signal strengths for the individual Higgs boson production modes are also measured and are shown in Fig. 6.12b. All measurements are consistent with the expectations from the SM.

6.7.3.3 $H \rightarrow WW^{(*)} \rightarrow 2 l 2\nu$ Decay Mode

The $H \rightarrow WW^{(*)}$ decay mode has a large branching fraction (~20%) and a relatively low-background final state. The study of this final state in which both W bosons decay leptonically is based on the signature with two isolated, oppositely charged, high p_T leptons (electrons or muons) and large missing transverse momentum, $E_T{}^{miss}$, due to the undetected neutrinos. The signal sensitivity is improved by separating events according to lepton flavor; into e^+e^-, $\mu^+\mu^-$, and $e\mu$ samples and according to jet multiplicity into 0-jet and 1-jet samples. The dominant background arises from irreducible non-resonant $WW^{(*)}$ production, and the dominant uncertainties arise from the estimation, using the data themselves, of the backgrounds from top quark pair, $WW^{(*)}$ and DY production.

The final states are categorized according to the number of associated jets, with the 0-jet category dominating the overall sensitivity. Events are selected that contain two leptons of either different or the same flavour. The large background from tt production, the different and same flavour final states are further categorized with 0, 1 and 2 associated jets. In the different-flavour final state, dedicated 2-jet categories are included to enhance the sensitivity to VBF and VH production mechanisms.

ATLAS has presented measurements of the inclusive cross section of Higgs boson production via the gluon–gluon fusion (ggF) and vector-boson fusion (VBF) modes [43], based on an integrated luminosity of 36.1 fb^{-1} recorded at $\sqrt{s} = 13$ TeV in 2015–2016. The combined transverse mass distribution for $N_{jet} \leq 1$ is shown in Fig. 6.13a. The ggF and VBF cross-sections times the $H \rightarrow WW^{(*)}$ branching ratio are measured to be 12.6 ± 1.0(stat) ${}^{+1.9}{}_{-1.8}$ (syst) pb and 0.50 ± 0.24 (stat) ± 0.18 (syst) pb, respectively, in agreement with the SM predictions, as illustrated in Fig. 6.13b.

CMS has published results on the decay mode $H \rightarrow WW^{(*)}$ using data corresponding to an integrated luminosity of 35.9 fb^{-1}, collected at $\sqrt{s} = 13$ TeV during 2016 [44]. The expected relative fraction of different Higgs boson production mechanisms in each category is shown in Fig. 6.14a, together with the expected signal yield. Combining all channels, the observed cross section times branching fraction is $1.28^{+0.18}{}_{-0.17}$ times the SM prediction for the Higgs boson with a mass of 125.09 GeV. The ratio of the observed and the predicted cross sections for the main Higgs boson production modes is shown in Fig. 6.14b. All are consistent with the predictions from the SM.

6.7.3.4 The $H \rightarrow \tau\tau$ Decay Mode

All of the decay modes discussed so far test the direct coupling of the Higgs boson to bosons, and only indirectly probe, through quantum loops, its coupling to fermions.

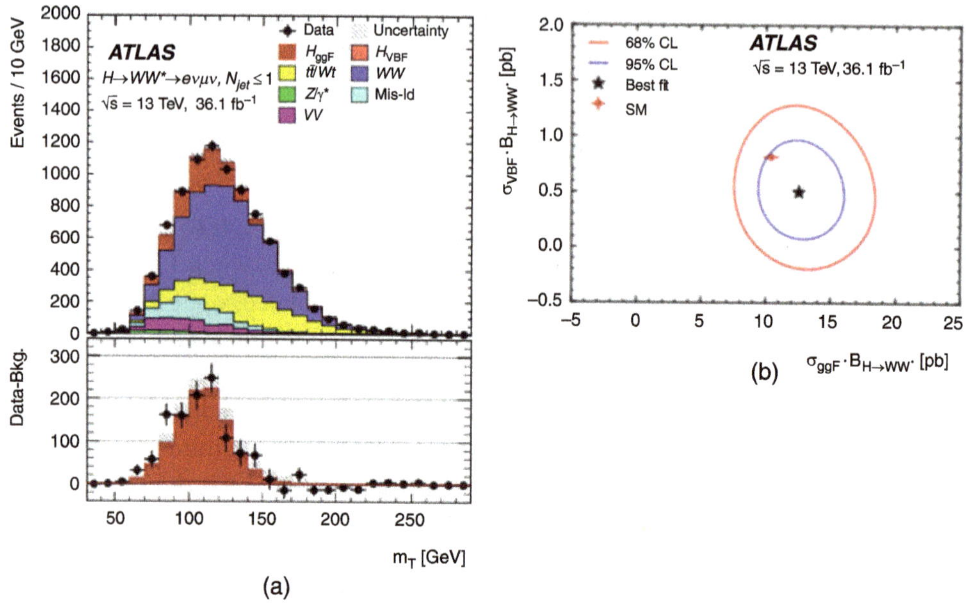

Fig. 6.13 (**a**) Post-fit combined transverse mass distribution for $N_{jet} \leq 1$. The bottom panel shows the difference between the data and the estimated background compared to the distribution for a SM Higgs boson with $m_H = 125GeV$. The H_{VBF} contribution is too small to be visible. (**b**) 68% and 95% confidence level two-dimensional likelihood contours of $\sigma_{ggF} \cdot B_H \to WW^{(*)}$ vs. $\sigma_{VBF} \cdot B_H \to WW^{(*)}$, compared to the SM prediction shown by the red marker

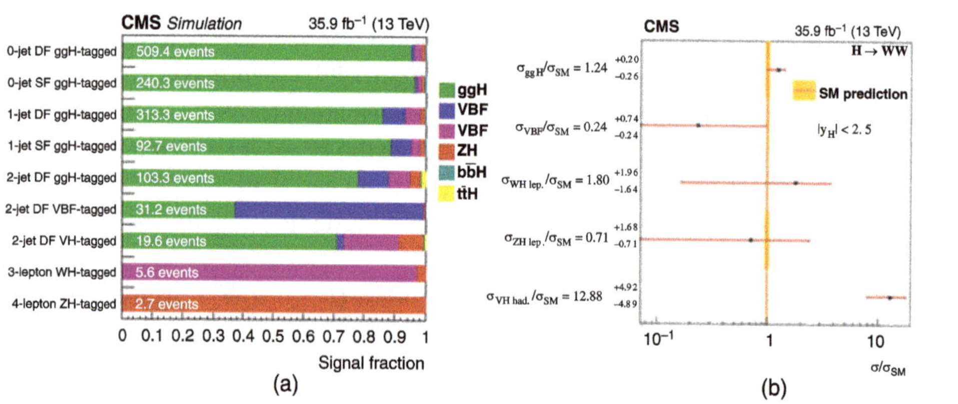

Fig. 6.14 (**a**) Expected relative fraction of different Higgs boson production mechanisms in each category included in the combination, together with the expected signal yield. (**b**) Observed cross sections for the main Higgs boson production modes, normalized to the SM predictions. The vertical line and band correspond to the SM prediction and associated theoretical uncertainty

The $H \to \tau\tau$ mode provides the best sensitivity for the direct measurement for Higgs boson coupling to fermions. It benefits from a relatively large branching fraction, a moderate mass resolution (~10–20%) and provides good sensitivity to both the ggH and VBF production processes.

The $H \rightarrow \tau\tau$ mode is studied via tau decays to $e\mu$, $\mu\mu$, $e\tau_h$, $\mu\tau_h$, $\tau_h\tau_h$, where electrons and muons arise from leptonic τ-decays and τ_h denotes a τ lepton decaying hadronically. Each of these categories is further divided into three sub-categories labeled 0-jet, boosted and VBF. The 0-jet category helps constrain background normalisation, identification efficiencies, and energy scales, and systematic uncertainties in the background model. The main irreducible background, $Z \rightarrow \tau\tau$ production, and the largest reducible backgrounds (W + jets, QCD multijet production, and top quark pair) are evaluated from control samples in data.

CMS has observed the $H \rightarrow \tau\tau$ mode using a data sample corresponding to an integrated luminosity of $35.9 fb^{-1}$ at $\sqrt{s} = 13$ TeV [45]. Figure 6.15a shows the distribution of the decimal logarithm of the ratio of the expected signal and the sum of expected signal and expected background in each bin of the mass distributions used to extract the results, in all signal regions. The background contributions are separated by decay channel. The inset shows the corresponding difference between the observed data and expected background distributions divided by the background expectation, as well as the signal expectation divided by the background expectation. The best fit of the product of the observed $H \rightarrow \tau\tau$ signal production cross section and branching fraction is $1.09^{+0.27}_{-0.26}$ times the SM expectation. The combination with the corresponding measurement performed with data collected by the CMS experiment at center-of-mass energies of 7 and 8 TeV leads to an observed

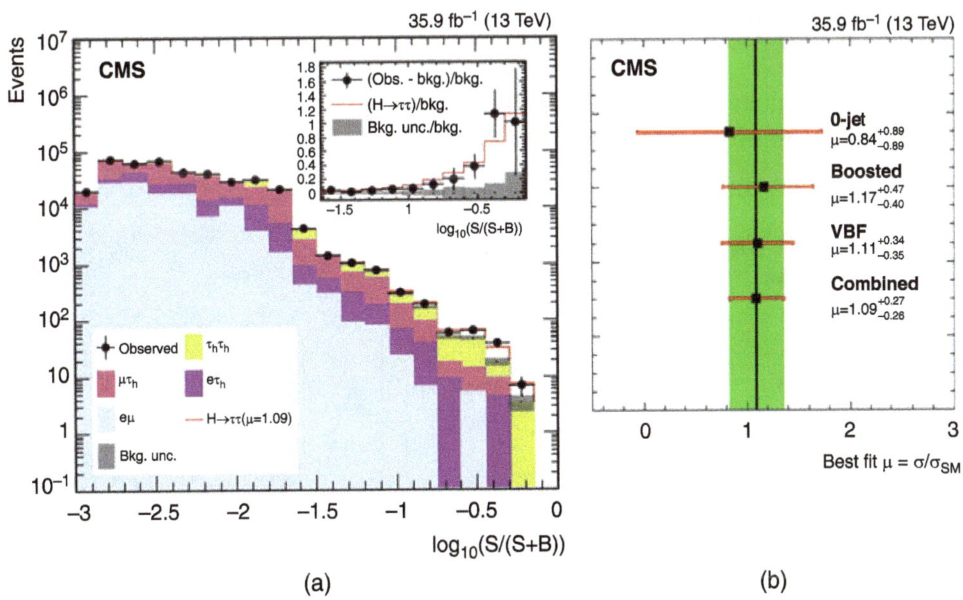

Fig. 6.15 (**a**) Distribution of the decimal logarithm of the ratio between the expected signal and the sum of expected signal and expected. The inset shows the corresponding difference between the observed data and expected background distributions divided by the background expectation, as well as the signal expectation divided by the background expectation. (**b**) Best fit signal strength per category

significance of 5.9 standard deviations, equal to the expected significance. Figure 6.15b right plots the signal strength per category for $m_H = 125.09$ GeV.

ATLAS has observed the $H \rightarrow \tau\tau$ mode using 36.1 fb^{-1} of data recorded at $\sqrt{s} = 13$ TeV [46]. All combinations of leptonic and hadronic tau decays were considered. Combining all data taken at $\sqrt{s} = 7$, 8 and 13 TeV, the observed (expected) significance is found to be 6.4 (5.4) standard deviations. Using the data taken at $\sqrt{s} = 13$ TeV, the total cross section, in the $H \rightarrow \tau\tau$ decay channel, is measured to be 3.71 ± 0.59 (stat) $^{+0.87}{}_{-0.74}$ (syst) pb, for $m_H = 125$ GeV, assuming the relative contributions of its production modes predicted by the SM. Total cross sections are determined separately for vector boson fusion production and gluon–gluon fusion production to be $\sigma(\text{VBF}, H \rightarrow \tau\tau) = 0.28 \pm 0{:}09$ (stat) $^{+0.11}{}_{-0.09}$ (syst) pb and $\sigma(\text{ggF}, H \rightarrow \tau\tau) = 3.0 \pm 1.0$ (stat.) $^{+1.6}{}_{-1.2}$ (syst) pb, respectively. The measured values for $\sigma(H \rightarrow \tau\tau)$, when only the data of individual channels are used, are shown in Fig. 6.16a, along with the result from the combined fit. The theory uncertainty in the predicted signal cross section is shown by the yellow band. Figure 6.16b shows the likelihood contours in the variables (ggF, $H \rightarrow \tau\tau$) and (VBF, $H \rightarrow \tau\tau$) for the combination of all channels. The 68% and 95% CL contours are shown as dashed and solid lines, respectively, for $m_H = 125$ GeV. The SM expectation is indicated by a plus symbol and the best fit to the data is shown as a star. All measurements are in agreement with SM expectations.

6.7.3.5 $H \rightarrow b\bar{b}$ Decay Mode

In the SM, fermions couple directly to the Higgs boson via the Yukawa interaction. A clear test of this hypothesis would be the measurement of the $H \rightarrow b\bar{b}$ coupling.

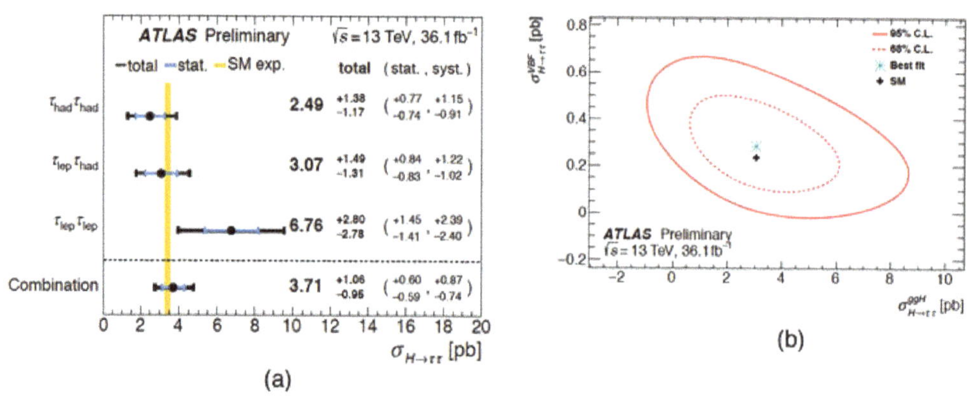

Fig. 6.16 (**a**) Measured values for $\sigma(H \rightarrow \tau\tau)$ when only the data of individual channels are used. Also shown is the result from the combined fit. The total $\pm$ 1σ uncertainty in the measurement is indicated in black, with the individual contribution from the statistical uncertainty in blue. (**b**) Likelihood contours for the combination of all channels in the [(ggF, $H \rightarrow \tau\tau$) v/s (VBF, $H \rightarrow \tau\tau$)] plane. The SM expectation is indicated by a plus symbol and the best fit to the data is shown as a star

The $H \rightarrow bb$ decay mode has by far the largest branching ratio (~58%). However, this is the most difficult decay channel to observe, since bottom quark pairs are prolifically produced by QCD processes and give rise to a formidable background. The cross section for b-quark pair production, σ_{bb}(QCD), is ~$10^7 \times \sigma(H \rightarrow bb)$. Therefore the search concentrates on Higgs boson production in association with a W or Z boson using the following decay modes: $W \rightarrow e\nu/\mu\nu$ and $Z \rightarrow ee$ or $\mu\mu$ or $\nu\nu$. The $Z \rightarrow \nu\nu$ decay is identified by the requirement of a large missing transverse energy. The Higgs boson candidate is reconstructed by requiring two b-tagged jets.

Events are selected in 0-, 1- and 2-charged lepton (e or μ) channels, to explore the $ZH \rightarrow \nu\nu bb$, $WH \rightarrow l\nu bb$, $ZH \rightarrow llbb$ signatures, respectively. Both experiments introduced several improvements since the initial searches including more efficient identification of b-jets, better dijet mass resolution and use of multivariate discriminants that better separate signal from background. Multivariate discriminants, built from variables that describe the kinematics of the selected events, are used to maximise the sensitivity to the Higgs boson signal. The signal extraction method is validated with, for example, the diboson analysis where the nominal multivariate analysis is modified to extract the VZ, $Z \rightarrow bb$ diboson process.

ATLAS has observed the mode $H \rightarrow bb$ by analyzing the combined data from Run 1 and Run 2 [47], corresponding to an integrated luminosity of 80fb^{-1} yielding an observed (expected) significance of 5.4 (5.5) standard deviations, thus providing direct observation of the Higgs boson decay into b-quarks. The signal strength is measured to be $1.01 \pm 0.12(\text{stat})\ ^{+0.16}{}_{-0.15}(\text{syst})$. Figure 6.17a shows the distribution of m_{bb} in data after subtraction of all backgrounds except for the WZ and $ZZ^{(*)}$ diboson processes using data taken at $\sqrt{s} = 13$ TeV. The contributions from all lepton channels, $p^V{}_T$ regions, and number-of-jets categories are summed and

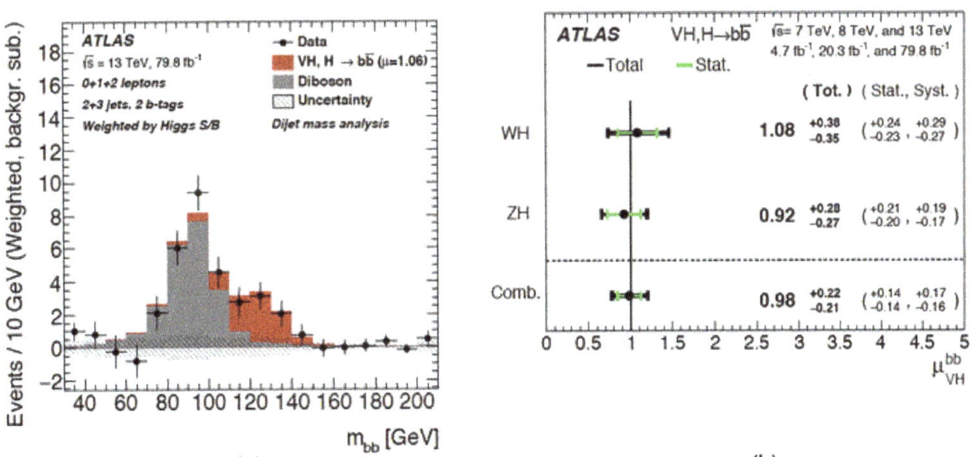

(a) (b)

Fig. 6.17 (**a**) Distribution of m_{bb} in data after subtraction of all backgrounds except for the WZ and ZZ diboson processes, as obtained with the dijet-mass analysis. (**b**) Fitted values of the Higgs boson signal strength, μ, for $m_H = 125$ GeV for the WH and ZH processes and their combination, using the 7 TeV, 8 TeV and 13 TeV data

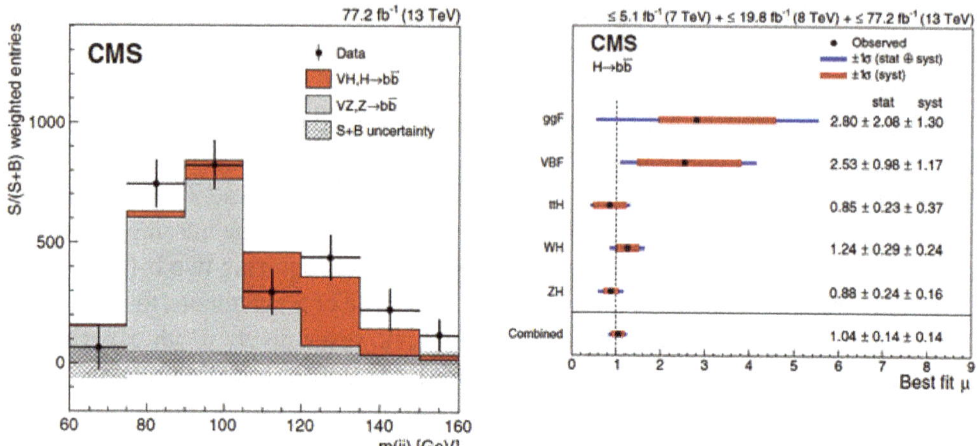

(a) (b)

Fig. 6.18 (**a**) Dijet invariant mass distribution for events weighted by S/(S + B) in all channels combined in the 2016 and 2017 data sets. The error bar for each bin represents the presubtraction 1σ statistical uncertainty on the data, while the gray hatching indicates the 1σ total uncertainty on the signal and all background components. (**b**) Best-fit value of the H → bb signal strength with its 1σ systematic (red) and total (blue) uncertainties for the five individual production modes considered, as well as the overall combined result. The vertical dashed line indicates the SM expectation

weighted by their respective S/B, with S being the total fitted signal and B the total fitted background in each region. The expected contribution of the associated WH and ZH production of a SM Higgs boson with $m_H = 125$ GeV is shown, scaled by the measured signal strength (μ =1.06). The size of the combined statistical and systematic uncertainty for the fitted background is indicated by the hatched band. Figure 6.17b shows the fitted values of the Higgs boson signal strength, where $\mu(\text{VHbb}) = 0.98\ ^{+0.22}{}_{-0.21}$ for $m_H = 125$ GeV for the WH and ZH processes and their combination, using the 7 TeV, 8 TeV and 13 TeV data.

CMS has observed the mode H → bb. Figure 6.18a shows the weighted dijet invariant mass distribution for events weighted by S/(S + B) in all channels combined in the 2016 and 2017 data sets [48]. The data (points), the fitted VH signal (red) and VZ background (grey) distributions, with all other fitted background processes subtracted, except that from dibosons are shown in Fig. 6.18a. Figure 6.18b shows the best-fit value of the H → bb signal strength for the five individual production modes considered, as well as the overall combined result. The vertical dashed line indicates the SM expectation. All results are extracted from a single fit with $m_H = 125.09$ GeV. CMS has made measurements, using data collected at $\sqrt{s} = 7$, 8, and 13 TeV, and observes an excess of events at $m_H = 125$ GeV with a significance of 5.6 standard deviations, where the expectation for the SM Higgs boson is 5.5, and a signal strength of 1.04 ± 0.14 (stat.) ± 0.14 (syst.).

6.7.3.6 H → μμ Decay Mode

The $H \rightarrow \mu^+\mu^-$ decay mode extends the test of the Higgs boson's coupling to the second generation of fermions. Several scenarios beyond the SM predict a higher branching fraction than the one predicted in the SM (2.2×10^{-4} at $m_H = 125$ GeV).

The dominant and irreducible background arises from the $Z/\gamma_* \rightarrow \mu\mu$ process that has a rate several orders of magnitude larger than that from the SM Higgs boson signal. However, due to the precise muon momentum measurement achieved by ATLAS and CMS, the dimuon mass resolution is excellent ($\approx$ 2–3%). A search is performed for a narrow peak over a large but smoothly falling background. For optimal search sensitivity, events are divided into several categories. Taking advantage of the superior muon momentum measurement in the central region events can be subdivided by the pseudorapidity of the muons, or by selections aiming at specific production processes. A category selecting the vector boson fusion process with its distinctive signature and relatively large cross section is particularly useful.

ATLAS has performed this search using data corresponding to an integrated luminosity of 36.1 fb^{-1} collected at $\sqrt{s} = 13$ TeV [49]. No significant excess is observed above the expected background. When combined with the data taken at $\sqrt{s} = 7$ and 8 TeV, the observed (expected) cross-section upper limit is 2.8 (2.9) times the SM prediction.

The search in CMS, using an integrated luminosity corresponding to 35.9 fb^{-1} recorded at $\sqrt{s} = 13$ TeV [50], and combining with data taken at $\sqrt{s} = 7$ and 8 TeV, yielded an observed (expected) cross-section upper limit is 2.92 (2.16) times the Standard Model prediction.

6.7.3.7 ttbar H Production Mode

As $m_t > m_H$ the Yukawa coupling of the Higgs boson to top quarks cannot be tested directly. However, it can be measured through the measurement in the pp. → ttH production process. The coupling of the Higgs boson to the top quark, the heaviest particle in the SM, could be very sensitive to the effects of physics beyond the SM.

Although the pp. → ttH production process only contributes around 1% of the total Higgs-boson production cross section, the top quarks in the final state offer a distinctive signature and allow many Higgs-boson decay modes to be accessed. Of these, the decay to two b-quarks, the Higgs boson decay mode with the largest branching fraction, also is sensitive to the b-quark's Yukawa coupling, the second largest in the SM.

A top quark decays almost exclusively to a bottom quark and a W boson, with the W boson subsequently decaying either to a quark and an antiquark or to a charged lepton and its associated neutrino. The Higgs boson has a rich spectrum of decay modes, and ttH production is studied using a wide variety of final state event topologies, with the Higgs boson decaying into bb, $WW^{(*)}$, $\tau\tau$, $\gamma\gamma$, and $ZZ^{(*)}$ pairs.

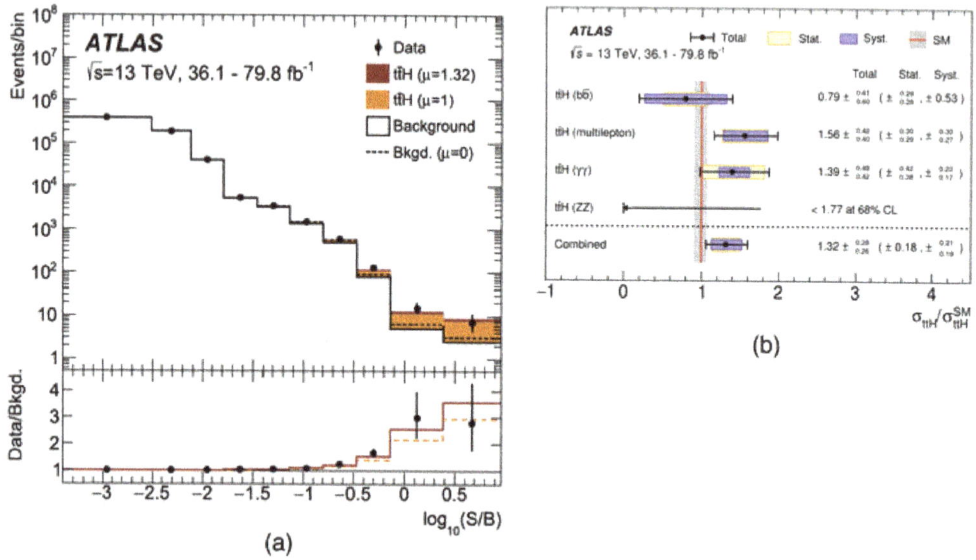

Fig. 6.19 (**a**) Observed event yields in all analysis categories in up to 79.8fb^{-1} of 13 TeV data. The lower panel shows the ratio of the data to the background estimated from the fit with freely floating signal, compared to the expected distribution including the signal assuming $\mu = 1.32$ (full red) and $\mu = 1$ (dashed orange). The error bars on the data are statistical. (**b**) Combined *ttH* production cross section, as well as cross sections measured in the individual analyses, divided by the SM prediction

ATLAS has observed this production mode using data taken at $\sqrt{s} = 7$ TeV, 8 TeV and 13 TeV corresponding to integrated luminosities up to 79.8 fb^{-1}. The Higgs boson decays included comprise *bb*, *WW**, $\tau + \tau -$, $\gamma\gamma$, and *ZZ**. The observed significance is 6.3σ, compared to an expectation of 5.1σ [51]. Assuming SM branching fractions, the total *ttH* production cross section at $\sqrt{s} = 13$ TeV is measured to be 670 ± 90(stat.) $^{+110}{}_{-100}$(syst.) fb, in agreement with the SM prediction. Figure 6.19a shows the observed event yields in all analysis categories. The background yields correspond to the observed fit results, and the signal yields are shown for both the observed results ($\mu = 1.32$) and the SM prediction ($\mu = 1$). The ranking of the discriminant bins is carried out by $\log_{10}(S/B)$, where *S* is the extracted signal yield and *B* the extracted background yield. Figure 6.19b shows the combined *ttH* production cross section, as well as cross sections measured in the individual analyses, divided by the SM prediction. The black lines show the total uncertainties, and the bands indicate the statistical and systematic uncertainties. The red vertical line indicates the SM cross-section prediction, and the grey band represents the PDF and α_S uncertainties and the uncertainties due to missing higher-order corrections.

CMS has observed ttH production in a combined analysis of data at $\sqrt{s} = 7$, 8, and 13 TeV, corresponding to integrated luminosities of up to 5.1, 19.7, and

35.9 fb^{-1}, respectively [52]. An excess of events is observed with an observed (expected) significance of 5.2 (4.2) standard deviations, over the expectation from

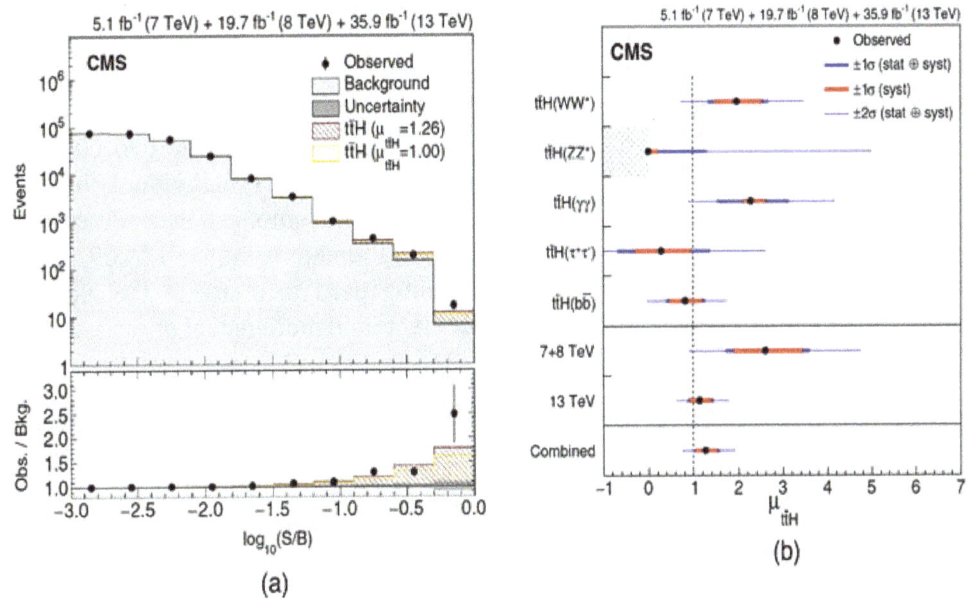

Fig. 6.20 (**a**) Distribution of events as a function of the decimal logarithm of S/B. The shaded histogram shows the expected background distribution. The two hatched histograms, each stacked on top of the background histogram, show the signal expectation for the SM ($\mu_{ttH} = 1$) and the observed ($\mu_{ttH} = 1.26$) signal strengths. The lower panel shows the ratios of the expected signal and observed results relative to the expected background. (**b**) Best fit value of the ttH signal strength modifier μ_{ttH}, with its 1σ and 2σ confidence intervals for upper section) the five individual decay channels considered, middle section) the combined result for 7 + 8 TeV alone and for 13 TeV alone, and lower section) the overall combined result. The SM expectation is shown as a dashed vertical line

the background-only hypothesis for $m_H = 125.09$. The combined best-fit signal strength normalized to the standard model prediction is $1.26^{+0.31}{}_{-0.26}$. Figure 6.20a shows the distribution of events as a function of the decimal logarithm of S/B, where S and B are the expected postfit signal (with $\mu_{ttH} = 1$) and background yields, respectively, in each bin of the distributions considered in this combination. The shaded histogram shows the expected background distribution. The two hatched histograms, each stacked on top of the background histogram, show the signal expectation for the SM ($\mu_{ttH} = 1$) and the observed ($\mu_{ttH} = 1.26$) signal strengths. The lower panel shows the ratios of the expected signal and observed results relative to the expected background. Figure 6.20b plots the ttH signal strength modifiers, μ_{ttH}, for the various selections and the overall combined result. The SM expectation is shown as a dashed vertical line.

6.7.4 *Combining the Results*

6.7.4.1 Mass of the Observed State

The mass of the Higgs boson is measured using the two decay channels that give the best mass resolutions namely $H \to \gamma\gamma$ and $H \to ZZ^{(*)} \to 4\ l$. ATLAS and CMS have combined their results from Run 1 [53]. The results were obtained from a simultaneous fit to the reconstructed invariant mass peaks in the two channels and for the two experiments. The measured masses from the individual channels and the two experiments were found to be consistent amongst themselves. The combined measured mass of the Higgs boson was found to be $m_H = 125.09 \pm 0.21$ (stat.) $\pm$ 0.11 (syst.) GeV, a value subsequently used in many of the analyses discussed above. The results of these measurements and more recent ones are shown in Fig. 6.21 [54].

The mass of the Higgs boson, combined with the measured top quark mass, has cosmological implications. The current measurement of m_H, along with that of the top quark mass [$m_t = 173.21 \pm 0.51$ (stat) ± 0.71 (syst)] indicate that our universe is in a metastable state, which eventually will tunnel through the potential barrier to the true vacuum in which space collapses, albeit over a period of time that is many orders of magnitude larger than the lifetime of the universe so far.

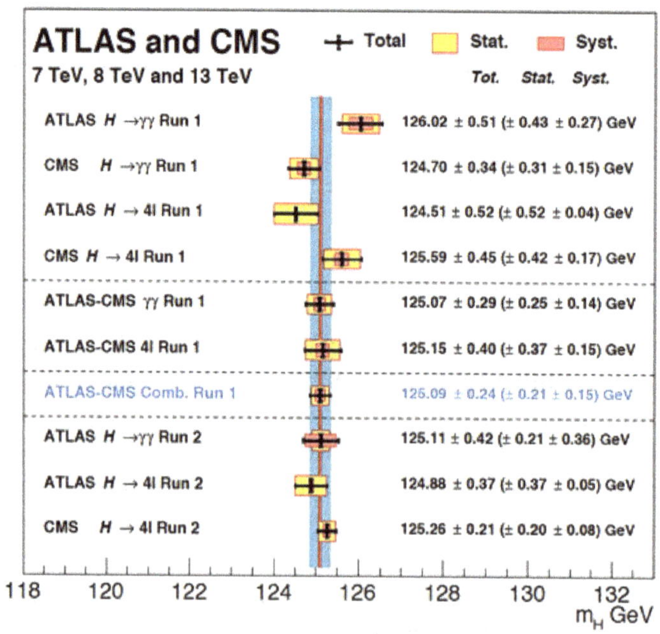

Fig. 6.21 Summary of the CMS and ATLAS mass measurements in the $\gamma\gamma$ and $ZZ^{(*)}$ channels in Run 1 and Run 2. Particle Data Group [54]

6.7.4.2 Compatibility of the Observed State with the SM Higgs Boson Hypothesis: Signal Strength

In the SM, the Higgs boson is a fundamental scalar particle with spin-parity $J^P = 0^+$, and couples to fundamental fermions as $m_f{}^2/v^2$ and to fundamental bosons as $m_V{}^4/v^2$ where $v = 246$ GeV. Several individual tests of compatibility with expectations from the SM have been discussed above.

Here we discuss the signal strength, μ, as determined by the combination of results from all channels. ATLAS and CMS combined their data from Run 1 [37] from the analysis of five production processes, namely gluon fusion, vector boson fusion, and associated production with a W or a Z boson or a pair of top quarks, and of the five decay modes $H \rightarrow \gamma\gamma$, $ZZ^{(*)}$, $WW^{(*)}$, bb, and $\tau\tau$. All results are reported assuming a value of 125.09 GeV for the Higgs boson mass. The Higgs boson production and decay rates measured by the two experiments are combined within the context of three generic parameterisations: two based on cross sections and branching fractions, and one on ratios of coupling modifiers. Several interpretations of the measurements with more model-dependent parameterisations are also given. The combined signal yield relative to the SM prediction is measured to be 1.09 ± 0.11. The error is broken down as ± 0.07 (statistical), ± 0.04 (experimental systematic), ± 0.03 (theoretical on background) and $\pm$ 0.07 (theoretical on signal).

The most recent measured values of μ, using the above-mentioned channels, are:

- ATLAS: $\mu = 1.13\,^{+0.09}{}_{-0.08}$, using 79.8 fb^{-1} at $\sqrt{s} = 13$ TeV [55],
- CMS: $\mu = 1.17 \pm 0.1$, using 35.9 fb^{-1} at $\sqrt{s} = 13$ TeV [56].

The error in the measurement from ATLAS has the following breakdown: ± 0.05 (statistical), ± 0.05 (experimental systematic), ± 0.03 (theoretical on background) and ($+0.05$, -0.04) (theoretical on signal). This should give a flavor of the possibilities for extrapolation into the future.

6.7.4.3 Compatibility of the Observed State with the SM Higgs Boson Hypothesis: Couplings

The 25 products, $\mu_i \times \mu_f$, where i (f) is the production (decay) index can also be considered as free parameters. This can be viewed as the measurements of cross sections times branching fractions, sigma×BR, by production mechanism and decay mode. The results from the ATLAS and CMS have been combined and are illustrated in Fig. 6.22 from the Particle Data Group [54] showing compatibility with the SM.

Figure 6.23a, from the ATLAS experiment [55], illustrates the dependence of the Higgs boson couplings on mass of the decay particles (μ, τ, b-quark, W, Z and t-quark). The plot is made for reduced coupling strength modifiers, κ, using $\kappa_F(m_F/v)$ for fermions and $(\sqrt{\kappa_V})(m_V/v)$ for vector bosons, where $v = 246.22$ GeV. The couplings modifiers κ_F and κ_V are measured assuming no BSM contributions

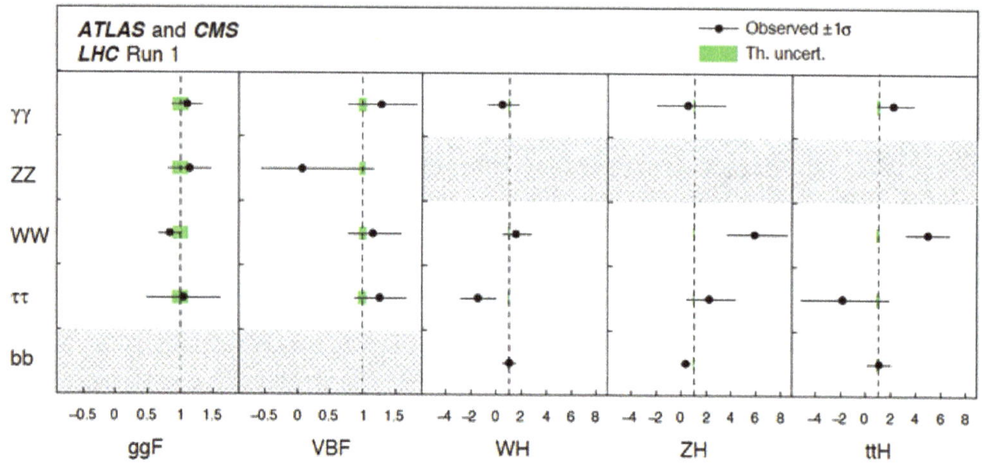

Fig. 6.22 Combined measurements of the products σ.BR for the five main production and five main decay modes. The hatched combinations require more data for a meaningful confidence interval to be provided

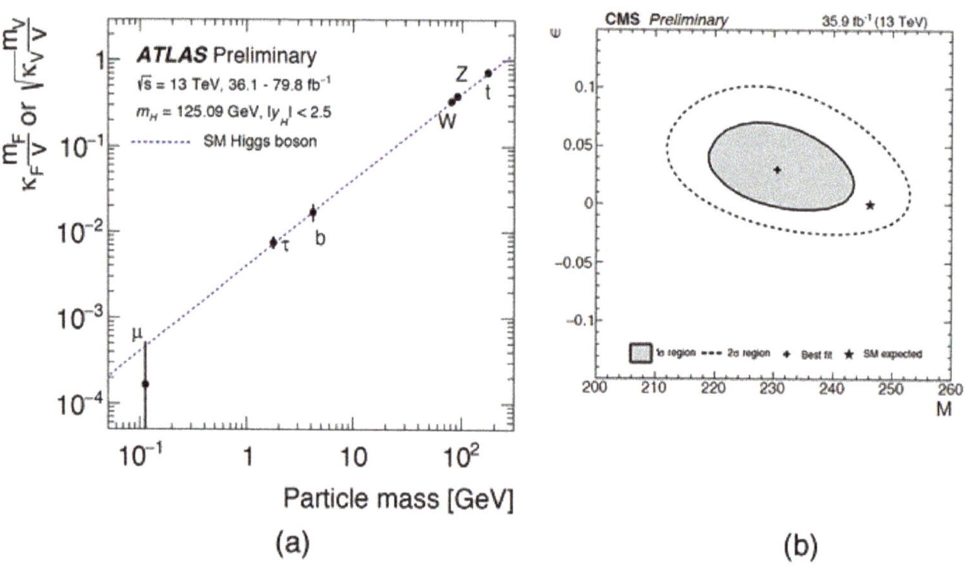

Fig. 6.23 (**a**) Reduced coupling strength modifiers $\kappa_F(m_F/v)$ for fermions (t, b, τ, μ) and $(\sqrt{\kappa_V})(m_V/v)$ for vector bosons (W, Z) as a function of their masses m_F and m_V, respectively. The SM prediction for both cases is also shown (dotted line). (**b**) Likelihood scan in the M-ε plane. The best-fit point and, 1σ, 2σ CL regions are shown, along with the SM prediction

to the Higgs boson decays, and the SM structure of loop processes. The line is the expectation from the SM.

CMS has made a similar fit, shown in Fig. 6.23b [56], using a phenomenological parameterization relating the masses of the fermions and vector bosons to the corresponding κ modifiers using two parameters, denoted M and ε. In such a model one can relate the coupling modifiers to M and ε as $\kappa_F = v\, m_f^{\epsilon}/M^{1+\epsilon}$ for fermions, and $\kappa_V = v\, m_V^{2\epsilon}/M^{1+2\epsilon}$ for vector bosons. The SM expectation, $\kappa_i = 1$, is recovered when (M, ε) = (v, 0).

Both Fig. 6.23a and b demonstrate good compatibility with the SM within the errors on the measurements.

6.7.4.4 Compatibility of the Observed State with the SM Higgs Boson Hypothesis: Quantum Numbers

Ascertaining the quantum numbers of the Higgs boson is essential to the understanding of its nature and its coupling properties. According to the Landau-Yang theorem the observations made in the diphoton channel excludes the spin-1 hypothesis and restricts possibilities for the boson to have spin-0 or -2. The diphoton decay mode also implies that the boson has charge conjugation has C-even.

To identify the spin-parity of the Higgs boson the production and decay processes are examined in several analyses. The angular distributions of the decay particles can be used to test various spin hypotheses.

Much can be gleaned from the decay mode $H \rightarrow ZZ^{(*)} \rightarrow 4\,l$, where the full final state is reconstructed, including the angular variables sensitive to the spin-parity, along with a very favourable signal/background ratio. CMS has used the information from the five angles (see Fig. 6.24a) and the two dilepton pair masses combined to form a discriminant based on the 0^+ nature of the Higgs boson [57].

ATLAS has also tested various J^P hypotheses, and in particular 0^+ and 0^-.

In all scenarios investigated by both CMS [57] and ATLAS [58] experiments, the data are compatible with the 0^+ hypothesis, excluding a pseudoscalar nature at CLS levels of 99.8% and 98.0%, respectively. The expected distribution in ATLAS of the test statistic for the SM hypothesis (in blue) and several alternative spin and parity hypotheses is compared in Fig. 6.24b. The combination of the three decay processes allows the exclusion of all considered non-SM hypotheses at amore than 99.9% CL in favour of the SM spin-0 hypothesis.

6.8 Conclusions and Outlook

In July 2012 the ATLAS and CMS experiments announced the discovery of a Higgs boson, confirming the conjecture put forward in the 1960s. Further results from the two experiments show that, within the current measurement precision, the Higgs boson has the properties predicted by the SM. However, several theories of physics beyond the SM (BSM) predict the existence of more than one Higgs boson, and one of these would only be subtly different from that predicted in the SM one, with signal strengths differing by between 0.5–5%, depending on the model in question, indicative of the required level of sensitivity to distinguish it from a SM Higgs boson.

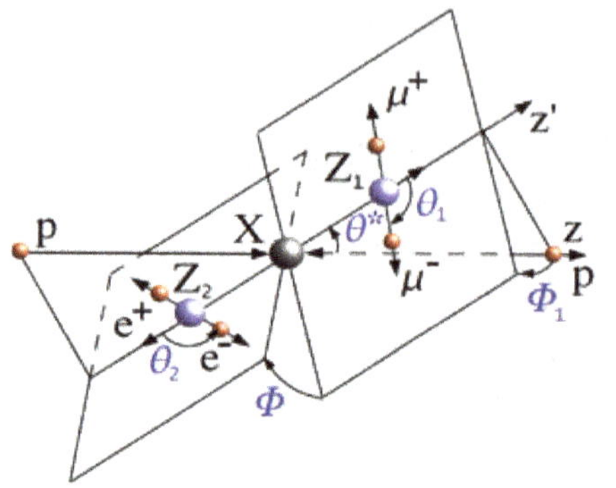

(a)

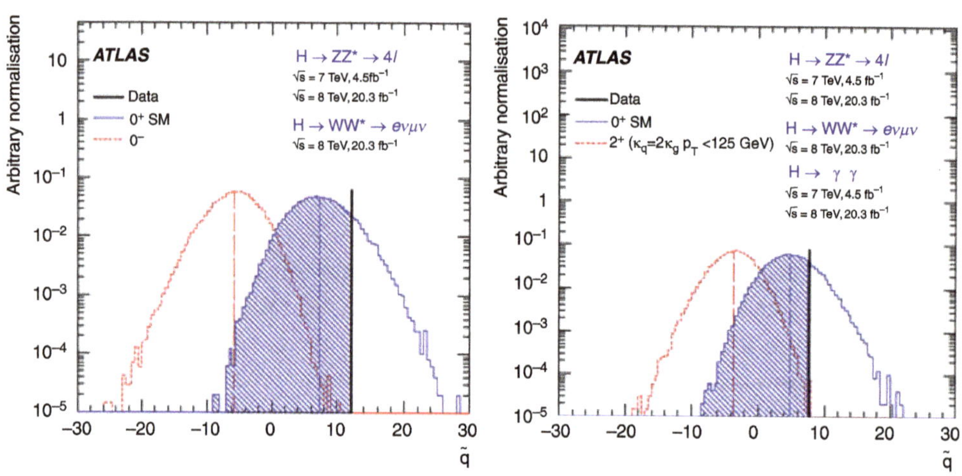

(b)

Fig. 6.24 (**a**) Illustration of the production and decay of a particle $X \to Z_1Z_2 \to 4\ l$ with the two production angles θ_* and Φ_1 shown in the X rest frame and three decay angles θ_1, θ_2, and Φ shown in the Z_i and X rest frames, respectively. (**b**) Examples of distributions of the test statistic (˜q) defined in for the combination of decay channels left) 0^+ versus 0^- right) 0^+ versus the spin-2 model. The observed values are indicated by the vertical solid line and the expected medians by the dashed lines. The shaded areas correspond to the integrals of the expected distributions used for the rejection of each hypothesis

In Run 2 (2015–2018) the LHC provided proton-proton collisions at $\sqrt{s} = 13$ TeV with a peak instantaneous luminosity of 2×10^{34} cm^{-2} s^{-1}, a factor of two beyond the design value. It is intended to operate the accelerator at $\sqrt{s} = 14$ TeV after the next long shutdown (2019–2020) and to integrate a luminosity corresponding to some 300 fb^{-1} by the end of Run 3 (2021–2024). More precise measurements of the properties of the new boson will be made, as well as a more extensive exploration of physics beyond the SM, for which many possibilities are conjectured including supersymmetry, extra dimensions, unified theories, superstrings etc.

The results presented in Chap. 6 are still mostly dominated by statistical errors. The ATLAS and CMS experiments continually update their results that can be found through their websites quoted in references [36, 37]. Much more data need to be collected to enable rigorous testing of the compatibility of the Higgs boson with the SM and to get clues to physics lying beyond the SM in case of a significant deviation. This is one of the main motivations for the high luminosity LHC project, labeled the HL-LHC.

Europe's topmost priority in particle physics calls for the exploitation of the full potential of the LHC, including the high-luminosity upgrade of the accelerator and detectors with a view to collecting ten times more data than in the initial design. It is planned to increase the instantaneous luminosity of the LHC to 5×10^{34} cm^{-2} s^{-1}, and record, by around 2035, an integrated luminosity corresponding to ~3000 fb^{-1} (ten times larger than the original design value). Such an integrated luminosity also requires substantial upgrades of the ATLAS and CMS experiments, now underway, to allow a very precise measurement of the properties of the Higgs boson the study of its rare decay modes and self-coupling, in addition to the search for physics beyond the SM. Many theories beyond the SM make different predictions for the properties of one or more Higgs bosons.

Based on the currently analysed data ATLAS and CMS experiments have recently made projections of the attainable sensitivity for such measurements by the end of the HL-LHC phase [59, 60]. As around 150 million Higgs bosons will be produced a search can also be made for exotic and rare decays of the boson. Figures 6.25 and 6.26 show two sets of projections; Scenario 1 (S1) using the current theoretical errors or Scenario 2 (S2) where the theoretical errors are halved. The extrapolations show the possibility of measuring the individual signal strengths with a precision of between 5–10% for an integrated luminosity of 300 fb^{-1}, and a few percent for a dataset corresponding to 3000 fb^{-1} per experiment, dominated by theoretical erros. The per-production mode signal strength parameters are projected to be measurable with uncertainties of between 3–6% for a dataset corresponding to

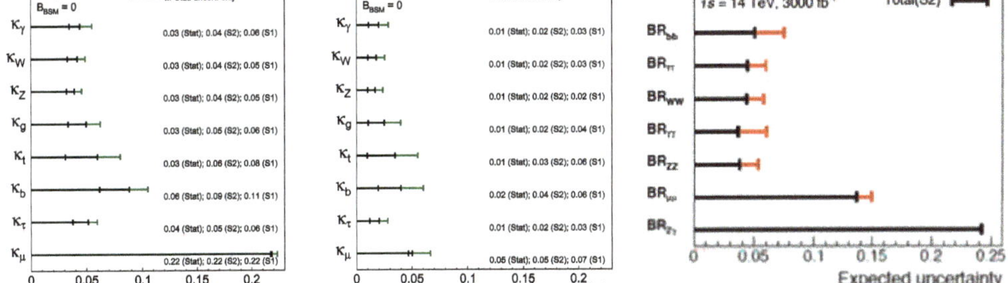

Fig. 6.25 Summary plot from CMS showing the total expected $\pm 1\sigma$ uncertainties in S1 and S2 on the per-decay mode signal strength parameters for 300 fb^{-1} (left) and 3000 fb^{-1} (centre). The statistical-only component of the uncertainty is also shown. (right) Expected uncertainty on the branching ratio measurements in ATLAS for the gg, ZZ$^{(*)}$, WW$^{(*)}$, tt, bb, $\mu\mu$ and Zγ decay channels normalized to their SM predictions assuming SM production cross section for scenarios S1 (red) and S2 (black)

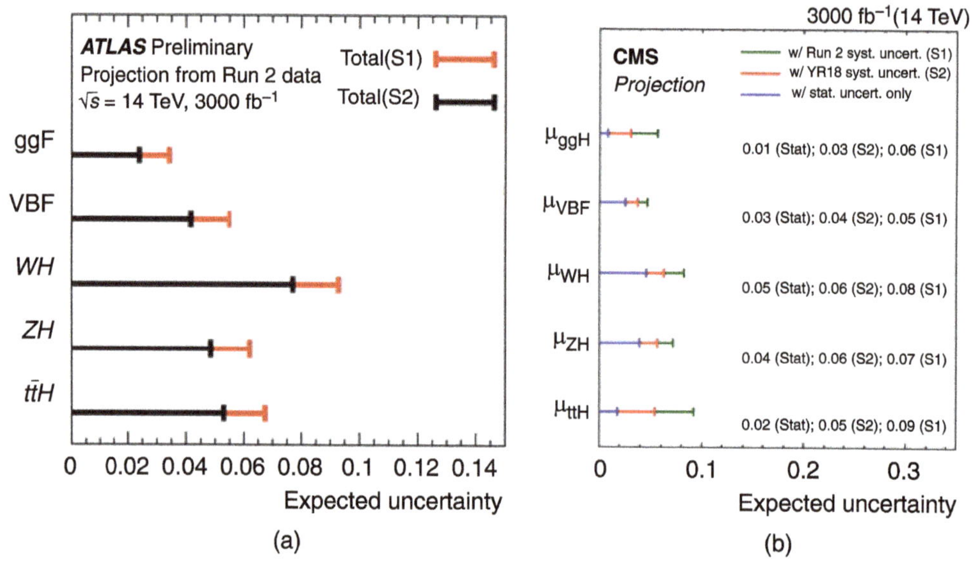

Fig. 6.26 (**a**) Expected uncertainty on the measurement of cross sections in ATLAS for the ggF, VBF, WH, ZH and ttH production modes normalized to their SM predictions assuming SM branching fractions for Scenarios S1 (red) and S2 (black). (**b**) Summary plot from CMS showing the total expected $\pm 1\sigma$ uncertainties in S1 and S2 on the cross section measurements for 300 fb^{-1} (left) and 3000 fb^{-1} (right). The statistical-only component of the uncertainty is also shown

3000 fb^{-1} per experiment. Of particular note, in view of a future electron-positron collider, is the projection for the measurement of the ttH coupling with a precision of ~5% per experiment.

The discovery of a Higgs boson implies the discovery of a fundamental scalar field that pervades the universe. Astronomical and astrophysical measurements point to the following composition of energy-matter in the universe: ~4% normal matter that "shines", ~23% dark matter, and the rest in the form of "dark energy." Dark matter is weakly and gravitationally interacting matter with no electromagnetic or strong interactions. These are the properties carried by the lightest supersymmetic particle. Hence the question: Is dark matter supersymmetric in nature? Fundamental scalar fields could well have played a critical role in the conjectured inflation of our universe immediately after the Big Bang, and in the recently observed accelerating expansion of the universe that, among other measurements, signals the presence of dark energy in our universe.

The discovery of the Higgs boson could turn out to be a portal to physics beyond the SM. Physicists at the LHC are eagerly looking forward to further running of the LHC, and the HL-LHC, and to establishing the true nature of the Higgs boson, to find clues or answers to some of the other fundamental open questions in particle physics and cosmology. The exploitation of the LHC is in its infancy, having recorded a small fraction of the finally anticipated integrated luminosity, and the expectations for other discoveries in the coming decades are high.

Acknowledgements The construction, and now the operation and exploitation, of the large and complex ATLAS and CMS experiments have required the talents, the resources, and the dedication of thousands of scientists, engineers and technicians worldwide. This paper is dedicated to all who have worked on these experiments. The superb construction, and efficient operation of the LHC accelerator and the WLCG computing are gratefully acknowledged.

The authors are greatly indebted to C. Seez for careful reading of this manuscript.

References

1. L. Evans (ed.), The Large Hadron Collider, a Marvel of Technology, EPFL Press, 2nd Edition, 2018;
2. L. Evans, P. Bryant (ed.) and LHC Machine, JINST 03 (2008) S08001.
3. F. Englert and R. Brout, Phys. Rev. Lett. **13** (1964) 321.
4. P.W. Higgs, Phys. Lett. **12** (1964) 132.
5. P.W. Higgs, Phys. Rev. Lett. **13** (1964) 508.
6. G.S. Guralnik, C.R. Hagen, and T.W.B. Kibble, Phys. Rev. Lett. **13** (1964) 585.
7. P.W. Higgs, Phys. Rev. 145 (1966) 1156.
8. T.W.B. Kibble, Phys. Rev. 155 (1967) 1554.
9. The Infinity Puzzle, Frank Close, Basic Books (2011).
10. S. Weinberg, Phys. Rev. Lett. 19 (1967) 1264.
11. A. Salam, Proceedings of the eighth Nobel symposium, ed. N. Svartholm, p. 367. Almqvist & Wiskell, 1968.
12. S.L. Glashow, Nucl. Phys. 22 (1961) 579.
13. G. t'Hooft and M. Veltman, Nucl. Phys. B44(1972)189.
14. F. J. Hasert et al., Phys. Lett. B46(1973)38.
15. D. Gross and F.Wilczek, Phys. Rev. Lett. 30(1973)1343.
16. D. H. Politzer, Phys. Rev. Lett. 30(1973)1346.
17. UA1 Collaboration, Experimental Observation of isolated large transverse energy electrons with associated missing energy at $\sqrt{s}$=540 GeV, Phys. Lett. B122 (1983) 103; Experimental Observation of Lepton Pairs of Invariant Mass Around 95 GeV/c2 at the CERN SPS Collider, Phys. Lett. 126B (1983) 398.
18. UA2 Collaboration, Observation of single isolated electrons of high transverse momentum in events with missing transverse energy at the CERN p-pbar collider, Phys. Lett. B122 (1983) 476; Evidence for $Z^0 \to e^+e^-$ at the CERN anti-p – p Collider, Phys. Lett. 129B (1983) 130.
19. ATLAS Collaboration, Observation of a new particle in the search for the Standard Model Higgs boson with the ATLAS detector at the LHC, Phys. Lett. B716 (2012) 1.
20. CMS Collaboration, Observation of a new boson at a mass of 125 GeV with the CMS experiment at the LHC, Phys. Lett. B716 (2012) 30.
21. M. Della Negra, P. Jenni and T.S. Virdee, Journey in the Search of the Higgs Boson: The ATLAS and CMS Experiments at eth Large Hadron Collider, Science Vol. 338 no. 6114 (2012) 1560.
22. P. Jenni and T.S. Virdee, The Discovery of the Higgs Boson at the LHC, in "60 Years of CERN Experiments", Ed. H. Schopper and L. Di Lella (World Scientific, 2015), 978-981-4644-14-3.
23. M. Della Negra, P. Jenni and T. S. Virdee, The Construction of ATLAS and CMS, Ann. Rev. Part. Sci. 68 (2018) 183.
24. ALEPH, DELPHI, L3, OPAL Collaborations, and LEP Working Group for Higgs Boson Searches, Phys. Lett. B 565 (2003) 61.
25. ALEPH, CDF, D0, DELPHI, L3, OPAL, SLD Collaborations, the LEP Electroweak Working Group, the Tevatron Electroweak Working Group, and the SLD Electroweak and

Heavy Flavour Groups, Precision electroweak measurements and constraints on the standard model, CERN PH-EP-2010-095, http://lepewwg.web.cern.ch/LEPEWWG/plots/winter2012/, arXiv:1012.2367, 2010, http://cdsweb.cern.ch/record/1313716.

26. CDF and D0 Collaborations, Evidence for a Particle Produced in Association with Weak Bosons and Decaying to Bottom-Antibottom Quark Pair in Higgs Boson Searches at the Tevatron, Phys. Rev. Lett. 109, 071894 (2012).
27. J.R. Ellis, M.K. Gaillard, D.V. Nanopoulos, Nucl. Phys. B 106 (1976) 292.
28. LHC Higgs Cross Section Working Group, https://twiki.cern.ch/twiki/bin/view/LHCPhysics/LHCHXSWG, http://arxiv.org/abs/1101.0593 (2011), http://arxiv.org/abs/1201.3084 (2012), http://arXiv:1307.1347 (2013), http://arXiv:1610.07922 (2016).
29. C.J. Seez, T.S. Virdee, L. Di Lella, R.H. Kleiss, Z. Kunszt, W.J. Stirling, in: G. Jarlskog, D. Rein (Eds.), Proceedings of the Large Hadron Collider Workshop, Aachen, Germany, 1990, p. 474, CERN 90-10-V-2/ECFA 90-133-Vol-2.
30. M. Della Negra, D. Froidevaux, K. Jakobs, R. Kinnunen, R. Kleiss, A. Nisati, T. Sjostrand, in: G. Jarlskog, D. Rein (Eds.), Proceedings of the Large Hadron Collider Workshop, Aachen, Germany, 1990, p. 509, CERN 90-10-V-2/ECFA 90-133-Vol-2.
31. Proceedings of the Large Hadron Collider Workshop, Aachen, Germany, 1990, CERN 90-10-V-2/ECFA 90-133.
32. G. Flugge, ed. Proceedings of the General Meeting on LHC Physics Detectors: Towards the LHC Experimental Programme, Geneva: ECFA/CERN (1992).
33. N. Ellis, T. S. Virdee, Ann. Rev. Nucl. Sci. 44 (1994) 609.
34. The LHC Detector Challenge, Physics World, Vol. 17, No. 9, (2004); Detectors at LHC, Phys. Rep. 403-404 (2004) 401.
35. ATLAS Collaboration, https://atlas.cern/discover/about, Letter of Intent, CERN-LHCC-92-004 (1992); Technical Proposal, CERN-LHCC-1994-043 (1994); JINST 3 (2008) S08003.
36. CMS Collaboration, https://cms.cern/, Letter of Intent, CERN-LHCC-92-003 (1992); Technical Proposal, CERN-LHCC-1994-038 (1994); JINST 3 (2008) S08004.
37. ATLAS and CMS Collaborations, Measurement of the Higgs boson production and decay rates and constraints on its couplings from a combined ATLAS and CMS analyss of the LHC pp collision data at $\sqrt{s}$=7 and 8 TeV, JHEP08 (2016) 045.
38. ATLAS Collaboration, Measurements of the Higgs boson properties in the diphoton decay channel using 80 fb^{-1} of pp collision data with the ATLAS detector, ATLAS-CONF-2018-028 (2018).
39. CMS Collaboration, Measurements of Higgs boson properties in the diphoton decay channel in proton-proton collisions at $\sqrt{s} = 13$ TeV, JHEP11 (2018) 185.
40. ATLAS Collaboration, Measurements of Higgs boson coupling properties in the $H \to ZZ^* \to 4l$ decay channel at $\sqrt{s} = 13$ TeV with the ATLAS detector, JHEP03 (2018) 095.
41. CMS Collaboration, Measurements of the properties of Higgs boson decaying in the four-lepton final state at $\sqrt{s} = 13$ TeV, CMS PAS HIG-18-001.
42. ATLAS and CMS Collaborations, Combined measurement of the Higgs Boson Mass in pp Collisions at $\sqrt{s}$=7 and 8 TeV with the ATLAS and CMS Experiments, Phys. Rev. Lett. 114 191803 (2015).
43. ATLAS Collaboration, Measurements of gluon–gluon fusion and vector-boson fusion Higgs boson production cross-sections in the $H \to WW^* \to e\nu\mu\nu$ decay channel in pp collisions at $\sqrt{s}$=13TeV with the ATLAS detector, Phys. Lett. B 789 (2019) 508.
44. CMS Collaboration, Measurements of the properties of Higgs boson decaying to a W boson pair in pp collisions at $\sqrt{s} = 13$ TeV, Phys. Lett. B 791 (2019) 96.
45. CMS Collaboration, Observation of the Higgs boson decay to a pair of τ leptons with the CMS detector, Phys. Lett. B 779 (2018) 283.
46. ATLAS Collaboration, Cross-section measurements of the Higgs boson decaying to a pair of tau leptons in proton-proton collisions at $\sqrt{s}$=13 TeV with the ATLAS detector, Phys. Rev. D 99, 072001 (2019).
47. ATLAS Collaboration, Observation of $H \to b\bar{b}$ decays and VH production with the ATLAS detector, Phy. Lett. B786 (2018) 59.

48. CMS Collaboration, Observation of the Higgs Boson decay to Bottom Quarks, Phys. Rev. Lett. 121, 121801 (2018).
49. ATLAS Collaboration, Search for the Dimuon Decay of the Higgs Boson in pp Collisions at $\sqrt{s}$=13 TeV with the ATLAS detector, Phys. Rev. Lett. 119, 051802 (2017).
50. CMS Collaboration, Search for Higgs boson decaying to two muons in proton-proton collisions at $\sqrt{s}$=13 TeV, Phys. Rev. Lett. 122, 021801 (2019).
51. ATLAS Collaboration, Observation of Higgs boson production in association with a top quark pair at the LHC with the ATLAS detector, Phys. Lett. B 784 (2018) 173.
52. CMS Collaboration, Observation of ttH Production, Phys. Rev. Lett 120, 231801 (2018).
53. ATLAS and CMS Collaborations, Precise determination of the mass of the Higgs boson and tests of compatibility of its couplings with the standard model predictions using collisions at 7 and 8 TeV, Eur. Phys. J. C (2015) 75:212.
54. C. Patrignani et al., Particle Physics Group, Chin. Phys. C, 40 100001 (2016)
55. ATLAS Collaboration, Combined measurements of Higgs boson production and decay in proton-proton collision date at $\sqrt{s}$=13 TeV, ATLAS-CONF-2018-031 (2018).
56. CMS Collaboration, Combined measurements of Higgs boson couplings in proton-proton collisions at $\sqrt{s}$=13 TeV, Eur. Phys. J. C 79 (2019) 421.
57. CMS Collaboration, Study of the mass and spin-parity of the Higgs boson candidate via its decays to Z boson pairs, Phys. Rev. Lett., **110** (2013) 081803.
58. ATLAS Collaboration, Study of the spin of the Higgs boson in diboson decays with the ATLAS detector, Eur. Phys. J C75 (2016) 476.
59. ATLAS Collaboration, Projections for measurements of Higgs boson cross sections, branching fractions, coupling parameters and mass with the ATLAS detector, ATL-PHY-PUB-2018-054.
60. CMS Collaboration, Sensitivity projections for Higgs boson properties measurements at the HL-LHC, CMS PAS FTR-18-011 (2018).

Permissions

All chapters in this book were first published in PPRLV1TE, by Springer; hereby published with permission under the Creative Commons Attribution License or equivalent. Every chapter published in this book has been scrutinized by our experts. Their significance has been extensively debated. The topics covered herein carry significant information for a comprehensive understanding. They may even be implemented as practical applications or may be referred to as a beginning point for further studies.

The contributors of this book come from diverse backgrounds, making this book a truly international effort. We would like to thank all the contributing authors for lending their expertise to make the book truly unique. They have played a crucial role in the development of this book. Without their invaluable contributions this book wouldn't have been possible. They have made vital efforts to compile up to date information on the varied aspects of this subject to make this book a valuable addition to the collection of many professionals and students.

This book was conceptualized with the vision of imparting up-to-date and integrated information in this field. To ensure the same, a matchless editorial board was set up. Every individual on the board went through rigorous rounds of assessment to prove their worth. After which they invested a large part of their time researching and compiling the most relevant data for our readers.

The editorial board has been involved in producing this book since its inception. They have spent rigorous hours researching and exploring the diverse topics which have resulted in the successful publishing of this book. They have passed on their knowledge of decades through this book. To expedite this challenging task, the publisher supported the team at every step. A small team of assistant editors was also appointed to further simplify the editing procedure and attain best results for the readers.

Apart from the editorial board, the designing team has also invested a significant amount of their time in understanding the subject and creating the most relevant covers. They scrutinized every image to scout for the most suitable representation of the subject and create an appropriate cover for the book.

The publishing team has been an ardent support to the editorial, designing and production team. Their endless efforts to recruit the best for this project, has resulted in the accomplishment of this book. They are a veteran in the field of academics and their pool of knowledge is as vast as their experience in printing. Their expertise and guidance has proved useful at every step. Their uncompromising quality standards have made this book an exceptional effort. Their encouragement from time to time has been an inspiration for everyone.

The publisher and the editorial board hope that this book will prove to be a valuable piece of knowledge for students, practitioners and scholars across the globe.

Index

Printed in the USA
CPSIA information can be obtained
at www.ICGtesting.com
LVHW082200201023
761596LV00009B/1511

9 798888 360149